MW00835378

FUNCTIONS

FUNCTIONS

New Essays in the Philosophy of Psychology and Biology

Edited by

ANDRÉ ARIEW, ROBERT CUMMINS, and MARK PERLMAN

OXFORD
UNIVERSITY PRESS

Great Clarendon Street, Oxford OX2 6DP

Oxford University Press is a department of the University of Oxford.
It furthers the University's objective of excellence in research, scholarship,
and education by publishing worldwide in

Oxford New York

Auckland Bangkok Buenos Aires Cape Town Chennai
Dar es Salaam Delhi Hong Kong Istanbul Karachi Kolkata
Kuala Lumpur Madrid Melbourne Mexico City Mumbai Nairobi
São Paulo Shanghai Singapore Taipei Tokyo Toronto
with an associated company in Berlin

Oxford is a registered trade mark of Oxford University Press
in the UK and in certain other countries

Published in the United States
by Oxford University Press Inc., New York

First published 2002

British Library Cataloguing in Publication Data
Data available

Library of Congress Cataloging in Publication Data
Functions: new essays in the philosophy of psychology and biology/edited
by André Ariew, Robert Cummins, and Mark Perlman.
p. cm.
Includes bibliographical references and index.
1. Psychology—Philosophy. 2. Biology—Philosophy. I. Ariew, André.
II. Cummins, Robert. III. Perlman, Mark, 1964–
BF38.F86 2002 124—dc21 2002020184

ISBN 0–19–925580–6

ISBN 0–19–925581–4 (pbk.)

1 3 5 7 9 10 8 6 4 2

Typeset by Hope Services (Abingdon) Ltd.
Printed in Great Britain
on acid-free paper by
T. J. International Ltd.
Padstow, Cornwall

ACKNOWLEDGEMENTS

Funding for this anthology came in part from the University of Rhode Island Council for Research Faculty Development Grant, and the National Science Foundation grant no. 0115500. We thank two anonymous reviewers for helpful comments. Peter Momtchiloff is an excellent editor. You could not ask for a better copy-editor than Hilary Walford. At least one of us would like to thank Deborah Kallman for vital support.

CONTENTS

Part Three
Teleosemantics

Part Four
Methodological Issues

Introduction

André Ariew and Mark Perlman

Humans often create objects with a purpose in mind, endowing the objects with particular functions. It is taken to be relatively unproblematic to have artifacts receive their functions from the intentions of their inventors. For example, can-openers are invented with the function of opening cans. But functions have a checkered philosophical history when applied to the natural world. In the past, they have been expected to carry a lot of explanatory weight, sometimes to the point of investing every event and every part of every object with either some cosmic function or some end-directing inner force. In the enlightenment and beyond, the mechanistic 'new science' was a sustained attempt to rid science of functions and other Aristotelian influences. Still, functions play a significant role in biology, and, recently, in psychology. We routinely explain how and why an organism is the way it is by citing the functions of its parts. Are we slipping back to a pre-enlightenment mode of explanation or are there other ways to ground functional explanation in biology and psychology? This is where philosophers can help the biologists, physiologists, and psychologists, or, if we are not careful, make everything even worse. We take the following to be the state of the art among philosophers of biology and psychology.

The statement 'hearts function to pump blood' indicates a causal role the organ plays within the circulatory system of vertebrates. No author in these pages would deny the legitimacy of causal role functional analysis. However, human hearts contribute to all sorts of 'systems' in virtue of their effects: a heart's pumping makes 'blub-blub' noises, which contributes to a doctor's vitality monitoring system or a parent's ability to soothe a distressed baby. Hearts weigh about four pounds, contributing to a human's total weight. And so on. There is nothing illegitimate about pointing out the multitude of causal roles hearts play, since it is true that normal hearts produce all sorts of effects besides pumping that contribute to all sorts of systems besides the circulatory system. In other words, there is some explanatory utility in pointing out the various ways the heart functions *as*. However, common sense tells us it is plain silly to assert that *the* function of the heart is, say, to weigh four pounds. Intuitively, pumping blood for the circulatory system is the 'proper' function, while the others are mere (sometimes fortuitous) 'side effects'. To stretch the intuition a little further, what distinguishes an item's proper function from the others is that the former explains (vaguely put) why the item exists or will continue to exist. Let us stipulate for the sake of labeling (admitting that there may be some controversy in the following claim): functional explanations that pick out an item's 'proper' function are 'teleological explanations' in virtue of the fact that they explain why an item exists or will continue to exist.

Perhaps our intuition that teleological explanations are legitimate is grounded by an analogy between natural items and human artifacts. The basket placed on the floor near our desks serves all sorts of functional roles in virtue of its multitude of effects (depending on what it is made of and its spatial configuration). We can set it mouth-side up to play office basketball; or we can lift it inverted above our heads to catch mice. If one is so inclined we can place it mouth-first over our heads to hide our head and faces. These are all things that the waste-basket functions *as*. But *the* function of the basket is temporarily to store unwanted items. It contributes to our society's waste-disposal system. Of course, one may deny that particular functional attribution and replace it with another one; after all, people purchase waste-baskets for all sorts of reasons. Whatever the case, the intuitions about waste-baskets are the same: the functional assignment depends on the *conscious intention* of the inventor, manufacturer, purchaser, user, etc. One picks out the item's 'purpose'. Here 'purpose' indicates proper function and answers the teleological question about the item's existence.

While it is debatable whether the analogy between natural items and human artifacts is philosophically fruitful, it is clear that a naturalist would reject the justification of applying teleological explanations to natural items on the basis of conscious intention. While hearts may be manipulated by humans, as in the case of using the blub-blub noise to soothe distraught babies, human manipulation does not determine an item's 'proper' function. However, what is not so clear—and hence is a somewhat contentious issue among philosophers—is whether there is a naturalistic, non-pragmatic way to sustain the distinction between how an item functions and *the* function of the item. In other words, it is debatable whether teleological explanations play a legitimate role in biology.

So we have come to a fork in the philosopher's road. Some argue that teleological explanations are illegitimate in biology. Others argue that natural selection theory grounds teleology—that is, some aspect of natural selection theory justifies the distinction of an item's 'proper' function. (This path has its own forks having to do with what about natural selection grounds teleology.) Others think that the distinction is grounded by attention to species typicality. (Perhaps there are others who think Developmental System's Theory grounds the distinction.)

Philosophers, including contributors to this volume, have made names for themselves blazing their own roads, putting up blocks in the road of others, or even creating their own forks by discussing issues related to functional or teleological explanation. For example, how does one individuate a trait to which we ascribe a function? Can natural selection pick out selected effects when several effects are fitness enhancing? The latter problem is particularly important for the philosopher of mind operating under the representational theory of mental content.

The old question in mind and meaning, 'What makes ideas mean what they do?', has recently been updated into the question: 'What gives mental representations their content?' Mental representations have intentionality: they are *about* other objects, or they *point to* things, they *mean* things. The difficult question is how they can have this unique characteristic. What is the source of intentionality? The issue is made more difficult by the possibility of mistakes—a representation may be *about* one thing yet sometimes be used to represent some other thing, or be about one thing yet represent it incorrectly. When a representation points to something not included in its representational content, this is known as *misrepresentation.* This familiar phenomenon is notoriously difficult to explain adequately. But we get a very promising answer from biology: perhaps representations get their content from their evolutionary function (or the function of the mechanisms that produce them or use them in cognition). The feature that makes functions an inviting source of mental content is that an object can have a function but not actually perform that function on all occasions. Even when it does not perform its function, it can still be said to have that function. This split between what an object actually does and what its function is makes functions seem ideal for providing content to mental representations, including their ability to misrepresent. But it depends crucially on specifying exactly what the function is.

The volume is divided into four themes. The contributors of this volume have all provided short abstracts preceding their essays. Here, we describe how the contributions fit into the four general themes. The first part is devoted to the history of teleology from Aristotle to Darwin and beyond. André Ariew and Michael Ruse are the contributors. The second and largest part includes comparisons between, and critiques and defenses of, one or another theory of functional and teleological explanation. The theories in contention include the etiological, the pragmatic, and the general goal-contribution analysis accounts of function (more on these theories and terminology below). In this volume, the compare/contrast essays come from Ruth Millikan and Valerie Gray Hardcastle. Millikan compares and contrasts her own etiological account with Robert Cummins's pragmatic account. Hardcastle compares etiological and pragmatic accounts in general. Robert Cummins provides a critique of the etiological account (or any account that attempts to ground functions to natural selection). Christopher Boorse defends his own general goal-contribution analysis of function. The defenders of etiological accounts include: William Wimsatt (in a revision of his 1972 paper), David J. Buller, and Peter H. Schwartz. Wimsatt's account is quite similar to Larry Wright's. Buller defends the idea that natural selection explains what a trait is 'designed' to do. Schwartz offers a difficult problem for etiological accounts and provides a possible solution. In the third part, three contributors critique teleosemantics: Mark Perlman, Berent Enç, and D. M. Walsh. Perlman provides a brief history of teleosemantics before offering his version of the so-called indeterminism problem. Enç articulates why teleosemantics cannot solve the indeterminancy problem. Walsh argues that the indeterminancy problem is rooted in a particular feature of etiological accounts. In the fourth and final part, three authors discuss methodological issues related to functional explanation. Mohan Matthen asks whether the function of rationality is to produce true beliefs. Colin Allen wonders what a *trait* is. Karen Neander defends the position that biological traits are individuated by their functions.

A note about terminology: the reader should be forewarned that terminology is not fixed in this area. Each contributor may seem wedded to his or her own terms though there is some overlap between authors. This reflects a common practice in the functions literature at large: there are some substantive theoretical disagreements over terms. Terminology is inevitably theory-laden, and so it is a mistake to try to cross-translate too closely. With that in mind, we leave the readers to fend for themselves.

Part One

History of Teleology and Functional Explanation: From Socrates to Darwin and Beyond

1. Platonic and Aristotelian Roots of Teleological Arguments

André Ariew

ABSTRACT

Aristotle's central argument for teleology—though not necessarily his conclusion—is repeated in the teleological arguments of Isaac Newton, Immanuel Kant, William Paley, and Charles Darwin. To appreciate Aristotle's argument and its influence I assert, first, that Aristotle's naturalistic teleology must be distinguished from Plato's anthropomorphic one; second, the form of Aristotle's arguments for teleology should be read as instances of inferences to the best explanation. On my reading, then, both Newton's and Paley's teleological arguments are Aristotelian while their conclusions are Platonic. Kant and Darwin's arguments are likewise Aristotelian while their conclusions are unique.

The authors contributing to this anthology focus on a variety of issues concerning 'functional explanations' in both biology and psychology. 'Functional explanation' is our chosen term because 'teleological explanation' is thought to imply backwards causation or bizarre ontological categories (for example, vital forces) attributable to the teleological theories of Plato and Aristotle. Functional explanation is not so imbued and hence, as opposed to teleology, is an appropriate topic for naturalistic analysis. However, scholars of ancient Greek science and metaphysics know that Aristotle's and Plato's teleologies are richer and more interesting than many of the writers on modern functional explanation realize. Between Aristotle's and Plato's writings are found several different categories of teleology, only some of which invoke bizarre metaphysical entities, and several powerful arguments for the legitimacy of teleological explanation in biology.

I would like to thank the following for comments on earlier drafts: Paul Bloomfield, Chris Stephens, David Buller, Julia Annas, Donald Zeyl, James Lennox, Mohan Matthen, Mark Perlman, Denis Walsh, John Beatty, Albert Silverstein, Galen Johnson, and the audience participating in the History of Science colloquium at the University of Minnesota, where this paper was presented in September of 2000. I thank John Beatty for inviting me to give the colloquium. Special thanks goes to Slobodan Zunjich for greatly influencing my views and providing me with extensive comments on many drafts. I could not have written this paper without his help.

There is considerable disagreement about the relation between teleological explanation and functional explanation. Consider Woodfield's influential remark that teleological explanations are part of the domain of purposive behavior while functional explanations are part of the larger domain of system analysis (Woodfield 1976). The distinction has no clear support, however, in the writings of Aristotle from which the modern concept of 'functional explanation' takes root. According to Aristotle's schema, functional explanations are a *subset* of teleological ones.

Most importantly, insofar as Aristotle's teleology pertains to explanations of natural items, it is misleading to cast off Aristotle's teleology as reading purposive behavior into natural events. This perception of Aristotle's teleology is the result of conflating Aristotle's naturalistic teleology with Plato's. As I will discuss in this essay, Plato's natural teleology is and Aristotle's is not creationist, anthropomorphic, and externally evaluative. Plato's natural teleology invokes the concept of the 'good'. Aristotle's does not. Aristotle's natural teleology is and Plato's is not naturalistic, immanent, and functional.

Aristotle's central mode of argument for both artifactual and natural teleology is an inference to the best explanation: teleology best explains facts about the organic world. We shall see arguments of this type in three separate discussions within Aristotle's *Physics*, book II. The distinction between Plato's teleology and Aristotle's natural teleology is worth revisiting for three reasons (which provide the three theses for this essay). 1. Both Aristotle's and Plato's teleology arguments are more sophisticated than historically and currently presented. 2. Many teleology arguments in post-eighteenth-century science are variants on Aristotle's inferences to the best explanations while the conclusions are *Platonic.* I will demonstrate this dualism in the works of Isaac Newton, Immanuel Kant, and William Paley. 3. Once we strip away Platonic teleology from Aristotle's inferences to the best explanation, the question 'To what extent was Darwin a teleologist?' can be answered plainly: Darwin endorsed a subset of Aristotle's teleology.

1. Aristotelian versus Platonic Teleology

1.1. Aristotle's Teleology

Teleological explanation in Aristotle pertains broadly to goal-directed actions or behavior. Aristotle invokes teleology when an event or action pertains to goals: 'that for the sake of which' (e.g. *Phys.* II. 194^b32). Following David Charles (1995; with some modification) we can distinguish two distinct conceptions of teleology in Aristotle's writings and at least two sets of subcategories:

I. Agency-centered teleology
   (i) *Behavioral.* Activities undertaken for the sake of something, which may be either a state or further action.
   (ii) *Artifactual.* Activities undertaken for the sake of producing an object of a certain sort (artifact).

II. Teleology pertaining to natural organisms
   (iii) *Formal.* Biological developmental processes that occur for the sake of self-preservation or preservation of the species (form).
   (ii) *Functional.* Parts of organisms that are present for the sake of the organism possessing them.

I and II are *distinct* notions of teleology: Aristotle should have used two words to distinguish them. Agent-specific teleology (I) is purposive, rational, and intentional, and represents an external evaluation. The goal is the object of an agent's *desire* or choice. In behavioral teleology (i) actions are done out of the desire to produce a goal—for example, walking is for the sake of health. In artifactual teleology (ii) the object (artifact) is produced for the sake of achieving some goal—for example, building a house to shelter oneself. Agents are *aware* of the means to fulfill the goal (Charles 1995: 107). To explain why a builder builds a house or to explain why doctors do what they do, we can cite their goal: a builder builds for the sake of shelter and a doctor doctors for the sake of health.

Teleology pertaining to natural organisms is distinct: *non*-purposive (though seemingly so), *non*-rational, *non*-intentional, and *immanent*—that is, an inner principle of change. The goal is *not* an object of any agent's desire. In formal teleology (iii) the *telos* is an inherent property of the process to complete the organism's developmental end state as seen in the form of the species (Zunjich, pers. com.). For example, plants require nourishment for self-fulfillment of the (species) form. So, roots extend downwards rather than upwards for the sake of nourishment (199^a29). In functional teleology (iv) the *telos* is inherent in the relationship between the part of the organism in question and the whole organism. For example, sharp teeth are in the front of the mouth for the sake of tearing (199^b24). Sharp teeth contribute to the flourishing of organisms possessing them. Put more strongly, carnivores flourish *because* they possess sharp teeth. This is consistent with the form of functional explanation that many authors in this volume accept: sharp teeth persist in nature among carnivores because they contribute to the flourishing of carnivores. In neither (iii) nor (iv) is the *telos* a conscious goal of the organism. Nor is the goodness of the process a part of the explanation for what occurs. Roots are not aware that it is good to grow downwards. Rather, a consequence of roots growing downwards is that plants flourish; those that do not grow downwards do not flourish. I summarize the differences in figure 1.1.

	Answer to: 'why is X there'?	*Awareness of means to a goal*	*Valuation*
Agents I(i), I(ii)	In terms of what's good	Agent is aware and flexible to change means	Deliberate, external
Organic development II(iii)	Contributes to the development of the organism according to its form (species)	Unaware and inflexible to change means	Goal is in the form of the species
Natural objects II(iv)	Contributes to self-flourishing	Unaware and inflexible to change means	Goal is property of relation between part and organism

Fig. 1.1. Summary of differences between Plato's and Aristotle's conception of *telos*
Source: adapted from Lennox (1992).

The roots of agency-based teleology (I) are found in the writings of Plato, while teleology pertaining to natural organisms (II) is Aristotle's own. Next, I compare Plato's and Aristotle's teleology. Then, I examine the three central Aristotelian arguments for teleology that I claim are at the core of many teleological arguments throughout history, despite the Platonic conclusions of many of them.

1.2. *Plato's Teleology*

In the *Phaedo* Plato recounts Socrates' criticisms of the Pre-Socratics for missing the real cause of the orderly arrangements of natural phenomena. Anaxagoras explains the orderly arrangements of the cosmos by means of mechanistic principles of motion of matter such as air, water, and ether. Simple material motions are what Anaxagoras takes to be the Reason for the motion in the cosmos. Socrates is unsatisfied. He expected Anaxagoras to explain how the natural order was the *best* of possible world orders. The difference is captured in asking the analogous question, 'Why does Socrates sit in prison?'. While facts about physiology, the composition of bones and sinews and their arrangements, offer a complete explanation of his current position in prison, the explanation is unsatisfactory, for it does not provide the *real* reason for Socrates' predicament. Socrates remains in prison because remaining rather than escaping is what Socrates deems the *best* course of action. Reference to the simple motion does not capture best intentions.

In the *Timaeus* Plato takes up Socrates' challenge to provide an account of why the cosmos was created for the best. Again, reference to simple motions provides a complete explanation of the orderly arrangement of the heavens. However, it is necessary to take account of the good if we want fully to understand the order of the heavens (Morrow 1950: 425). The true cause is agency working for the best.

This ordered world is of mixed birth; it is the offspring of a union of Necessity and Intellect. Intellect prevailed over Necessity by persuading it to direct most of the things that come to be toward what is best, and the result of this subjugation of Necessity to wise persuasion was the initial formation of this universe. (*Timaeus*, 48a, trans. Zeyl)

Agency is constrained both by goodness and in the materials available. For example, while spheres may be the finest shape possible, other shapes are used in nature because the sphere has already been used for the body of the cosmos (Strange 1985: 28).

It is important to note the distinctions between Plato's teleology and Aristotle's. Each is influenced by a different cosmology. For Plato the universe is an artifact, as are the living organisms within (thus subsuming Aristotle's III and IV into I(ii)). The demiurge is the general cause of all motion, including motion on earth. Aristotle fundamentally distinguishes between the cause of motion in the heavens and the cause of motion on earth. The heavens are incorruptible. The primary motion of the sun, stars, and planets is circular—the natural tendency of the distinctly heavenly element, ether. In contrast, earth (or rather the sublunar realm) is corruptible, with motions described in terms of the natural tendencies of the distinctly earthly elements, fire, air, water, and earth. None of these elements tends to circular motion. However, the distinction between heavenly and earthly motion has a caveat: the circular movement of the sun, stars, and planets causes the earth's seasons, which exert a general influence upon growth on earth (Balme 1987: 277).

For Plato's teleology, the striving towards good depends on a standard of excellence in the forms. The artifacts of the universe (including the living organisms therein) are created after the model of the forms. Hence the standard of excellence that drives the striving towards the good is external to the object itself. Aristotle's teleology is immanent, not external to the object. Organic development is an activation of a particular potentiality as seen in the form of the species to which the individual belongs. That activation is not external to the individual but is an inner principle of change (*Phys.* II, ch. 1). Consequently, on Aristotle's account, while humans are sensitive to the means by which they attain their particular goals, there is no explicit requirement that the goal is *best* for the individual's requirements. Plato's and Socrates' teleology is stronger in that actions are always for the best (Annas 1982: 314). I summarize the distinction between Aristotle's and Plato's teleology in figure 1.2.

	Cosmology	*Source of change*	*Valuation*
Plato	Creator (demiurge) governs all motion	External model	Action is for the *best* (from a cosmological point of view)
Aristotle (both agent and organismal)	Distinct motions for heavens and earth	An inner principle of change (Immanent)	Action or part is *useful* to individual

Fig. 1.2. Summary of the differences between Plato's and Aristotle's teleology

2. Aristotle's Arguments against the Materialist

2.1. Aristotle's Argument from Flourishing

All of Aristotle's arguments for teleology we will consider are pitted against the materialist conception that, roughly, materials and their necessary causes are sufficient to explain all physical events. Material necessity refers to a physical event that is the result of the nature of the matter involved as opposed to being interfered with by some external force (Cooper 1987: 260). Aristotle agrees with the materialist that citing the materials and their causal interactions suffice for the explanation for *some* physical events. For example, the reason why the sky rains is due to the material necessity of sky and water: 'what is drawn up must cool, and what has been cooled must become water and descend' (198^b19–20). The 'must' refers to the natural unimpeded regularities of the sky.[1] Further, in the case that the rain spoils a man's crop ('on the threshing floor'), the spoilage comes as a result of rain's natural tendency to come out of the sky. Rain does not fall for the sake of spoiling a man's crop; the result is due to both the material necessity of rainfall and the unfortunate placement of the crop. In other words, Aristotle accepts the abductive inference from the *observational fact* (O) of the particular occurrence of rain spoiling a man's crop to the *hypothesis* (H (materialist)) that the observation is a coincidence.

O: Rain falls.

H (materialist): nature and motion of simple bodies.

According to Aristotle H (materialist) sufficiently explains O. Further, in the case of:

[1] *Pace* Furley (1985). Furley believes that Aristotle rejects the materialist view at all levels. I prefer the interpretation presented here, which is probably more mainstream. However, my main concern is Aristotle's argument for teleology so not a lot rides on this controversy for me.

O′: Rain falls and spoils a man's crop.
H (materialist)′: coincidence.

Aristotle agrees that the H (materialist)′ sufficiently explains O′. That is, coincidence sufficiently accounts for the relationship between rainfall and spoilage of a man's crop.

However, Aristotle contends that coincidence is not sufficient to explain all events. Consider the dental arrangement of humans and some animals: sharp teeth grow in front and broad molars in the back. Aristotle asks what accounts for the fact that carnivores possessing this particular dental arrangement (nearly) invariably prosper 'whereas those which grew otherwise perished and continue to perish'?.

Since the materialist denies that natural events occur for the sake of some end, he or she would have to accept that the usefulness of the dental arrangements occurs as a coincidental result of the material necessity of dental matter. The situation is essentially no different from the man's crop spoiling because of rainfall.

By Aristotle's lights, there is a difference: evidence that particular dental arrangements are *useful* to the organism comes from the fact it is a *regular occurrence in living nature.* It happens nearly invariably; organisms with different dental arrangements nearly always die. So, the proper explanation is that sharp teeth grow in front and broad molars in the back *for the sake* of an organism's flourishing. The 'goal' is inherent in the nature of growth.

Unlike the case of rain, where Aristotle accepted as coincidence the relation between rain and crop spoilage, Aristotle cannot accept as coincidence the fact that organisms possessing sharp teeth in front and broad molars in the back invariably flourish while organisms possessing alternative arrangements invariably die. The latter phenomenon is better explained by teleology: possessing sharp teeth in front and broad molars in the back occurs for the sake of the organism flourishing. That is, teleology can be abductively inferred from the fact that the dental arrangements regularly contribute to the flourishing of individuals possessing them. To schematize (redefining the variables O and H accordingly):

O: Sharp teeth growing in front and broad molars in the back regularly lead to the flourishing of carnivores possessing that arrangement.

O′: Alternative dental arrangements lead to the death of carnivores possessing the alternative.

H (materialist): What does not occur by simple movement occurs by chance. The difference between O and O′ is by chance.

H (teleologist): The difference between O and O′ is that, for carnivores, a particular dental arrangement (sharp teeth in front, broad molars in the back) *occurs for the sake* of flourishing.

Aristotle argues that H (materialist) insufficiently accounts for the difference between O and O′ while H (teleologist) sufficiently accounts for the difference between O and O′.

I will refer to this Aristotelian argument as 'the argument from flourishing'.

2.2. *Aristotle's Argument from Regularity*

In *Physics* II.9 Aristotle strengthens his argument against the materialist by providing an alternative explanatory scheme. In addition to the nature and movement of simple bodies (material necessity) and chance, Aristotle offers a third mode of explanation: hypothetical necessity.[2]

What is hypothetical necessity? Take eyelids, for example. Eyelids are flaps of skin that protect eyes from easy external penetration. According to Aristotle, the eyelid material—the flaps of skin—is *necessary* for the sake of eye protection. The necessity referenced here is called 'hypothetical necessity': it is a constraint on materials given the specific purpose for which the part will be used. Not any material will do for the sake of eye protection, only eyelid material, given the specific form of eye protection that humans and other animals require. This is meant to be taken strongly: the actual materials that compose an organ are required for the completion of the process where completion is the goal of development. Put differently, if there had not been a need for eye protection, there would not have been materials present to form eyelids (Cooper 1987: 255).

The concept of hypothetical necessity makes clear the relationship between functional (iv) teleology and formal (iii) teleology. Consider the example: eyelid material is present for the sake of eye protection (that is the function of eyelid material). So, eyelid material has a functional role (iii) to play in the growth of eye protection. Further, eye protection is necessary for seeing, and seeing occurs for the sake of the organism's growth (iv). The necessity is granted to matter, eyelids, and is conditional in that it contributes to the goal of natural growth. Eyelid material contributes to natural growth by affording eye protection, which itself is crucial for the function of seeing (Cooper 1987).

Hypothetical necessity is inherent in actions pertaining to deliberate agents as well (Charles 1988: 119). In such a case, hypothetical necessity explains why some action has been taken or why some object has been created. These occur because of the agent's goal. In this case, the agent is aware of the goodness of the action or object as a means to the goal.

I follow John Cooper (1987) in viewing Aristotle's argument for hypothetical necessity in terms of an inference to the best explanation for regularities of

[2] One might argue that hypothetical necessity refers to a teleology that does not invoke the final *aitia* (Zunjich, pers. com.).

processes. To make the case stronger let us switch examples to the development of a newborn from sperm, egg, and the usual background developmental conditions. Accordingly, the materialist cannot account for how these materials conspire to produce fetuses (nearly) every time. In other words, by appeal to simple motion and material cause, materialists cannot fundamentally distinguish between:

1. physical forces that are unconstrained to produce a range of different possible outcomes; and
2. physical forces that (nearly) always result in the same product—a newborn.

The materialist's only recourse is an appeal to *coincidence*. Aristotle's reply is that coincidence is insufficient to account for the regularity of the conjugation seen in organic development because chance operates only in unusual circumstances (198^{b}35–199^{a}3). The principle of hypothetical necessity better explains the regularity of development: the materials are there for the sake of producing the conjugation that leads to the development of newborns.

O: sperm and egg invariably conjugate to produce newborns.

H (materialist): accident.

H (teleologist): hypothetical necessity.

H (teleologist) better explains the regularity by which we observe organic development because accidents are rare in nature.

On Aristotle's account, materials are, so to speak (in Cooper's words) 'the seat of the necessity' (Cooper 1987: 255). However, these material arrangements are conditional on the production of newborns being something that occurs in nature. In this way, goal is prior to matter (Charles 1995: 121 n.). That is, sperm and egg do cause the goal of producing newborns; however, the goal of newborns is not there because of the sperm and egg. Quite the opposite: if newborns are to exist (and they do by nature), then sperm and egg and the process that leads them have to exist. That is what it means that sperm and eggs have to exist *for the sake of newborns*.

We will refer to this argument for hypothetical necessity as 'the argument from regularity'.

2.3. *Aristotle's Argument from Pattern*

Finally, it is worth considering a third inference to the best explanation Aristotle employs to support the irreducibility of teleology in explanation. This time Aristotle recognizes that the same teleological scheme applies to explain a particular sort of organization that regularly occurs both within human action and in the non-human natural world. The organization he has in mind is exemplified in the following cases: housebuilding, leaves growing to shade fruit, roots descending for nourishment (rather than rising), nestbuilding in birds, and webmaking in spiders. In all of these cases we recognize a certain *pattern of arrangement and sequential order*. For example, in development of an artifact (such as housebuilding) or in nature (as in roots descending downwards), all the steps of development occur in sequence, which leads up to the final state. Further, parts of an object that contribute to some whole effect are situated to contribute to the whole effect (Charles 1995: 115). These patterns do not happen by accident. Rather they occur in every instance where the relevant organization is found—for example, in the intentional production of artifacts (housebuilding) or the non-deliberate formation of natural objects (webmaking, nestmaking, roots descending, leaves shading fruit). It is in this respect that Aristotle famously remarks that 'as in art, so in nature' (*Phys.* 199ᵃ9–10) and 'as in nature, so in art' (199ᵃ15–16). The same pattern that explains certain organizations found in nature also explains the same organizations found in artifacts (Charles 1995: 115). This 'certain organization' is just goal-directed activity. Aristotle infers teleology from patterns of order and arrangement. We will call this the 'argument from pattern'.

To strengthen this argument, Aristotle presents the first instance where teleology preserves a distinction between function and accident, except for Aristotle the term is a 'mistake'.[3] Mistakes occur when one of the stages required to achieve the goal has failed to complete its role in the production of the goal. Mistakes occur, for example, when a doctor pours the wrong dosage or when a man miswrites or when monstrosities such as 'man-headed ox-progeny' or 'olive-headed vine-progeny' develop. The same teleological pattern whereby each stage of development occurs in order for the sake of the goal allows us to explain the difference between what occurs by art or nature, on the one hand, or by mistake, on the other. What occurs by art or nature follows the pattern successfully, while mistakes or the creation of monsters feature a failed developmental stage.

On the face of it, Aristotle *presupposes* teleology in order to explain it. If so, Aristotle is guilty of circular reasoning. However, on closer inspection, Aristotle does not commit the fallacy. Teleology is not part of the explanandum; orderliness and functional relationships are. Contrast orderliness among the normal beings with disorder found in monstrosities. The difference is explained teleologically. The inference is something like the following:

O (nature): Orderly developmental patterns occur by nature.
O (nature)′: Disorderly developmental patterns lead to mistakes or monstrosities.

[3] The function/accident distinction is crucial for modern-day teleology.

Analogously,

O (artifact): Orderly creative procedures lead to functional artifacts.
O (artifact)′: Disorderly creative procedures lead to mistakes.
H (materialist): All phenomena are explained according to the same materialistic principles. There is no essential difference in their explanation.
H (teleologist): Orderly patterns occur *for the sake of* the form while monstrosities do not.

The teleological explanation better explains what distinguishes O from O′. According to Aristotle, the materialists cannot explain what goes wrong when mistakes occur or what goes right when developed or created things work.

So far I have spent much of this essay explicating Aristotle's teleology arguments and distinguishing between Aristotle's 'localized' teleology from Plato's 'global', divine agent-centered teleology. After Aristotle, a pattern in teleological arguments emerges: a variation of one of Aristotle's three types of teleological argument is put forward in support of the existence of a Platonic divine agent. Aquinas exemplifies the melding of two teleologies whereby regularity of pattern is offered as evidence of design. As Ron Amundson so aptly puts it, 'In Aquinas's time it was easy to move from *always acts the same* to *acts for an end*, and thence to *achieves the best result*' (Amundson 1996: 16). The distinguishing Aristotelian feature is the move from 'always acts the same' to 'acts for an end'. The extra inference is Platonic and explains why the end 'achieves the best result'. Later on we see this same pattern in teleological arguments from Newton, Whewell, Paley, Kant, and Darwin.

Commentators have failed to appreciate this pattern in their interpretation of post-Aristotelian teleological arguments for many reasons. First, they often interpret what I have been calling 'teleological arguments' as 'arguments *from* design'. The latter argument infers the existence of a creator to explain purpose in nature. While I do not doubt that such arguments have been offered in history, we should recognize that the inference from purpose to agent is poor. Teleological explanations are supposed to be contrasted with material explanations. A materialist thinks that there are no purposes in nature to explain. So, an inference *from purpose* begs the question that purpose exists and requires an explanation. None of Aristotle's three arguments, the argument from flourishing, the argument from regularity, and the argument from pattern, begs the question against the materialist. In each, Aristotle infers teleology from the relevant factor, flourishing, regularity in development, or patterns of order and arrangement. For one example (to refresh our memories), Aristotle argues that dental arrangements are *useful*, and hence grow for the sake of their usefulness, because those carnivores that possess the particular dental arrangement tend to prosper, while those that do not, die. I hope to demonstrate that the more careful teleologists follow Aristotle's lead.

A second reason why commentators fail to appreciate the sophistication of teleological arguments is that some appear to hold that the key feature of a teleological argument is the analogy between human artifacts and natural 'designs' (e.g. Hurlbutt 1965: 14). As we have seen, such analogies are neither a distinguishing feature—there are other sorts of teleological arguments—nor, in the case that such arguments are presented, primarily an argument from analogy. Aristotle's argument that features the analogy is an inference to the best explanation and not an argument from analogy. In arguments from analogy, a feature ascribed to a target subject is ascribed to an analog. The strength of the analogy depends on the degree to which the analog *resembles* the target. For example, we might think that, since dog biology resembles human biology, and since humans have a circulatory system, dogs do too. We do not evaluate an inference to the best explanation in the same way. There is no comparison between targets and analogs. Instead, an inference to the best explanation begins with an observation and considers which hypothesis offered might explain the observation. Again, Aristotle's three arguments for teleology feature the inference to the best explanation schema.

Finally, many commentators dismiss Aristotelian teleology as it is purported to ascribe fishy vital forces or bizarre backwards causation to nature. However, I think this is the biggest misreading of Aristotle. First, as I mentioned above, there are good reasons to think that Aristotle's final aitia are not *causes* but *reasons* or *explanations.* Of course, on this reading, there is an open question whether Aristotle thought that his final aitia corresponded to irreducible ontological properties of the world above that of material causes, or whether he viewed them as useful forms of explanations.[4] Nevertheless, even if one holds that final *aitia* are ontologically irreducible to material causes, it does not follow that these irreducible properties are forward-'looking', intentional 'vital forces'. A 'vital force' is a force that drives a causal process. Picking one out would be to pick out the source of motion or developmental change. However, in Aristotle's account of explanation, to attribute this role to final *aitia* would be to collapse the distinction between final *aitia* and *efficient causes* (Gotthelf and Lennox 1987: 201). It is the latter, *causal aitia* that pick out the source of change.

[4] I follow many commentators in thinking that final aitia do pick out an ontological category distinguished from material causation. Aristotle most likely would have thought that human intentionality was not reducible to material causes. And, likewise, organic development (growth) is irreducible to causal laws of motion.

3. Cosmological Teleology

3.1. Newton

Centuries later Aristotle's teleological arguments reappear in inferences to explain the order that govern the motions in the cosmos. Ironically, Isaac Newton, who is best known for his mechanistic physics, employs an Aristotelian teleology inference. Truth is, Newton was not a thoroughgoing proponent of a mechanical universe. In a letter to Richard Bentley, Newton lists a number of questions that he thinks the mechanical sciences *cannot* answer, including:

> What is there in places almost empty of matter, and whence is it that the sun and planets gravitate towards one another, without dense matter between them? To what end are comets; and whence is it that planets move all in one the same way in orbs concentrick, while comets move all manner of ways in orbs very excentrick; and what hinders the fixed stars from falling upon one another? How came the bodies of animals to be contrived with so much art, and for what ends were their several parts?' (*Opera Omnia*, IV. 237)

Newton presents evidence for teleology in both the motions of the solar system—'cosmological teleology'—and in the adaptability of living organisms to their environments—'biological teleology'.

In a revealing passage, Newton remarks on the ontology of gravity: 'Gravity must be caused by an agent acting constantly according to certain laws; but whether this agent be material or immaterial, I have left to the consideration of my readers' (Amundson 1996: 15). This clearly leaves room for teleology in Newton's cosmology. But so far these passages are negative; they state the limitations of mechanical sciences (Hurlbutt 1965: 7). A hint of a positive teleological argument comes later in Newton's letter to Bentley:

> To make this system, therefore, with all its motions, required a cause which understood, and compared together, the quantities of matter in the several bodies of the sun and planets, and the gravitating powers resulting from thence; the several distances of the primary planets from the sun, and of the secondary ones from *Saturn, Jupiter*, and the Earth; and the velocities, with which these planets could revolve about those quantities of matter in the central bodies; and to compare and adjust all these things together in so great and variety of bodies, argues that cause to be not blind and fortuitous, but very well skilled in mechanicks and geometry'. (*Opera Omnia*, IV. 431–2; quoted in Hurlbutt 1965: 7)

Newton's argument is similar to Aristotle's inference from pattern. Accordingly, the stable motions of the heavens depend on a singular arrangement of planet sizes, distances, number, and position. Implied here is that, if the system were arranged haphazardly—that is, by blind chance—its balance would have been compromised. Hence blind and fortuitous causes do not explain the origins of the universe's stable motions. Rather, the delicate balance we see in the solar system suggests a creative origin: an act of intelligent design.

O: The solar system exhibits a balanced arrangement of variously sized planets.

O′ (counterfactual): Had the arrangement been haphazard, the balance would not exist.

The best inference is teleological: the arrangements exist for the sake of the balance. However, as we see, Newton goes a step further and postulates the existence of a skilled designer. So, while Newton's argument resembles Aristotle's, the conclusion is Plato's. The telos is the intention of a skilled designer.

Newton knew that the harmony and stability of the solar system have exceptions in the orbital speeds of Jupiter and Saturn. Jupiter's speed was accelerating while Saturn's was decelerating. Newton argued in the *Optics* that the solar system would fall apart, the stability compromised, unless the orbital speeds of Jupiter and Saturn were adjusted. Perhaps, Newton hypothesized (despite the fact that Newton famously despised hypotheses), comets played the adjustment role (Amundson 1996: 18). If so, the eccentric motion of the comets would be explained: they restore stability in the solar system. So, there is an interesting difference between Plato's demiurge and Newton's divine creator. While Plato's demiurge created in a single act, Newton's God intervened with its creation.

3.2. The Death Knell of Cosmological Teleology

LaPlace eventually solved the problem of the exceptional orbits of Saturn and Jupiter, demonstrating that the orbits would eventually reverse, creating an oscillation that is stable in the long run (Amundson 1996: 20). Consequentially, there was less of a motivation to think that the eccentric orbits of the comets had the purpose to adjust the solar system, since the exceptional orbits of Saturn and Jupiter were self-correcting. Worse for Newton's teleology, LaPlace put forward the hypothesis that the solar system coalesced from nubular clouds. If correct, this would explain the origins of the solar system without reference to an intelligent designer.

Yet, cosmological teleology dies hard. As Whewell argued, LaPlace's Nebular Hypothesis for the origins of the solar system merely forced the issue of origins back a step. Accordingly, we are left with an open question of what accounts for the laws that govern the coalescing of nubulae. This opens the door again for teleology: 'What but design and intelligence prepared and tempered this previously existing element, so that it should by its natural changes produce such an orderly system?' (Whewell 1836; quoted in Amundson 1996: 21).

Spinoza despised such arguments, calling them arguments *ad ignorantiam*. Underhill (1904: 224) captures the spirit of Spinoza's disdain nicely:

a tile falls from a roof on a man's head and kills him: the tile, they argue, must have fallen on purpose to kill him. Otherwise, if it had not been God's will, how could all the cicumstances have concurred just then and there? You may answer: It happened because the wind blew and the man was passing that way. They will urge—Why did the wind blow and why did the man pass that way just at that time? If you suggest fresh reasons, they will ask similar questions, because there is no end of such questioning, until you take refuge in that *ignorantiae asylum*, the will of God'.

As a consequence of LaPlace's work, the popularity of cosmological teleology waned while the popularity of biological teleology waxed. The cosmos lacked a means/end patterning from which teleology could be inferred. Recall Aristotle's argument from pattern whereby evidence for teleological principles is found in particular orderly arrangements of developmental phenomena such as in organic development of adapted organisms or the creation of human artifacts. The last gasp of cosmological teleology seized on that pattern in the correlations between organic cycles and astronomical time period (Amundson 1996: 21). Just as Aristotle considered thousands of years before, Whewell argued that the correspondence between the solar year and the vegetative growth cycle suggested, not chance, but 'intentional adjustment' (Whewell 1836: 26).

4. Biological Teleology

The remaining figures we will consider, Kant, Paley, and Darwin, apply teleology to biological explanations as opposed to cosmological explanations. Again, the Aristotelian influences on these figures are striking. Of the three, only Paley will endorse a Platonic *telos*.

4.1. Kant

Kant distinguishes two sorts of causation, mechanical and teleological. Mechanical causes exhibit a progressive series of causes preceding their effects. Teleological causes exhibit both a progressive and a *regressive* series of causal chains whereby effects *both* precede and proceed from their causes. An effect can be the *cause* of its preceding cause. Regressive cause-and-effect chains are most clearly represented in purposive human behavior. For example, the existence of a house is the cause of rental income, yet the 'representation' of the income is the cause of building the house in the first place (Butts 1990: 5).

Kant concludes (a sketch of the argument is below) that the processes of nature can *be understood only* teleologically. Interestingly, the *telos* Kant ascribes to nature is meant to be distinct from the *telos* ascribed to human purposive behavior. Kant is being careful to avoid the Platonic conclusion that natural processes serve useful ends as evaluated from 'on high'. Rather, natural *telos* is immanent, *in rerum natura*, very much in the mold of Aristotle's natural teleology. So, when final causes are ascribed to human behavior, they refer to 'utility'—as in iron is useful to shipbuilding; when final causes are ascribed to natural processes, they refer to 'internal', biological ends.

What is Kant's argument for the existence of these biological internal ends? Kant's answer is consistent with his general epistemology: to understand nature we must view it 'as if' nature is rational and acts for practical ends. That is to say not that nature *is* rational but that nature acts as a rational analog to a living being (Butts 1990: 7).

I take Kant's argument so far to be similar to half of Aristotle's argument from pattern: the same pattern that explains organizations found in human activity is the same pattern that explains organization in nature. I say this is 'half' of Aristotle's argument from pattern because Aristotle's argument works both ways, 'as in art, so in nature' and 'as in nature, so in art'. The pattern Kant sees 'as in art, so in nature' is the progressive and regressive causal series. Where in nature is that pattern evident? The answer: in *self-preserving activity*. Kant considers three ways in which a tree may be 'regarded as an end to itself or internal end' (quoted in Underhill 1904: 226).

1. *Phylogenetic.* Reproduction begets organisms that resemble a generic kind (i.e. species). The kind is both the effect of continued generic existence and the cause of reproduction.
2. *Individual growth.* Growth is more than increase in size according to mechanical laws, for individuals deviate from their generic form to secure their own self-preservation under particular circumstances. This leads to originality in individual design unequalled in art.
3. *Functional part/whole relations.* Parts of animals form in a way that the maintenance of any one part depends reciprocally on the maintenance of the rest.

Note, Kant's conclusion is stronger than the argument from pattern presented above. According to Kant, we must *necessarily* think of nature as designed. That is what Kant means when he remarks that it would be absurd to expect that 'another Newton will arise in the future who will make even the production of a blade of grass understandable by us according to natural laws which no design has ordered' (quoted in Beatty 1990: 54). (I find the reference to Newton ironic, given our discussion above.) The remaining steps of Kant's argument, I think, are unAristotelian hence beyond the scope of this chapter.

My point has been to point out the Aristotelian kernel of Kant's biology. First, his distinction between external and internal ends reflects nicely

Aristotle's own distinction between what I called 'Agent-centered' and 'Natural' teleology. Second, Kant's argument for ascribing *telos* in nature resembles Aristotle's argument from pattern. Finally, reflect on Kant's remarks on growth (above). Kant recognized that mechanical principles are necessary to understand some parts of animal formation but mechanical principles *alone* cannot explain the individuality of growth. We explain the latter by reference to the self-preserving (teleological) activities of an individual organism. Aristotle would have been proud.

4.2. Paley and Darwin

Darwin is often thought to have brought the demise of teleological thinking in biology (Ghiselin 1969; Mayr 1988). But, since the concept of telos is so packed with different meanings, it is unclear what sort of teleology Darwin's theory of evolution by natural selection rejected. Darwin himself unabashedly utilized the concept of a final cause in his *Species Notebooks* and even in the *Origin of Species* itself (Lennox 1993: 410–11). Elsewhere, in an illuminating exchange between Asa Gray and Darwin, Gray commented on 'Darwin's great service to Natural Science in bringing back to it Teleology: so that instead of Morphology versus Teleology, we shall have Morphology wedded to Teleology' (quoted in Lennox 1993: 409). In response, Darwin wrote, 'What you say about Teleology pleases me especially and I do not think anyone else has ever noticed the point' (quoted in Lennox 1993: 409). The issue here is to what extent did Darwin reject teleology and to what extent did he support it? The key is to distinguish clearly between what I have been calling 'Platonic' and 'Aristotelian' teleology. A Platonic *telos* is an agent that operates or creates purposively for the sake of the best.[5] An Aristotelian telos is a property of an individual's functioning—its contribution to its own sustainability. One way to interpret the significance of Darwin's theory to teleology is to view Darwin's theory as a rejection of Platonic agency as the cause of natural phenomena. Rather than appealing to a divine creator with a good plan, Darwin appealed to facts about nature. This interpretation arises when we view Darwin's theory as an answer to, in particular, William Paley's argument from design. As we shall see, both Paley's argument and Darwin's response are plausibly seen as applications of Aristotle's teleology.

4.3. Paley

William Paley (1828) asks us to consider what we would infer about the presence of a watch lying on a heath. How did the watch come to exist? Had we

[5] Kurt von Baer thought Platonic teleology was misleading and even 'silly'. He blamed the association of Platonic teleology (he called it 'theological teleology') for much of the 'teleophobia'.

found a stone rather than a watch, it would suffice to infer that the stone had lain on the heath forever. However, that answer is not applicable to the watch, for watches, as opposed to stones, exhibit a particular organization, a singular order in the way their component parts are put together such that the hands move in accordance to time:

For this reason, and for no other, viz. That, when we come to inspect the watch, we perceive (what we could not discover in the stone) that its several parts are framed and put together for a purpose, e.g. that they are so formed and adjusted as to produce motion, and that motion so regulated as to point out the hour of the day . . . (Paley 1828: 5)

If the watch were composed in any other manner, had its parts been shaped or sized differently, the hands would not move in the same way (or not at all). Many commentators ignore this Aristotelian component of Paley's argument—the contrast between the functioning and malfunctioning watch depends on the arrangements of the parts of the object:

that if the different parts had been differently shaped from what they are, of a different size from what they are, or placed after any other manner, or in any other order, than that in which they are placed, either no motion at all would have been carried on in the machine, or none which would have answered the use that is now served by it. (Paley 1828: 5)

Paley infers purpose in the watch's intricate composition. This purpose is the rational intention of a creator.[6] Schematically, Paley's inference is as follows:

O: A particular assemblage produces motion.

O′: Deviations of the assemblage results in no motion.

H (designer): The assemblage is purposeful (put together by a designer).

It is important to note that Paley's inference does not depend on prior observations of watches being made by watchmakers.[7] Rather, Paley infers the existence of an intentional designer from the watch's complexity, arrangement, intricacies, and well-suitedness to the completion of certain tasks. The key point, what we will call 'the inferential step', is that certain patterns in artifacts suggest design and the existence of a designer. The pattern is exhibited when

[6] I realize that the passage above, where Paley writes, 'we *perceive* . . . that its several parts are framed and put together for a *purpose* . . .', suggests a different reading from the one I offered—namely, that Paley observes a purpose rather than inferring one. However, I think Paley's supporting examples are meant to be taken as evidence for what Paley 'perceives'. If I am wrong, that is, if the proper way to read the passage is that Paley infers a creator *from* the purpose he perceives rather than from the pattern of assemblage, then Paley's argument begs the question against the materialist who 'perceives' no purpose in nature.

[7] Paley (1828: 42) writes: 'Nor would it, I apprehend, weaken the conclusion, that we had never seen a watch made—that we had never known an artist capable of making one . . . Ignorance of this kind exalts our opinion of the unseen and unknown artist's skill, if he be unseen and unknown, but raises no doubt in our minds of the existence and agency of such an artist, at some former time and in some place or other.'

the artifact's effect requires a particular arrangement or order in its parts. Had the artifact exhibited any other order, it would probably not have produced its wondrous effects. As Paley (1828: 10) writes, 'Arrangement, disposition of parts, subserviency of means to an end, relation of instruments to a use, imply the presence of intelligence and mind'.

Next, Paley applies the same inference to an object with extraordinary complexity, a watch that, in addition to ability to track time, is capable of self-replication. Now, the parts are more complex, and the order of parts more crucial. The most obvious inference is, according to Paley, a designer with extraordinary abilities. Let us call this additional step in the argument, the 'increasing order of complexity': the more complex the organization, the more complex the design, the more cunning the creator.

Finally, Paley applies both the 'inferential step' and the 'increasing order of complexity' to account for living things. As with the existence of the watch and the self-replicating watch, creatures and organs of the natural world demonstrate a super-extraordinary complexity, order, and arrangement. The only rational inference is a designer of sufficient intelligence and purpose. That designer must be God, according to Paley. Much of the rest of Paley's book is an ode to the complexity and intricacies found in nature.

Many commentators take Paley to infer a creator *from purpose found in objects.* Paley writes, above, that we perceive that parts of the watch are put together for a purpose. I do not read the inference the same way. If it were so, then the argument would be question begging—presupposing purpose to infer teleology. Rather, I take the inference to be similar to Aristotle's analogy between artifacts and nature (Sober 2000). Paley's 'inferential step' is similar to Aristotle's analogy between artifacts and nature. Recall, Aristotle inferred teleology from patterns of order and arrangement. The end states depend on previous parts in an appropriate position to contribute to the whole. If these arrangements are not present, either development shuts down, or the organization fails to produce a particular effect ('mistakes'). Paley's teleological inference runs the same way, from particular order and arrangements for both nature and artificial contrivances.

Yet, Paley's *telos* is an agent, a designer with intentions to create the *best* possible world while Aristotle's is immanent and relative—relative to organisms in their surroundings. Paley's goodness is global. Purpose is in the good intentions of a creator that has created the best possible cosmos. Paley's *telos* is clearly in the mold of Plato's demiurge. Since Paley's teleology is much stronger than Aristotle's both in the concept of striving for the best and in globalizing the perspective, Paley's inference requires the additional inference, 'the increasing order of complexity'. While Aristotle's inference recognizes a pattern in both art and nature, Paley's inference recognizes a pattern in art that is more exquisite in nature.

4.4. Darwin

Perhaps Darwin's answer to Paley's design argument was to demonstrate how the good designs in nature could be explained differently from the good designs of artifacts; replacing a Platonic creator with the blind forces of natural selection. Richard Dawkins (1986: 5) has popularized this reading of Darwin's contribution to biology:

> A true watchmaker has foresight: he designs his cogs and springs, and plans their interconnections, with a future purpose in his mind's eye. Natural selection, the blind, unconscious, automatic process which Darwin discovered, and which we now know is the explanation for the existence and apparently puposeful form of all life, has no purpose in mind. It has no mind and no mind's eye. It does not plan for the future . . . If it can be said to play the role of watchmaker in nature, it is the *blind* watchmaker.

On Dawkins's reading the *explanandum* is the same for both Paley and Darwin: the existence of highly complex and intricate creatures well suited for the task of reproduction. Darwin himself invites this reading of the explanatory role the theory of natural selection plays, since much of the *Origin* compares natural selection with artificial selection.

On this approach, Darwin's task is steep. He has to explain how natural forces could conspire to assemble products displaying intricate and complex orders and arrangements that are so well suited to the environmental conditions. Recall that most important for Paley's creator is that its intentions and powers for creation are for the *best* in a global sense. Darwin then needs to demonstrate how a non-intentional physical force (or set of forces) could produce creations that match the global standard. Some historians think that Darwin's theory succeeds: 'Here we have nature selecting, in that we have a deliberate metaphor that has nature doing what man familiarly does, but doing it much better' (Hodge 1991: 214). But there are two problems with this approach. First, by regarding the living world as full of good designs, Darwin's theory is not clearly a better explanation for their existence than the creation theory. While a Darwin supporter might succeed in showing how certain natural processes could produce good designs, there always remains a nagging doubt as to whether the blind forces of natural selection could produce so many different perfections. Second, and more importantly, this interpretation ignores one of two components of Darwin's evolutionary theory, the 'tree-of-life' hypothesis. Darwin viewed all species as sharing a history, all evolving from a single common ancestor. Darwin's theory of natural selection, the second component, explains how species evolve from ancestral species; how modifications lead new species to branch out of old ones.

The proponent of Paley's natural theology most clearly opposes Darwin's tree of life and hence sees no motivation for the theory of natural selection.

Accordingly, creationists view each species as the unique creation of an all good God and thereafter immutable and eternal. How could Darwin demonstrate the superiority of his tree-of-life hypothesis? Should Darwin infer evolution from the same perfections and intricacies that Paley viewed as evidence for God's handiwork? No. As S. J. Gould (1980: 20) puts it, 'ideal design is a lousy argument for evolution for it mimics the postulated action of an omnipotent creator. Odd arrangements and funny solutions are proof of evolution—paths that a sensible God would never tread but that a natural process, constrained by history, follows perforce'. Darwin's argument against a creator and for a non-intentional force of nature is found in the awkwardness of developmental patterns, and the seemingly poor designs of nature. Baleen whales develop teeth in neotony only for them to be reabsorbed into the baleen structure that they use to feed on krill. Why would an omnibenevolent God bother to allow whales to develop teeth that won't be used later in life? Pandas get at the tender shoots of bamboo through the inefficient process of running the stalks along an inflexible spur of bone that juts out like a thumb. Why did God give pandas this clumsy design feature?

Paley (1828: 6–7) argued that design is evident in mishaps as well, for the purpose is clear even if the system does not achieve it. However, Paley is referring to instances of failed development—for example, deformed individuals. Darwin's mishaps are flaws of type—'design' flaws from a creator's point of view. Darwin writes: 'Rudimentary organs may be compared with the letters in a word, still retained in the spelling, but become useless in the pronunciation, but which serve as a clue in seeking for its derivation' (quoted in Gould 1980: 27). To illustrate (something close to) Darwin's language example, the fact that Spanish, French, and Italian assign similar names to numbers is evidence that the words did not arise *de novo* for each language (Sober 1993: 42). See Table 1.1 for an example.

Given the data, compare a 'creationist' hypothesis with an 'evolutionary' one:

H (creationist): each language is the result of an *independent* act of creation by a wise creator.

Table 1.1. Similarities between languages for words indicating numbers

Number	French	Italian	Spanish
1	*un*	*uno*	*uno*
2	*deux*	*due*	*dos*
3	*trois*	*tre*	*tres*
4	*quatre*	*quattro*	*cuatro*

H (evolutionist): the different languages are derived from modification of a common language.

In this case, the evolutionary hypothesis is clearly the better inference from the data.

Analogously, the reabsorbtion of a whale's teeth in its mother's womb is evidence that whale development is not a separate act of creation but survives as a remnant and modification (by natural selection) of an ancestral developmental pattern. In other words,

H (evolutionist): Organic traits are derived and modified from the traits of their ancestors.

better explains the evidence from 'poor' design than does:

H (creationist): Each species is the result of an independent act of creation.

Let us take stock of the importance of Darwin's answer to Paley's argument to the issue of teleology. Darwin, in gathering evidence for his 'tree-of-life' hypothesis, debunks Paley's Platonic teleology whereby organic traits are intentional designs of a supreme creator. However, by debunking Platonic teleology, it does not follow that Darwin has debunked natural teleology altogether. Platonic teleology is only one sort, Aristotelian teleology is an entirely different sort. Evidence from vestigial and 'odd arrangements' suggests that organic traits are not derived from a purposeful act of creation but rather organic traits are derived and modified from the traits of their ancestors through natural selection. That is, Darwin replaces the hand of creation with a non-intentional 'force', natural selection.

How is natural selection a teleological 'force'? I see remnants of two sorts of teleology operating in Darwin. The key to seeing both is within Darwin's concept of natural selection, which can be summed up as follows: as a result of individuals possessing different heritable abilities striving to survive and reproduce in local environments, comes an explanation for changes in trait composition of populations through time. Traits become prevalent in populations because they are useful to organisms in their struggle to survive. Aristotle's *functional* teleology is preserved through the idea that an item's existence can be explained in terms of its *usefulness* (Lennox 1993). What makes a trait *useful* is that it provides certain individuals an advantage over others in their own struggle to survive and reproduce. Secondly, the concept of individual striving to survive and reproduce plays the fundamental role in Darwin's explanation for the origins of organic diversity. The same concept reminds us of Aristotle's *formal* teleology—the striving for self-preservation. Usefulness is not a global valuation, a 'for the best' in Plato's sense, but an immanent feature of the relation between developing organism and their local environmental conditions (including their competitors). Traits that allow the organisms possessing them to be 'better suited' to survive the struggle will be better represented in future populations. Likewise, Aristotle's 'usefulness' is a property of the individual's relation to the local environmental conditions. Recall the example: sharp teeth are in front for the sake of tearing. Sharp teeth contribute to the flourishing of organisms possessing them, whereby the flourishing depends on the carnivore's local environment.

There are significant differences between Aristotle's formal teleology and Darwin's. Compare Darwin's view of the *source* of trait variations that organisms come to possess with Aristotle's idea that the origin of traits exist for the sake of the flourishing of organisms possessing them. In Darwin's view, variants arise by 'chance'—that is, variants develop *independently* from any relation to the environment. Darwin's theory of the source of variation is distinctly unAristotlelian. In Aristotle's view, traits develop *for the sake* of the individual's self-preservation. In fact, Karl von Baer critiques Darwin on this very point (Lenoir 1982: 270). According to this critique, if 'blind necessity' is the only force operating, then the fundamental questions of biology—development, adaptation, and the like—will remain unintelligible. An explanation that strings together mechanical processes lacks the fundamental principle that connects the processes to a particular end (Lenoir 1982: 271). I interpret von Baer's criticism to be close to Aristotle's argument from regularity: the materialist lacks the principle that distinguishes one material process from any other. Consequentially, what distinguishes developmental processes that lead to living newborn from one that fails?

Another difference between Aristotle's and Darwin's teleology concerns Aristotle's concept of hypothetical necessity. Recall that, for Aristotle, an item's usefulness *constrains* the necessity of the materials. That is, *because* eyes are useful for seeing, the organic ingredients coalesce. The need to see necessitated the existence of eye materials (fluid, lids, and so on). For Darwin, this is exactly backwards: the materials constrain function. Natural selection operates on the materials (the variants) that are available to it. That's why pandas possess such an awkward mechanism for manipulating bamboo shoots. The panda's thumb is a modification of the enlarged radial sesamoid that the ancestors of pandas and its cousin species (bears and raccoons) possessed (Gould 1980: 23). The panda's thumb is a 'contraption', modified from the anatomy of what was available for selection to operate upon.

This last point, I think, begins to explain Asa Gray's remark (which I quote again): 'Darwin's great service to Natural Science in bringing back to it Teleology: so that instead of Morphology versus Teleology, we shall have Morphology wedded to Teleology' (quoted in Lennox 1993: 409; see also Amundson 1996: 32). The reference to 'Morphology' refers to a school of thought that advanced a 'unity-of-plan' theory of organic diversity. Accordingly, members of a taxonomic group are accounted for in terms of resemblances between members of the same and other taxonomic groups. Traits that resemble each other across taxonomic groups are called 'homologues' and indicate a 'common plan' throughout nature.

Morphologists thought that picking out homologous structures constituted picking out essential categories in nature. That is, the existence of homologous structures indicates the fundamental laws of body plans.[8] However, Darwin wondered how to explain the prevalence of variants to the 'common plans'? To this he invokes natural selection. Natural selection operates over pre-existing structures competing for limited resources in a common environment. So, while structures pre-exist their adaptive uses, it is the process that produces adaptations that explain morphological change. (Note, mutations, migrations, genetic recombination, all explain the existence of variants to the common plan, but it is natural selection that makes some of these variants prevalent in certain populations.)

5. Conclusion

When we see appeals to teleology in science, it is crucial to identify what *kind* of teleology, and which kind of argument for it. While many scientists and philosophers of science have rightly rejected the Platonic *telos* with its arcane metaphysical trappings, other teleology in science is not wedded to such metaphysics. If biology has an ineliminable teleology, this is not so bad as long as it is one of the more restrained Aristotelian versions of teleology.

References

Amundson, Ron (1996), 'Historical Development of the Concept of Adaptation', in Michael Rose and George V. Lander (eds.), *Adaptation* (Boston: Academic Press, 11–51.

Annas, J. (1982), 'Aristotle on Inefficient Causes', *Philosophical Quarterly*, 32: 311–26.

Ayala, F. J. (1970), 'Teleological Explanations in Evolutionary Biology', *Philosophy of Science*, 37: 1–15.

Balme, D. M. (1987), 'Teleology and Necessity', in A. Gotthelf and J. Lennox (eds.), *Philosophical Issues in Aristotle's Biology* (Cambridge: Cambridge University Press), 9–20.

Beatty, John (1990), 'Teleology and the Relationship between Biology and the Physical Sciences in the Nineteenth and Twentieth Centuries', in F. Durham and R. D. Purrington (eds.), *Some Truer Method: Reflections on the Heritage of Newton* (New York: Columbia University Press), 113–44.

[8] Some modern-day morphologists still hold this view for quite persuasive reasons. See the work of Goodwin.

Butts, Robert E. (1990), 'Teleology and Scientific Method in Kant's Critique of Pure Reason', *Nous*, 24: 1–16.

Charles, David (1988), 'Aristotle on Hypothetical Necessity and Irreducibility', *Pacific Philosophical Quarterly*, 69: 1–53.

——(1995), 'Teleological Causation in the *Physics*', in Lindsay Judson (ed.), *Aristotle's 'Physics': A Collection of Essays* (Oxford: Oxford University Press).

Cooper, John M. (1987), 'Hypothetical Necessity and Natural Teleology', in A. Gotthelf and J. Lennox (eds.), *Philosophical Issues in Aristotle's Biology* (Cambridge: Cambridge University Press), 243–74.

Dawkins, Richard (1986), *The Blind Watchmaker (New York:* W. W. Norton & Co.).

Furley, David (1985), 'The Rainfall Example in the "Physics II.8"', in Allan Gotthelf (ed.), *Aristotle on Nature and Living Things* (Pittsburgh: Mathesis Publications Inc.).

Gayon, Jean (1998), *Darwinism's Struggle for Survival* (Cambridge: Cambridge University Press).

Ghiselin, Michael (1969), *The Triumph of the Darwinian Method* (Berkeley and Los Angeles: University of California Press).

Gotthelf, Allan (1987), 'Aristotle's Conception of Final Causality', in A. Gotthelf and J. Lennox (eds.), *Philosophical Issues in Aristotle's Biology* (Cambridge: Cambridge University Press), 204–22.

——and Lennox, J. (1987) (eds.), *Philosophical Issues in Aristotle's Biology* (Cambridge: Cambridge University Press).

Gould, Stephen Jay (1980), *Panda's Thumb: More Reflections in Natural History* (New York: W. W. Norton & Co.).

Hodge, J. (1991), 'Darwin, Whewell and Natural Selection', *Biology and Philosophy*, 6: 457–60.

Hurlbutt, R. H. (1965), *Hume, Newton, and the Design Argument* (Lincoln, Neb.: University of Nebraska Press).

Lennox, James G. (1985), 'Plato's Unnatural Teleology' in Dominique O'Meara (ed.), *Platonic Investigations* (Washington: Catholic University of America Press; repr. in James G. Lennox, *Aristotle's Philosophy of Biology: Studies in the Origins of Life Science* (Studies in Biology and Philosophy series) (Cambridge: Cambridge University Press), Ch. 13.

——(1992), 'Teleology' in Evelyn Fox Keller and Elizabeth A. Lloyd (eds.), *Keywords in Evolutionary Biology* (Cambridge, Mass.: Harvard University Press), 324–33.

——(1993), 'Darwin *was* a Teleologist', *Biology and Philosophy*, 8: 409–21.

Lenoir, Timothy (1982), *The Strategy of Life: Teleology and Mechanics in Nineteenth Century German Biology* (Chicago: University of Chicago Press).

Mayr, Ernst (1988), *Toward a New Philosophy of Biology* (Cambridge, Mass.: Harvard University Press).

Morrow, G. (1950), 'Necessity and Persuasion in Plato's *Timaeus*', *Philosophical Review*, 59: 147–63.

Paley, W. (1828), *Natural Theology*, 2nd edn. (Oxford: J. Vincent).

Ruse, M. (1979), *The Darwinian Revolution: Science Red in Tooth and Claw* (Chicago: University of Chicago Press).

Sober, Elliott (1993), *Core Questions in Philosophy: A Text with Readings* (Upper Saddle River, NJ: Prentice Hall).

——(2000), *Philosophy of Biology*, 2nd. edn. (Boulder, Colo.: Westview Press).

Strange, Steven K. (1985), 'The Double Explanation in the *Timaeus*', *Ancient Philosophy*, 5: 25–39.

Underhill, G. E. (1904), 'The Use and Abuse of Final Causes', *Mind*, 13/50: 220–41.

Whewell, W. (1836), *Astronomy and General Physics, Considered with Reference to Natural Theology* (Bridgewater Treatise 3), a new edn. (Philadelphia: Carey, Lea & Blanchard).

Woodfield, Andrew (1976), *Teleology* (Cambridge: Cambridge University Press).

Zeyl, D. (2000), *Timeaus* (Cambridge, Mass.: Hackett Publishing Company).

2. Evolutionary Biology and Teleological Thinking

MICHAEL RUSE

ABSTRACT

Teleological thinking and language are things that fully permeate evolutionary biological science, even though they are absent from the physical sciences. Through an analysis looking back at the history of biology, I find the source of the difference to lie in the way(s) in which evolutionary biology, unlike the physical sciences, resembles the world of humans and their artefacts. I defend evolutionary biology from a number of criticisms suggesting that this teleology is a sign of weakness, arguing rather that the evolutionist faces problems different from the physical scientist and that teleological thinking is important and powerful.

There is something distinctive about biological language, particularly evolutionary biological language. There one finds talk of 'purposes' or 'functions' or 'ends' in a way missing from the physical sciences. In biology, it makes perfectly good sense to ask a question like: 'What function does the sail on the back of Dimetrodon serve?' Or: 'What is the purpose of the appendix in the human being?' Or: 'Do vertebrates have sex in order to spread new mutations?' None of this forward-looking language, commonly known as 'teleological' language—language where one is trying to understand the present or the past in terms of the future—seems generally appropriate in the physico-chemical sciences. No one would ask what 'purpose' or 'end' the planet Mars serves. Nor would one say that hydrogen combines with oxygen 'in order to' make water. Hydrogen does indeed combine with oxygen to make water, but it does not do so 'in order to make' water. Nor would one say that the 'function' of the sliding continental plates is to create earthquakes and volcanoes. Sliding plates do create earthquakes, but that is not their function. They do not have a function.

In this chapter, it is my intention to look at the teleological thinking of evolutionary biology, to ask about its nature. I want to see why teleological language seems appropriate in evolutionary biology, but not in the physico-chemical sciences; to ask whether in some sense such language is eliminable, and if indeed any such elimination would be desirable; and to find out if, in some sense, evolutionary biology is fundamentally and interestingly different from the physico-chemical sciences—perhaps superior, perhaps inferior—precisely because it relies on such a way of thinking. (Good background reading includes Nissen 1997; Allen *et al.* 1998; Buller 1999).

1. Explications

Teleological language is not peculiar to evolution and biology. It occurs elsewhere, most particularly in the context of human beings: that is to say, human beings *qua* human beings, rather than human beings *qua* products of evolutionary processes. Here, in the context of human beings, we find two main kinds of teleology. The first is that which centres on individual humans themselves. The second is that which centres or focuses on humans as a whole taken over history: humans considered over long periods of time. It is the first kind of teleology, that focusing on individual humans, that will be the main point of my discussion; although I shall make a few remarks about the second, more historically oriented kind of teleology. With respect to the first kind of teleology, again one finds that there is a division: first, between the teleology of humans themselves, and then, second, the teleology of the objects or artefacts created by humans. Let us take these two categories in turn.

By the teleology of humans themselves, I mean the intentionality and subsequent actions of humans, as they strive to achieve goals that they have set or desire (Dennett 1987). Let us take a student who wants to be a professor of philosophy. He or she realizes that, in order to gain such a position or status, it is necessary to have a Ph.D. degree. The student therefore sends off for university graduate catalogues and makes an application and so forth. Then, if qualified, he or she gains acceptance to a department, does the required coursework, examinations, and thesis, and achieves the Ph.D. I take it that this is a teleological state of affairs: the student's behaviour, now or in the past, is being guided by the future achievement of the gaining of the Ph.D.

Or, to be a little more precise, by the thoughts or intentions of gaining the Ph.D. in the future. The reason and need for such precision are simple. If one says merely that the future-gaining-of-the-Ph.D. is in fact influencing the student-now-writing-for-university-catalogues, one is committing oneself to causes in the future affecting events in the present. Unfortunately, this runs into some classic problems, notably that of the 'missing-goal object' (Mackie 1966). What happens if the student fails the thesis oral and never gets the Ph.D.? This means that the non-existent Ph.D. is supposedly affecting the present behaviour of writing for university catalogues. Surely this cannot be, and so one has to invoke something else as the cause of the present writing. But then it would seem that one has two students doing identical things, namely sending for university catalogues. But, whereas the one's behaviour is being causally affected by a successful Ph.D. thesis defence some five or six years hence, the other student's behaviour cannot be so explained. And this is the case even if neither student has any responsibility for the difference in success. (Suppose one student is killed in a car accident.) In order to get around this problem—a problem that obviously does not exist with regular causation, since this is always past or present—it is preferable to think of the student's behaviour now being affected by the student's hopes and intentions, now or in the past. One still has a teleological situation, in as much as the hopes or intentions make reference to desired states of affairs in the future, whether or not these states actually come to pass.

The second kind of teleology centring on individual humans themselves is that of the design or function that one puts into, or uses, when creating objects or artefacts. I have a loaf of fresh bread and I want to cut it up. A normal knife does not seem to work very well, for, however sharp, the bread gets pushed around and squished. Any slice coming off the loaf looks very sorry for itself, indeed. I think about how to get around this problem and decide that, perhaps, a serration along the cutting edge of the knife will do the trick. I therefore make, or have made, a knife with a serrated edge, and find now it works just fine, cutting even the freshest loaf of bread. I have here a designed object, an artefact, that I have made 'in order to' cut bread satisfactorily. The object has teleology imputed or given to it through my intentions: that is to say, it has a derivative teleology parasitic on the teleology of the first kind, discussed just above. I wanted the end result of a nicely cut loaf of bread, and therefore I made a knife with a serrated edge in order to achieve that end. Again, we have a teleological explanation in as much as the knife designed and manufactured now or in the past is being understood in terms of future effects, namely neatly cut loaves of bread. Since the teleology is parasitic, there is no need to worry about the missing goal-object problem: although I have made the knife in order to cut the bread, even if the bread does not get cut, it is still the case that my intentions now are the causal fuel of the creation of the knife, rather than some future event that may or may not obtain.

Notice incidentally that what is distinctive about this artefact kind of teleology is that we would use the word 'design' in this context. I suppose we could use the word 'design' in the first context, as in 'my design is to become a university professor'. But usually we would say rather that 'my intention is to become a university professor'. Conversely, although one might well say that the bread knife is serrated with the intention of cutting bread, or the intent is that it cuts bread, one would not normally say the serrated knife has the intention of cutting bread. Intentions apply to conscious beings whereas design is what conscious beings do to things. (Hence 'my design is to become a university professor' plays on the fact that we might have an intention that involves making something of ourselves—we are the product of our own intentions.)

To complete this discussion of meanings, let me make very brief mention of historical teleology. Here, one is trying to understand the sweep of history in terms of its ends, rather than in terms of its initial foundations or conditions (Ruse 1996). One is trying to understand what humans do and how they behave in terms of what will happen rather than what did happen. Here, we also have two kinds of teleology, albeit divided rather differently from the division above in the individual human case. On the one hand, most obviously we have the religious or spiritual kind of teleology. Human history is related to the divine intentions and goals and aims. Things that happened in the past or present are understood in terms of what is going to happen in the future. For the Christian, for instance, humans were created by God in order that they might share in everlasting ecstasy with Him (Ruse 2001). Jesus came in order that we might be saved through His grace and thus achieve salvation. The whole sweep of history is understood, not just in terms of God's love in the past in creating us humans in his own image, but also in terms of what will presumably happen to us in the future. The keys here therefore are God's Providential intentions and interventions.

The second kind of historical teleology is more secular. Although it comes in various forms, usually it is related to some kind of belief in progress (Bury 1920; Wagar 1972). This was an idea that emerged in the eighteenth century (the Age of the Enlightenment), often in conscious opposition to Christian Providential teleology. The progressionist sees history as in some sense providing a long sweep up from primitive beginnings, and climbing ever higher: not through God's Providence, but through human effort. Finally, one achieves some sublime state, which may or may not be succeeded by yet a higher state. August Comte, the nineteenth-century French positivist, saw a sweep of history from the religious through to the metaphysical, and then on—thanks to Comte himself particularly!—to the scientific. Everything is to be seen in this context, with movement up from one form to another yet higher.

2. Organic Design

Enough by way of definition. Let us turn to biology, evolutionary biology in particular. For the moment, I will leave the question of historical teleology on one side and will also do the same with the kind of teleology that deals directly with intentions. Whether or not intentions are important in evolutionary biology, other than with humans, is a matter that I shall discuss later; although, fairly obviously, even if such intentions exist, they are not going to be quite as prominent as they are in the case of humans. Even those who would assert the reality of chimpanzee intentions do not want to talk about hymenoptera (ants, bees, and wasps) having intentions. Certainly no one thinks that vegetables have ends or goals of this kind.

But what kinds of ends or goals do vegetables or hymenoptera or any other organism have? The sort of question that puzzled us is: 'Why does Dimetrodon (a Permian reptile) have a sail on its back?' Or rather, it is the answer that puzzles us: 'The sail exists in order to—serves the function or end of—scaring off predators.' (More likely, the end was heat regulation, but no matter here.) Why is such teleological or functional language thought appropriate? Why are we allowed to say that the sail (existing now) is explained in terms of the scared predator (existing in the future)? In the light of the discussion in the last section, the answer can be given readily. Such language is thought appropriate because organisms or the parts of organisms (characteristics) are taken to be design-like: they are taken to be as if they were artefacts, or parts of artefacts, created by conscious intelligences in order to serve certain ends. Just as the serration on the knife exists in order to cut bread, so the sail on Dimetrodon exists in order to frighten off predators. The Dimetrodon sail and all like organic characteristics—commonly known as 'adaptations'—seem as if they were created or designed by a thinking being, and, because objects that have been created or designed by thinking beings are teleological (that is, they have teleology imputed to them), it is appropriate to use such forward-looking language (Wright 1973, 1976; Lewontin 1978).

Moreover, the end or purpose of organic characteristics is obvious. Whereas, although human artefacts have all sorts of different immediate ends, ultimately they serve human satisfaction, so analogously, just as organic characteristics have all sorts of immediate ends (the Dimetrodon sail has the immediate end of frightening off predators), the overall organic characteristics have a shared end, namely the satisfaction of their possessors' needs. However, whereas humans have all sorts of different things that satisfy them, not just life and survival but also happiness and pleasure and so forth, in the case of organisms taken generally it is life and survival and reproduction alone that counts. In other words, adaptations serve the end of survival and reproduction of their possessors, and, because they are artefact-like—as if they were designed—teleological language is appropriate.

Now, thus far I have rather stated my case rather than argued for it. You ask: What justification can be given for claiming that teleological language is appropriate? I reply: Because organisms are as if designed. But am I right? Here, as a good evolutionist, I turn to history for help. It is a simple and true fact that, down through the ages, people in the West interested in organisms have seen them as if designed, thinking indeed that they seem as if designed because they were in fact designed! This is an argument that has its origins in the work of Plato, the *Phaedo* and the *Timaeus* specifically, and was taken up by the Christian theologians, notably Aquinas. It had great staying power, and right up through the beginning part of the nineteenth century, for all that philosophers sometimes argued otherwise, it was accepted as a general truth that the organic world has all the evidences of construction, of having being made with intention, that is of having been designed (Woodfield 1976). Indeed, this sense of design was so strong that everyone concluded that design must indeed be genuine and that therefore there had to be a conscious intelligence behind the creation of organisms—such a conscious intelligence being identified with the Christian God. Thus, to take the classic statement of the teleological argument—the argument from design—put forward by Archdeacon William Paley in his *Natural Theology* (1802), the eye is very much akin to a telescope: just as telescopes have telescope-makers, so eyes have to have eye-makers. This therefore proves the existence of the Great Optician in the Sky. The teleological language we use in the case of organisms is appropriate, because organisms were created and designed: they are God's artefacts.

Now, of course, something important and pertinent happened in the middle of the nineteenth century. In 1859 Charles Darwin published his great work *On the Origin of Species by Means of Natural Selection*, in which he argued that all organisms (including ourselves) are the results of a long, slow, natural process of development from a few simple forms. Moreover, Darwin argued that the main mechanism or cause fuelling this change is what he called 'natural selection': the differential reproduction of organisms brought about by a struggle for existence; such a struggle being a consequence of the Malthusian pressures on all organisms: there is only a limited amount of space and food supplies, and therefore, given the geometric potential of organisms to increase in number, there necessarily has to be a population pressure and subsequent winnowing. Some organisms survive and succeed, and others fail. Since the successful tend to differ from the unsuccessful, over time there will be change or evolution (Ruse 1979).

Darwin's *Origin* did not necessarily disprove God as designer. It was still possible after *The Origin* to think that everything ultimately goes back to God and to His intentions. In fact, it seems pretty clear that this was the case for Darwin himself when he first published *The Origin*. Later in life his faith fell away entirely, but this was probably more because he could not reconcile belief in a Creator with the problem of evil than because in any sense he felt troubled by organic design. However, after Darwin and today especially, there are many who would deny absolutely and completely the existence of any Creator of any kind. Most notably, the Oxford evolutionist Richard Dawkins (1995) has been loud and articulate and ardent in his forceful atheism: in Dawkins's oft-stated opinion, even if Darwin did not make belief in a designing God impossible, he made possible the denial of such a belief in a designing God. After Darwin, one could put everything down to a natural process, that is to say, to the law-bound process of natural selection. There is no need to invoke God's creative intervention in the world.

Does this then mean that the language of intention or design was no longer used after *The Origin*, or that it was at once considered inappropriate? Far from it! No one was more committed to a teleological way of thought about organisms than was Charles Darwin (Lennox 1993). He was educated in the late 1820s at the University of Cambridge, where standard fare would have been the works of Archdeacon Paley. Darwin read them, *Natural Theology* in particular. As a consequence, Darwin accepted completely Paley's premiss that the eye and the hand and all other organic characteristics—that is to say, all other organic adaptations—seem as if designed. The point is that Darwin wanted to explain this 'as if design' through natural selection (whether or not God stands ultimately behinds this selection), rather than through a directly intervening God of miraculous intention and design. Nothing in Darwin's thought denied that organisms have an artefact-like appearance. What Darwin challenged was not the premiss, but the conclusion. He wanted to argue that natural selection can do the job without reference to a deity. He wanted to argue, in fact, that in as much as one is going to give a scientific explanation, it is necessary to appeal to something like natural selection. Reference to a deity is simply not what we find or expect in good science.

History bears out that Darwin was right in his intuitions and successful in his proposals. It took some considerable time for natural selection to be accepted as a fully adequate mechanism, but everyone agreed with Darwin that the world is as if designed. And, gradually, it became accepted that 'as if design' is a consequence of the Darwinian mechanism of natural selection: which is how matters rest to this day. Evolutionary biologists use teleological language because organisms through their adaptations are artefact-like: they seem as if designed. Today, however, following Darwin, evolutionary biologists explain this 'as if designed' through the law-bound process of natural selection: the differential survival of organisms in the struggle for existence and reproduction. Teleological language in evolutionary biology therefore refers to entities that have been produced by natural selection. But note that the reason why the teleological language is appropriate is not because natural selection was the causal factor as such. Rather, it is because natural selection produces entities, adaptations, which seem as if designed and teleological language is appropriate in such contexts. If organisms were not 'as if designed', teleology would not be appropriate. But since they are, it is.

The evolutionary biologist George Williams, in *Adaptation and Natural Selection*, one of the most influential books of the past half century, wrote as follows:

> Whenever I believe that an effect is produced as the function of an adaptation perfected by natural selection to serve that function, I will use terms appropriate to human artifice and conscious design. The designation of something as the *means* or *mechanism* for a certain *goal* or *function* or *purpose* will imply that the machinery involved was

fashioned by selection for the goal attributed to it. When I do not believe that such a relationship exists I will avoid such terms and use words appropriate to fortuitous relationships such as *cause* and *effect*. (Williams 1966: 9)

(I note, incidentally, that others who like me stress the significance of design for the proper understanding of biological teleology are precisely those who, like me, take seriously the history of evolutionary biology. See Kitcher 1993*a* on history and 1993*b* on teleology; Rudwick 1972 on history and 1964 on teleology; and Beatty 1990 on both.)

3. Metaphor

Let me run through some consequences or implications of the history-based argument that I have just given. First, note that what we have here is a metaphor. Organisms are being treated as if they were objects of design. Even if, ultimately, one thinks that they are literally objects of design, the point is that in the biological, post-Darwinian, evolutionary context one is treating them as if they were designed, without any implication that they are in fact designed. So, one is taking the concept of design, particularly the concept of artefact-design, and applying it to the organic world. One is—as theorists of metaphor claim—looking at the organic world through the lens of human artefact making (Lakoff and Johnson 1980). Therefore, at a certain level, we have a kind of Kantian situation: we are seeing the world according to our structure or categories, rather than finding something actually in the world. (Kant dealt with this issue in his third critique, the *Critique of Judgement*, arguing that, although teleology does not exist as an objective fact, we must treat organisms as if they are teleological. As a matter of fact, he thought that this precluded evolution, but even though his science is outdated his philosophy lives on.)

Of course, the world has to be of a certain kind or character for the metaphor to work properly and fruitfully. But the point is that the whole question of design, or design-likeness, is ours, rather than that of the world. At a basic level, therefore, this means that there is neither reason nor cause to think that the organic world is in some sense different with respect to its ultimate stuff, as it were. There is nothing in the teleological language of evolutionary biology to imply that organisms incorporate vital non-material forces or anything like that. We are just seeing things ultimately made of molecules, organized in a particular sort of way, in the context of a particular framework or metaphor. There is certainly no cause to talk about entelechies or *élans vitaux* striving to achieve certain ends or anything like that. In the language of philosophers, there is no denial whatsoever of 'ontological reductionism'. No one is saying that the teleology of organisms commits one to non-material entities (Ayala 1970).

Nor, obviously, is there any reason to think that we are ultimately trying to explain the present in terms of the future. We are, of course, trying to explain the present in terms of what we think the future will be like. And since, so often, the future does come out to be this way, we tend to drop or even forget the fact that it is the future as we now think it will be. But no one is saying that the sail on the back of Dimetrodon is actually being explained in terms of future frightening of predators. At a strictly causal level, one is indeed implying that the sail exists because earlier Dimetrodons with sails frightened predators. But this is another matter. The point is that there is no implication that organic causation does strange and weird things, from the viewpoint of time. It is true, however, that through the metaphor we are introducing value considerations into biology. Plato knew that teleology means values:

> I once heard someone reading from a book, which he said was by Anaxagoras, and saying that Mind was what arranged everything and caused everything. I was delighted with this 'cause', and it seemed to me to be somehow rather appropriate that Mind should be responsible for everything, and I supposed that if this were so, the ordering Mind ordered everything and placed each thing severally as it was best that it should be; so that if anyone wanted to discover the cause of anything, how it came into being or perished or existed, he simply needed to discover what kind of existence was *best* for it, or what it was best that it should do or have done to it. (*Phaedo* 97 b–c)

In the post-Darwinian age, we are dropping the reference to mind, but still we are judging features according to the roles they play in—their contribution towards—the well-being of their possessors. Failure to recognize this point is what vitiates analyses that simply try to unpack teleology in temporal terms without acknowledging design and value. (For instance, someone like Larry Wright wants simply to talk of means and ends, and of ends explaining means; but this lays him open to non-living, cyclical counter-examples, like the rain causing lakes causing clouds causing rain, where functional language would not be considered appropriate. See Bedau 1992.)

4. The Physico-Chemical World

The second point is that, without need of further explanation or discussion, we have the reason why teleological language is generally considered inappropriate in the physico-chemical realm. It is simply that, without natural selection, entities tend not to look very design-like. Hence, one does not use such language or feel that it is appropriate. The simple fact of the matter is that Mars does not seem as if it is an artefact, and the same is true of the plates that carry continents around the earth. It just does not seem appropriate to talk, for instance, of the purpose of Mars in circling the sun, or of the intention of the plates to cause earthquakes or volcanoes. Therefore, we do not use such language.

One might note, incidentally, that history supports the point just made. Before Darwin, natural theologians were having a great deal of trouble with the inorganic world. Look at the classic statements of natural theology published in the 1830s, the so-called *Bridgewater Treatises.* The authors (particularly William Whewell, who was asked to write on astronomy) showed and admitted that there is far less evidence of design in the inorganic world than there is in the organic world (Whewell 1833, 2001; Ruse 1977). This was a paradox for them, because they believed that the inorganic world was no less designed than the organic world. But they could not get away from the fact that the inorganic world does not show such evidence—unless one gets into such tortured claims that God put coal under the British Isles so that the Industrial Revolution might succeed (Buckland 1836). Everyone knew, however, that this kind of argument was doing the cause of natural theology more harm than good. Basically, it was seen that, because the inorganic world is not explicitly design-like, the human-based language of function and intention does not seem appropriate.

I would point out, however, that since here we are using metaphors nothing is absolute. There is no ontological or other reason why, in some sense, the inorganic world at some time might not appear as if designed: either through fortuitous circumstances or through some kind of inorganic equivalent of organic natural selection. If, therefore, one can find design-like or quasi-design-like phenomena in the inorganic world—and it is sometimes argued that one does find this, for instance, in the theory of light with the principle of least distance—then I feel no tension at the use of the language of function and intention. This does not threaten what has been said about organic teleology. Indeed, if anything it strengthens the case, because, as is generally the case with the use of metaphors, one does not expect to find any exact or tight fit.

5. Ambiguity

This brings me to my third point, which is equivalent to what has just been said, but applying now to the organic world. Because we are using a metaphor, we are going to expect some unevenness or ambiguity in its use around the edges in the organic world. For instance, there will perhaps be cases where things are adaptive in the first generation, without really having been caused by natural selection. In such a situation, things happen rather by chance and then selection picks them up. Perhaps a characteristic has been perfected by selection for one end, but then (because of changed circumstances) it turns out to be very adaptive for some other completely different end. Does one want to talk in terms of 'design' and 'intention' and 'function' in such a case? I really do not think that there is an ultimate answer here. Some people's intuitions will go one way. Other people's intuitions will go the other way. But this does not really affect the overall metaphor of design. One expects indeed that some people will be happy to use a metaphor more extensively than others. Where some will be expansive, others will be restrictive. So, if one finds one biologist prepared to use the language of design in a particular case, but another feeling uncomfortable about this, this is no more than one would expect when metaphorical language is at stake. It is confirmation rather than refutation of the basic claims. (It is for this reason that I deny being insensitive or ignorant in brushing aside those critics who would separate function and design, for instance Allen and Bekoff 1995. I simply do not think the borderline cases are enough to warrant such a separation, any more than the existence of Siamese twins stops me from talking about single individuals and about pairs of humans.)

6. Adaptationism

This leads straight into the fourth point, about whether or not the metaphor of design is appropriate right through the organic world. Some people, ultraDarwinians, think that it is. Sociobiologists such as Richard Dawkins (1976, 1986), John Maynard Smith (1958, 1981) and Edward O. Wilson (1975, 1994), not to mention myself (Ruse 1982, 1986, 1989) and some other philosophers like Daniel Dennett (1995), fall very much into this camp. We do not necessarily want to say that everything has immediate design-like features. But we usually want to argue that, even if something does not have artefact-like properties now, it had them in the past, or is in some way connected to such artefact making. The classic case, discussed by John Maynard Smith (1981), is of the four-limbedness of vertebrates. Why do we have four limbs rather than six, like insects, or eight, like arachnids? It is obvious why we do not have five, since this would lead to difficulty in walking. (Starfish, which do have five limbs, do not walk.) But why four rather than six? There seems no reason or point at all. However, Maynard Smith points out that, even if there is no point at the present, in the past when the vertebrates evolved they were aquatic creatures. Having two limbs fore and two limbs aft was the perfect way to rise or fall in the water, just as planes use their two wings forward and two wings behind in order to rise and fall through the air. So, even though there may not be an immediate purpose at this point, at some level purpose-like thinking is always useful and relevant and appropriate.

There are others who, although keen evolutionists and certainly admitting some level of design or intention (no one wants to deny that the eye is 'as if designed'), do want to argue that the metaphor is being used altogether too much, to the detriment of evolutionary biological research. Prominent here have been the paleontologist Stephen Jay Gould and his associate, the molecular geneticist Richard C. Lewontin. They argue that many organic characteristics are simply without end or purpose (Gould 1980, 1982; Gould and Lewontin 1979). Such features might be produced by random factors, like so-called genetic drift, or simply be constraints on development or some such thing. There is no reason to think that there is design or 'as if design' here. Not everything is as adaptive as ultraDarwinians claim.

As a matter of historical fact, this anti-adaptationism is a stance that goes back a long way (as Gould himself well realizes). It goes right back at least to the beginning of the nineteenth century, when German morphologists—the so-called *Naturphilosophen*—argued that the chief distinctive characteristic of the organic world, or indeed of the world generally, was isomorphism or symmetry (Gould 1977; Lenoir 1982; Richards 1992; Ruse 2000). Thus, the *Naturphilosophen*—and following on them people like Gould and Lewontin—make much of the similarities (known now as 'homologies') between the bones of the forelimbs of the human, the horse, the mole, the porpoise, the bird, and the bat: similarities, even though the limbs are all used for different functions. Such isomorphisms or homologies, argue Gould and Lewontin, are evidence that the world is nothing like as tightly design-like as Darwinian evolutionists argue. These critics claim that we are often misled by the metaphor of design, which is truly a throwback to a Christian style of thinking that most biologists do not now share. Hence, in many respects, we would be better off without the metaphor.

Although I myself have strong opinions on this matter (Ruse 1982), this is not my chief point of concern here. What I would want to say is that this difference between evolutionary biologists over this question of design is no more than one would expect, given such an analysis of mine. Sometimes people think that metaphors are appropriate; other times people think that metaphors are inappropriate; and often one gets considerable debate about whether a particular metaphor is of value in a particular case. Take the use of metaphor in the religious sphere. God is often thought of as a Father, although everybody knows that God is not literally a father, in the sense of the person who had sexual intercourse with one's mother, thus passing on 50 per cent of one's genes. Rather, God has father-like characteristics: He is loving and caring and so on and so forth, and also in some sense a Creator. Yet, as is well known today, many feminists and others would argue that this kind of metaphor used of God is altogether inappropriate and misleading. In as much as one thinks of God as a father, one also thinks of Him as being patriarchal, or as being stern and judgemental, as being dominant, and so on and so forth—all things that one might well want to question of God. These critics would prefer to use some other metaphor: perhaps God as Mother, to counter the God as Father metaphor. Or, perhaps, one should just drop entirely this whole parental metaphor scheme. Here, I am obviously not going to take a stand on the theological issues or the use of metaphor in this domain. It is enough that I draw attention to the fact that one frequently has clashes over metaphor, and that one expects to find (as one does indeed find) clashes over metaphor in evolutionary biology.

7. Natural Selection

The fifth point is that nothing that I have said in my analysis denies the fact that, often, biologists will use teleological language without much thought at all about natural selection. For instance, molecular biologists use all sorts of telelogical notions, even though they are probably totally ignorant of Darwin and his achievements. One finds talk about the genetic code being of such a form and nature in order to convey certain facts of information, and relatedly of the need to crack the code. One finds speculation about the purpose of the DNA and RNA, and of all the junk stuff that seems jammed in between the active parts. And so on and so forth. Here we have people who are using the language of artefacts, and yet who are probably totally ignorant of natural selection. Indeed, if reminded or educated, quite possibly contemptuous of it. Then at the other end of the spectrum of the biological world, we have the most old-fashioned of biologists, the comparative anatomists and morphologists. They also use teleological language even though they too are probably indifferent to natural selection, again virtually if not completely ignorant of it. They will talk about the 'function' of various bodily parts in invertebrates and the like, even though they have little or no interest whatsoever in what the selective virtues of such parts might be (Cummins 1975; Amundson and Lauder 1994).

Neither of these cases at all affects what I have said. Even though some molecular biologists may be ignorant of evolutionary biology—less so today, I suspect, than in the 1950s—this is not to say that the entities they consider were not produced by evolution. This is not to say that natural selection had no part in putting together the genetic code, or the DNA and RNA molecules. It is just that molecular biologists have tended to have no interest in origins: they just want to see how things work or how things relate to other things. The same is true also of morphologists. It may well be true that morphologists, and fellow travellers like systematists, are basically uninterested in evolution. Certainly they tend to have little time for natural selection, a tradition—if that is the right term—that flourished before and after *The Origin* was published (Ruse 1979, 1996, 1999). But the fact that morphologists are uninterested in what put their organisms together does not mean that natural selection did not play a role. And, in as much as this is true, functional language is appropriate.

8. Heuristic Virtues

My sixth and final point is about the usefulness of teleological thinking in evolutionary biology. Every so often one sees calls for intellectual or scientific purity. It is argued that, since teleological language is a throwback to vitalistic or Christian thinking, one should purge one's biology of any such talk (Ghiselin 1983; Kramer 1984). One should eschew talk of the function of the sail on the back of Dimetrodon. Rather, in a stern and rigorous manner one should speak only of natural selection on the sails of earlier Dimetrodons leading to the sail on the Dimetrodon one is now studying. Or, at least, one should be more cautious than one is at present (Godfrey-Smith 1999).

Against this point, I will make two responses. First, negatively, as we have seen, there is nothing ontologically or spiritually or otherwise offensive in the teleological language used in evolutionary biology. One is not thereby committed to a God, whether or not one wants to believe in one. One is not committed to vital forces or any such thing. And one is certainly not plunged into depths caused by the problem by the missing goal object or anything like that. So in these respects, teleological language is perfectly pure and appropriate. Nor, incidentally, does the introduction of values (as in ends being judged desirable or good) alter this conclusion. Admittedly, the logical positivists and their successors the logical empiricists were adamant that science is and must be value free, but we are much more relaxed about those sorts of things these days (Ruse 1999). Most importantly, the kinds of values that teleology introduces are not absolute values—an adaptation may do a parasite some good but no one is going to say that the parasite itself is good. Teleological values are more akin to relative values requiring a kind of evaluation, judging that something will help an organism do something. Even the logical empiricists agreed that such values as these are a central part of science and not to be denied or eliminated.

Secondly, positively and more importantly in response to those who would eliminate telelogical language and thinking from evolutionary biology, let me point out that this mode of thought plays an absolutely vital role. It is of key heuristic value. Without the teleological language, without thinking of organisms as if they were designed, much evolutionary biology would simply grind to a halt (Dennett 1995). Let me make my point by reference to a recent, already-classic study: that of the Harvard entomologist and sociobiologist Edward O. Wilson on the social insects, in particular on the division of labour amongst the ants.

Wilson (1980*a*, *b*, 1983*a*, *b*) looked at the caste system amongst the ants. He found, sometimes in the same species, four, five, or even six different forms of worker. The questions he asked were: 'Why do these different forms exist?'

'What is the purpose or function of the forms?' The answer that Wilson gave, relating ultimately back to natural selection, was that the different forms exist for different ends. Some ants are very large, and therefore serve as fighters or defenders of the nest. Some ants are midsized, and they are involved in cutting up leaves and the like, and bringing the pieces back to the nest. Some ants are even smaller: they are involved in chewing up the leaves and creating the gardens in which they grow a kind of fungus. And some ants, of minute size, are used for harvesting the fungus and for feeding the larvae: they look after the queen and the eggs and so on and so forth. All ants with different shapes, and with different ends and different functions. Without the metaphor of design, without thinking in terms of ends or purposes or functions, Wilson would have been absolutely stymied when he came to consider these highly sophisticated and complex social insects. But, using the metaphor of design as his conceptual heuristic tool, Wilson was able to cut right through the problem of caste differentiation.

What I am suggesting is that, with the metaphor of design, one has one of the most prized epistemic virtues that one can have in any scientific theory: predictive fertility (Ruse 1999). Given the artefact way of thought—regarding organisms as if they were put together consciously—evolutionary biologists have a vitally important tool for looking into the organic world. Rather than being thrown back at once on impossible questions about how natural selection operated in the distant past, evolutionists begin by asking how things work at the moment. When once they have got a handle on this question, they are able next to talk in terms of natural selection, trying to relate their studies back to what happened in the past. But this second stage can occur only if first they have used their design metaphor to ask pertinent questions about function. Who knows what caused the Dimetrodon sail? Just saying 'natural selection' is true and completely unhelpful. You are and will always remain completely in the dark about particulars. But ask first about what the sail could do now or in the future, ask about function, and you are on the way to finding out about selection in the past.

In other words, although perhaps possibly one might purge evolutionary biology of its teleological language, one would be doing so only at the greatest of all costs. Even if philosophers may think that such intellectual purity is desirable, fortunately practising evolutionary biologists are nothing like as foolish. In this sense, although evolutionary teleology does nothing to disturb ontological reduction, it makes one very wary about some other claims about reduction: especially those claims about theoretical reduction, where supposedly all biological theories are going to be shown at some point to be special instances of physico-chemical theories. Organisms are different from planets and rocks and continental plates, and they call for different modes of understanding. In this sense, evolutionary biology is autonomous.

9. Intentionality

I move along. Let me turn now briefly to the other forms of teleology that are listed at the beginning of this chapter.

First we have the teleology of human intentionality (Dennett 1984). I want to get a job as a philosophy professor, so in order to do so I set out to get a Ph.D. My end or goal is getting the Ph.D., and to get this I send away for university catalogues and so on and so forth. The teleology comes about through my planning and having a purpose and an end, and trying to achieve it. There are no causes and future affecting the present. What is affecting the present are my current intentions. But there is a forward-looking direction to everything that is going on. The question is whether or not we find anything analogous in the evolutionary world.

I have admitted already that we probably do, if only because humans are themselves part of the evolutionary world. So here we have got a teleology that is more than just a teleology of artefacts: except perhaps as we might want to say that we are designing ourselves in some artefact sense, but I will leave this alone as rather misleading. I have admitted also that probably it is appropriate to use such intentional language in the case of higher organisms like the chimpanzees. Today we have far more than mere anecdotes of chimpanzees and other higher organisms setting out to achieve certain ends and often achieving them, or at least showing that they did have such ends in view. From the work of Franz De Waal and others on chimpanzees (both captive and wild), we know that males prize being the head of the group (De Waal 1982; Goodall 1986). We know also that, in order to achieve such dominance, a male will have to forge alliances with certain key females in the group. Without such alliances, the male cannot hope to achieve and certainly not to maintain his status. It seems very clear that, at some level, males are aware of this, and indeed that the females are aware of this also. Moreover, an awful lot of planning goes on as males attempt to move up the pecking order, and as females decide to help—or sometimes decide not to help or to change allegiances. In cases like this it is not only tempting, but surely appropriate, to use intentional language. (Primatologists themselves have thought long and hard about using such language, deciding that it is entirely appropriate.)

For vegetables, and even for so-called lower organisms, such intentional language is not appropriate. Take the case of wasps, specifically the 'sphex' which is described in some detail by Daniel Dennett in several of his works. This wasp has a certain goal, but if for some reason it is disrupted the wasp simply goes through the motions all over again.

> When the time comes for egg laying, the wasp *Sphex* builds a burrow for the purpose and seeks to out a cricket which she stings in such a way as to paralyze but not kill it.

She drags the cricket into the burrow, lays her eggs alongside, closes the burrow, then flies away, never to return. In due course, the eggs hatch and the wasp grubs feed off the paralyzed cricket, which has not decayed, having been kept in the wasp equivalent of deep freeze. To the human mind, such an elaborately organized and seemingly purposeful routine conveys a convincing flavor of logic and thoughtfulness—until more details are examined. For example, the Wasp's routine is to bring the paralyzed cricket to the burrow, leave it on the threshold, go inside to see that all is well, emerge, and then drag the cricket in. If the cricket is moved a few inches away while the wasp is inside making her preliminary inspection, the wasp, on emerging from the burrow, will bring the cricket back to the threshold, but not inside, and will then repeat the preparatory procedure of entering the burrow to see that everything is all right. If again the cricket is removed a few inches while the wasp is inside, once again she will move the cricket up to the threshold and re-enter the burrow for a final check. The wasp never thinks of pulling the cricket straight in. On one occasion this procedure was repeated forty times, always with the same result. (Dennett 1984: 11, quoting Wooldridge 1963: 82)

It is hard to think that the wasp is anything but genetically determined and that all talks of intentionality are surely inappropriate. The wasp is not thinking about what is going on or anything like that. It is appropriate to think of the wasp in teleological terms, but only in terms of the artefact model discussed in earlier sections. The wasp (or its parts) is certainly adapted towards certain ends, and goes through its motions because it is thus adapted. But it is inappropriate to think that there is anything else going on here. There is certainly not the intentionality that we usually associate with humans—that is to say, with conscious, thinking beings.

Is this the end to the matter then? I am not quite sure that it is. Let us push our discussion a little further. What is the chief mark of human intentionality? A good case can be made for saying that humans are 'goal-directed', or 'directively organized'. By this I mean that, if (for some reason) human plans are disrupted they will regroup and try again. I am not sure that this is either a necessary or a sufficient condition of human intentionality, but it certainly does seem to be a strong feature of it. Thus, for instance, why would we want to say that I really do have in mind the goal of getting a university professorship? Because, if my aims of getting the Ph.D. are disrupted in one way, then I will regroup and try again in another way. In other words, a strong sign that one has intentionality is that one has the kind of goal-directedness that one associates with feedback mechanisms, such as were developed for torpedoes and rockets during the Second World War. If the target moves, one can respond to—redirect oneself towards—the target. Or, if the target stays where it is but environmental circumstances in some way knock one off course, one can regroup and bring oneself back on course again (Sommerhoff 1950; Braithwaite 1953; Nagel 1961).

Now, I do not introduce this notion of goal-directedness purely by chance. At one point back in the 1950s and 1960s, when the Second World War was a much larger part of everyone's imagination, the general approach towards biological teleology was made almost exclusively in terms of such goal-directedness. It was said that it is appropriate to use teleological language of organisms because they (and they alone) are goal-directed. This was a mistake (Ruse 1973). Because one speaks of something as being adapted—where teleology does enter—it does not follow that this thing is in any sense capable of responding to change. Perhaps the Dimetrodon with its sail can respond to change. Perhaps not. But it is still an adaptation all the same, and for this reason we have teleology. The mix-up was between 'adaptation' and 'adaptability' (Waddington 1957). To speak of an organism as being adapted makes no reference whatsoever to whether or not it is goal-directed. However, to speak of an organism as being adaptable is to say that it is adapted, but it is also to say that it is goal-directed. If, for some reason, the organism is disrupted from its ends, it can regain these ends: not necessarily always, but at least in some reasonable number of occasions.

All conscious intention involves adaptability. If the would-be alpha male cannot forge his allegiances with one female, then in order to build his path to power and fame he switches to another female. In this sense, he is being adaptable. But not all adaptability involves conscious intention, and what about these cases? What about the classic example of sweating and shivering? Mammals sweat and shiver in order to maintain a constant body temperature. If an organism gets too hot, it starts to sweat, and if an organism gets too cold, it starts to shiver: thereby usually, although not necessarily, bringing the organism back to the original temperature. There is no question of conscious intention here. There is teleology, because considered overall sweating and shivering are adaptations. But would one want to say that, in some sense, even though adaptability is not to be confused with adaptation, the adaptability confers an extra dimension of teleology? Would one want to say that, although one is granted an extra dimension of teleology in the case of conscious intention, and although one does not want to say that conscious intention always occurs with adaptability, nevertheless adaptability always implies some dimension of teleology over and above that of mere adaptation?

You can anticipate what I am about to say. The answer is somewhat fuzzy! One is dealing here with a metaphor: this time a metaphor brought from human intentionality, rather than from human artefacts. But the point I made above about metaphors still holds good. The whole thing about metaphors is not that they are right and wrong. Rather, they are appropriate or inappropriate and what one person might find appropriate another person might find inappropriate. In other words, metaphors have a subjective dimension that one does not find in other forms of thinking like mathematics. So what I would say is that, whilst I feel fairly confident that one can claim that there is an extra dimension of teleology when one is talking of conscious intention, even of non-human organisms, in other cases where one has adaptability there is no final or fixed answer. Some people, and I think I would want to include myself here, are probably inclined to say that goal-directness does introduce more teleology over and above the teleology of adaptation (Ayala 1970). Others, whilst recognizing that we do have goal-directedness (and that it can be important), would prefer not to think of any more teleology other than the teleology of adaptation. One has adaptations of a particular kind, perhaps adaptations of a particularly interesting kind, but not adaptations that are more teleological than any other adaptations. The teleology comes through the fact that shivering and sweating are adaptations, therefore just as design-like as rockets and torpedoes. The teleology does not come through their goal-directedness, but through the fact that goal-directed adaptations are simply human artefact-like.

My personal feeling is that this is all that can or indeed should be said about this topic. There is little point in trying to argue further on the matter. Everybody can and should recognize that there are some particularly interesting adaptations—namely, adaptations that involve adaptability over and above straight adaptation. Whether or not one wants to regard adaptability as being in some sense an extra dimension of teleology is purely a matter of taste. Some do and some do not, and that is all there is to be said about this. Some people like spinach, others do not, and we just have to learn to live with our differences.

10. Historical Teleology

We come now to the question of teleology considered over the long range of history. Again you will not be surprised to find that metaphor plays a key role and that this has the implications and consequences sketched in earlier sections. There are two main possible ways of viewing history teleologically. One is based on Christian Providentialism manner, where the whole of history is seen as a drama involving God and his favourite creation, humankind. History is the story of God's intentions, leading to our possible ultimate salvation and eternal ecstasy with Him. The other picture is secular, based on Progressionism. History is seen as one of progress, with humans creating an ever-better world through their own unaided effort.

It is fair to say that the Christian Providentialist view of history has little or no immediate relevance to the evolutionary scenario. It is certainly the case that the relationship between Christianity and evolutionary biology, Darwinian evolutionary biology, is a lot closer than many believe or want to believe (Ruse 1996). Indeed, apart from the already much-discussed metaphor of design, I would argue that the whole way of thinking in an evolutionary manner, that is to say in terms of origins, is something that goes back directly to Judaeo-Christian thought. The influential ancient Greeks believed that the universe is eternal: therefore, they had no cause to ask about origins of life or anything like that. The evolutionist, although he or she rejects the original Christian position on origins (as do most Christians today), nevertheless thinks that the question of origins is important. However, this said, there is really no question of the evolutionist trying to interpret evolution in any way analogous to or metaphorically through the lens of Christian Providentialism.

The story with progress is very different. The case can be made that the whole history of evolutionary theory—virtually up to the present—is the story of an epiphenomenon on the sociocultural idea of progress. The origins of evolutionism, in the late eighteenth and early nineteenth centuries, were a direct function of progressionist thinking in the sociocultural world. Early evolutionists, such as Erasmus Darwin (the grandfather of Charles Darwin) and the Frenchman Jean Baptiste de Lamarck, were ardent social progressionists, reading their secular ideology directly into the organic world: there finding this ideology transformed into evolutionism (Darwin 1794–6; Lamarck 1809; Burkhardt 1977; McNeil 1987)—a progressive evolutionism, from the ultimate simple form, the 'monad', to the most complex and sophisticated form, the 'man'. Naturally, they immediately read this evolutionism back into the cultural world, as justification for their original beliefs!

Then, as we come down through the ages, we find that progressivist beliefs remained at a high level: if anything, climbing yet higher. Darwin himself, as we now realize, was a ardent social progressionist, who read progress right into his evolutionary biology (Ospovat 1981). He thought that humans come out at the top, because of a kind of evolutionary 'arms race' between competing lines. Prey and predator get faster in tandem, and thus adaptations are improved. Ultimately, intelligence emerges and we humans have won the evolutionary stakes. Likewise, Darwin's contemporary and far more influential evolutionary theorist Herbert Spencer (1857) saw the whole of life as a progress up from the undifferentiated to the highly complex, from the blob to the mammal, from the savage to the European, and indeed to the Victorian gentleman of which he considered himself the epitome.

Progress was the ideology of evolutionists into the twentieth century. The great population geneticist Sir Ronald Fisher introduced his so-called fundamental theorem of natural selection as a kind of counter to the decaying effects of the second law of thermodynamics: he believed that, thanks to natural selection, in the organic world we find an ever-progressive climb upwards (Fisher 1930; Box 1978). And finally, even as we come up to the present, we learn that eminent evolutionists, not excluding Richard Dawkins himself, believe strongly in progress (Dawkins and Krebs 1979). John Maynard Smith argues that there are transitions in life's history, with ever greater upward jumps (Maynard Smith and Szathmary 1995). Dawkins believes in Darwinian arms races, ending with organisms with the largest onboard computers. And Edward O. Wilson (1992: 187) is virtually lyrical in his passion for progressionism.

> Biological diversity embraces a vast number of conditions that range from the simple to the complex, with the simple appearing first in evolution and the more complex later. Many reversals have occurred along the way, but the overall average across the history of life has moved from the simple and few to the more complex and numerous. During the past billion years, animals as a whole evolved upward in body size, feeding and defensive techniques, brain and behavioral complexity, social organization, and precision of environmental control—in each case farther from the nonliving state than their simpler antecedents did. More precisely, the overall averages of these traits and their upper extremes went up. Progress, then, is a property of the evolution of life as a whole by almost any conceivable intuitive standard, including the acquisition of goals and intentions in the behavior of animals.

11. Valid Thinking?

Now what can one say about this kind of thinking? Does one have a dimension of teleology at the historical level, over and above any teleology one might find at the level of individual organisms? I think one does and, although I myself feel somewhat uncomfortable with such thinking, I am not sure that one can say absolutely that it is wrong. However, if one is inclined to think teleologically in this way, then the following comments seem pertinent.

First, remember that once again one has a metaphor, a subjective entity very much dependent on the person using it. Not everyone accepts that social or cultural progress is a given. Apart from Christians, who are (or should be) more committed to Providence, many western intellectuals think that, in the light of two world wars, the development of atomic weapons, the overpopulation of the earth and much, much more, talks of progress in any simplistic way are not simply false, but almost obscene. They would therefore deny that progress is an adequate philosophy of history. My own suspicion is that scientists generally do not share this pessimism: they are used to success in their own realm and they are inclined to see it in the social and cultural realm also. But there certainly are some scientists, including some eminent evolutionary biologists, who are by no means enamoured with notions of progress. For instance, the important English sociobiologist Geoffrey Parker believes that social progress is probably not anything like as inevitable or likely as most people think. For this reason, he is not only inclined to deny social progress, but very unwilling to read any kind of progressionist philosophy into his evolutionary biology (Ruse 1999). Conversely, there are those (both within and without the scientific community) who are eager to deny that the course of biological evolution is in any wise progressive. Stephen Jay Gould (1989, 1996) is one who has argued long and ardently on this theme. He thinks that all claims about biological progress are ultimately doomed to failure.

The second point to be made is that, even amongst those who do believe in biological progress, there is usually a reluctance to allow or argue that such teleology or progressionism is in any sense inevitable. The Christian Providentialist presumably believes that there is an inevitability to history, even though most Providentialists would not argue that there is an inevitability to individual salvation. But, amongst progressionists, particularly biological progressionists, one generally finds a reluctance to argue that progress must occur. Often, of course, it does not. One gets all sorts of cases of degeneration—for instance, the evolution of parasites: these often lose many characteristics, including the sexuality that they had previously. But, more than just specific cases of degeneration (which, as in the parasite case, have good adaptive reasons), generally the feeling amongst evolutionary biologists is that there is no guarantee to progress (Simpson 1949). And few would argue that humans absolutely had to appear. Most would argue that, even if social progress is possible on an ongoing basis, there is certainly no guarantee or expectation that humans will evolve into something bigger and better in any kind of way that we would appreciate.

This denial is not universal. There are evolutionists who see a virtual built-in inevitability to their biological progress. One such was Herbert Spencer. In this century, Julian Huxley, the grandson of Thomas Henry Huxley and author of *Evolution: The Modern Synthesis* (1942), looked upon evolution as something that moved inexorably upwards. I am not sure that Huxley thought that we humans are going to evolve beyond the state that we have reached, but certainly he saw a necessary path leading up to humans. Today, Edward O. Wilson subscribes to something very close to an inevitability thesis: 'the overall average across the history of life has moved from the simple and few to the more complex and numerous. During the past billion years, animals as a whole evolved upward in body size, feeding and defensive techniques, brain and behavioral complexity, social organization, and precision of environmental control—in each case farther from the nonliving state than their simpler antecedents did' (Wilson 1992: 187). Arguing that this is no chance, but a necessary outcome of the evolutionary process, Wilson concludes: 'Progress, then, is a property of the evolution of life as a whole by almost any conceivable intuitive standard, including the acquisition of goals and intentions in the behavior of animals.' (I am noting this thinking not because I agree with it, but rather to show that, in the writings of some evolutionists, one gets a stronger sense of historical teleology than one does in the writings of others—so strong a sense, in fact, that it is surely quite appropriate to speak of some evolutionists' philosophy of history as being teleological in a very strong and positive sense.)

The third point I would make is that, even if one subscribes to some form of biological teleology at the historical level, this does not in any way commit one to some overall teleology or cosmic teleology of the universe. In fact, as I have hinted, often evolutionists put forward their own progressionism in the face of and despite what they see as the inevitable degeneration of the universe as a whole. R. A. Fisher is a case in point, and the same is true of others, Spencer in particular (Richards 1987). As it happens, Fisher was a Christian, thinking that God has given us our task on earth (eugenical preservation of the talented) corresponding to His task on earth (the production through evolution of humans). So presumably at some ultimate level, Fisher would have reverted to a Providential world-picture. But Spencer did not do so, and neither did others like Julian Huxley. Because one is a biological teleologist, this is not at all to say that necessarily one is going to be a teleologist in some broader cosmic perspective.

Three points of amplification are enough. As I have said, not everybody wants to accept the same metaphors. It is more a question of metaphors being appropriate or suitable than true or false. If someone wants to deny this kind of historical teleology or argues that one can see no real heuristic value to it—and I must confess that these are the directions of my inclinations—then it is surely open for them to do so. The case for some sort of overall historical teleology certainly seems significantly weaker than the case for teleology at the individual organismic level.

12. Conclusion

My survey and analysis of teleology as it occurs in evolutionary biology comes to an end.

I have tried to show how it arises, and why it persists. I have shown that it comes in various forms, and I have defended its use, particularly in the context of organic adaptation. There are reasons why evolutionists use teleological language, why they think teleologically, and why they would be very reluctant to speak and think otherwise. There are also reasons why such teleology is more or less peculiar to biological science and does not come into the physico-chemical sciences. This is not an ontological matter, but it is a question of perspective (Beckner 1969; Ruse 1989). Thanks to the evolutionary process, organisms invite teleological thought in a way that inanimate objects do not.

Finally, to speak to the one question that has not been addressed: I see no reason whatsoever why one should argue that teleological understanding is inferior to any other kind of scientific understanding. Nor do I see why, in consequence, one should say that evolutionary biology is thereby weaker or less adequate than the physico-chemical sciences. Evolutionary biology is different because the challenges it faces are different. Different modes of understanding are required, and, thanks to teleology, different modes of understanding are supplied. It is as simple as that. Difference among the sciences does not imply inequality, any more than difference among humans implies inequality.

References

Allen, C., and Bekoff, M. (1995), 'Biological Function, Adaptation, and Natural Design', *Philosophy of Science*, 62: 609–22.

—— ——and Lauder, G. (1998), *Nature's Purposes: Analyses of Function and Design in Biology* (Cambridge, Mass.: MIT Press).

Amundson, R., and Lauder, G. V. (1994), 'Function without Purpose: The Uses of Causal Role Function in Evolutionary Biology', *Biology and Philosophy*, 9: 443–69.

Ayala, F. J. (1970), 'Teleological Explanations in Evolutionary Biology', *Philosophy of Science*, 37: 1–15.

Beatty, J. (1990), 'Teleology and the Relationship between Biology and the Physical Sciences in the Nineteenth and Twentieth Centuries' in F. Durham and R. D. Purrington (eds.), *Some Truer Method: Reflections on the Heritage of Newton* (New York: Columbia University Press), 113–44.

Beckner, M. (1969), 'Function and Teleology', *Journal of the History of Biology*, 2/1: 151–64.

Bedau, M. (1992), 'Where's the Good in Teleology?', *Philosophy and Phenomenological Research*, 52: 781–806.

Box, J. F. (1978), *R. A. Fisher: The Life of a Scientist* (New York: Wiley).

Braithwaite, R. (1953), *Scientific Explanation* (Cambridge: Cambridge University Press).

Buckland, W. (1836), *Geology and Mineralogy* (Bridgewater Treatise, 6; London: William Pickering).

Buller, D. J. (1998) (ed.), *Function, Selection, and Design: Philosophical Essays* (Albany, NY: SUNY Press).

Burkhardt, R. W. (1977), *The Spirit of System: Lamarck and Evolutionary Biology* (Cambridge, Mass.: Harvard University Press).

Bury, J. B. (1920), *The Idea of Progress; An Inquiry into its Origin and Growth* (London: Macmillan).

Cummins, R. (1975), 'Functional Analysis', *Journal of Philosophy*, 72/20: 741–65.

Darwin, C. (1859), *On the Origin of Species* (London: John Murray).

Darwin, E. (1794–6), *Zoonomia; or, The Laws of Organic Life* (London: J. Johnson).

Dawkins, R. (1976), *The Selfish Gene* (Oxford: Oxford University Press).

——(1986), *The Blind Watchmaker* (New York: W. W. Norton & Co.).

——(1995), *A River out of Eden* (New York: Basic Books).

——and Krebs, J. R. (1979), 'Arms Races between and within Species', *Proceedings of the Royal Society of London, B* 205: 489–511.

De Waal, F. (1982), *Chimpanzee Politics: Power and Sex among Apes* (London: Cape).

Dennett, D. C. (1984), *Elbow Room* (Cambridge, Mass.: MIT Press).

——(1987), *The Intentional Stance* (Cambridge, Mass.: MIT Press).

——(1995), *Darwin's Dangerous Idea: Evolution and the Meanings of Life* (New York: Simon & Schuster).

Fisher, R. A. (1930), *The Genetical Theory of Natural Selection* (Oxford: Oxford University Press).

Ghiselin, M. (1983), 'Lloyd Morgan's Canon in Evolutionary Context', *Behavior and Brain Sciences*, 6: 362–3.

Godfrey-Smith, P. (1999), 'Adaptationism and the Power of Selection', *Biology and Philosophy*, 14: 181–94.

Goodall, J. (1986), *The Chimpanzees of Gombe: Patterns of Behavior* (Cambridge, Mass.: Harvard University Press).

Gould, S. J. (1977), *Ontogeny and Phylogeny* (Cambridge, Mass.: Harvard University Press).

——(1980), 'Is a New and General Theory of Evolution Emerging?', *Paleobiology*, 6: 119–30.

——(1982), 'Darwinism and the Expansion of Evolutionary Theory', *Science*, 216: 380–7.

——(1989), *Wonderful Life: The Burgess Shale and the Nature of History* (New York: W. W. Norton & Co.).

——(1996), *Full House: The Spread of Excellence from Plato to Darwin* (New York: W. W. Norton & Co.).

——and Lewontin, R. C. (1979), 'The Spandrels of San Marco and the Panglossian Paradigm: A Critique of the Adaptationist Programme', *Proceedings of the Royal Society of London, Series B: Biological Sciences*, 205: 581–98.

Huxley, J. S. (1942), *Evolution: The Modern Synthesis* (London: Allen & Unwin).

Kant, I. (1951), *Critique of Judgment*, trans. J. H. Bernard (New York: Haffner; 1st edn., 1790).

Kitcher, P. (1993*a*), *The Advancement of Science: Science without Legend, Objectivity without Illusions* (New York: Oxford University Press).

——(1993*b*), 'Function and Design', *Midwest Studies in Philosophy*, 18: 379–97.

Kramer, P. J. (1984), 'Misuse of the Term Strategy', *BioScience*, 3417: 405.

Lakoff, G., and Johnson, M. (1980), *Metaphors We Live By* (Chicago: University of Chicago Press).

Lamarck, J. B. (1809), *Philosophie zoologique* (Paris: Dentu).

Lennox, J. G. (1993), 'Darwin *was* a Teleologist', *Biology and Philosophy*, 8: 409–21.

Lenoir, T. (1982), *The Strategy of Life: Teleology and Mechanics in Nineteenth Century German Biology* (Dordrecht: Reidel).

Lewontin, R. C. (1978), 'Adaptation', *Scientific American*, 239/3: 213–30.

Mackie, J. (1966), 'The Direction of Causation', *Philosophical Review*, 75: 441–66.

McNeil, M. (1987), *Under the Banner of Science: Erasmus Darwin and his Age* (Manchester: Manchester University Press).

Maynard Smith, J. (1958), *The Theory of Evolution* (Harmondsworth: Penguin).

——(1981), 'Did Darwin Get it Right?', *London Review of Books*, 3/11: 10–11.

——and Szathmary, E. (1995), *The Major Transitions in Evolution* (New York: Oxford University Press).

Nagel, E. (1961), *The Structure of Science: Problems in the Logic of Scientific Explanation* (New York: Harcourt, Brace & World).

Nissen, L. (1997), *Teleological Language in the Life Sciences* (Lanham, Md.: Rowman & Littlefield).

Ospovat, D. (1981), *The Development of Darwin's Theory: Natural History, Natural Theology, and Natural Selection, 1838–1859* (Cambridge: Cambridge University Press, reissue 1995).

Paley, W. (1819), *Natural Theology* (*Collected Works*, iv; London: Rivington); 1st edn., 1802.

Richards, R. J. (1987), *Darwin and the Emergence of Evolutionary Theories of Mind and Behavior* (Chicago: University of Chicago Press).

——(1992), *The Meaning of Evolution: The Morphological Construction and Ideological Reconstruction of Darwin's Theory* (Chicago: University of Chicago Press).

Rudwick, M. J. S. (1964), 'The Inference of Function from Structure in Fossils', *British Journal for the Philosophy of Science*, 15: 27–40.

——(1972), *The Meaning of Fossils* (New York: Science History Publications).

Ruse, M. (1973), *The Philosophy of Biology* (London: Hutchinson).

——(1977), 'William Whewell and the Argument from Design', *Monist*, 60: 244–68.

——(1979), *The Darwinian Revolution: Science Red in Tooth and Claw* (Chicago: University of Chicago Press).

——(1982), *Darwinism Defended: A Guide to the Evolution Controversies* (Reading, Mass.: Benjamin/Cummings Publishing Company).

——(1986), *Taking Darwin Seriously: A Naturalistic Approach to Philosophy* (Oxford: Blackwell).

——(1989), *The Darwinian Paradigm: Essays on its History, Philosophy and Religious Implications* (London: Routledge).

——(1996), *Monad to Man: The Concept of Progress in Evolutionary Biology* (Cambridge, Mass.: Harvard University Press).

——(1999), *Mystery of Mysteries: Is Evolution a Social Construction?* (Cambridge, Mass.: Harvard University Press).

——(2000), *The Evolution Wars* (Santa Barbara, Calif.: ABC-CLIO).

——(2001), *Can a Darwinian be a Christian? The Relationship between Science and Religion* (Cambridge: Cambridge University Press).

Simpson, G. G. (1949), *The Meaning of Evolution* (New Haven, Conn.: Yale University Press).

Sommerhoff, G. (1950), *Analytical Biology* (Oxford: Oxford University Press).

Spencer, H. (1857), 'Progress: Its Law and Cause', *Westminster Review*, 67: 244–67.

Waddington, C. H. (1957), *The Strategy of the Genes* (London: Allen & Unwin).

Wagar, W. (1972), *Good Tidings: The Belief in Progress from Darwin to Marcuse* (Bloomington, Ind.: Indiana University Press).

Whewell, W. (1833), *Astronomy and General Physics* (Bridgewater Treatise, 3; London: William Pickering).

——(2001), *Of the Plurality of Worlds: An Essay*, ed. M. Ruse (Chicago: University of Chicago Press).

Williams, G. C. (1966), *Adaptation and Natural Selection: A Critique of Some Current Evolutionary Thought* (Princeton: Princeton University Press).

Wilson, E. O. (1975), *Sociobiology: The New Synthesis* (Cambridge, Mass.: Harvard University Press).

——(1980*a*), 'Caste and Division of Labor in Leaf Cutter Ants (hymenoptera formicidae, Atta) I. The Overall Pattern in Atta sexdens', *Behavioral Ecology and Sociobiology*, 7: 143–56.

——(1980*b*), 'Caste and Division of Labor in Leaf Cutter Ants (hymenoptera formicidae, Atta). II. The Ergonomic Optimization of Leaf Cutting', *Behavioral Ecology and Sociobiology*, 7: 157–65.

——(1983*a*), 'Caste and Division of Labor in Leaf Cutter Ants (hymenoptera formicidae, Atta) III. Ergonomic Resiliency in Foraging by Atta cephalotes', *Behavioral Ecology and Sociobiology*, 14: 47–54.

——(1983*b*), 'Caste and Division of Labor in Leaf Cutter Ants (hymenoptera formicidae, Atta) IV. Colony Ontogeny of Atta cephalotes', *Behavioral Ecology and Sociobiology*, 14: 55–60.

——(1992), *The Diversity of Life* (Cambridge, Mass.: Harvard University Press).

——(1994), *Naturalist* (Washington: Island Books/Shearwater Books).

Woodfield, A. (1976), *Teleology* (Cambridge: Cambridge University Press).

Wooldridge, D. (1963), *The Machinery of the Brain* (New York: McGraw-Hill).

Wright, L. (1973), 'Functions', *Philosophical Review*, 82: 139–68.

——(1976), *Teleological Explanation* (Berkeley and Los Angeles: University of California Press).

Part Two

Analysis: Functional Explanations Today

3. A Rebuttal on Functions

Christopher Boorse

ABSTRACT

This chapter answers most major objections to a general goal-contribution (GGC) analysis of functions as causal contributions to goals of a goal-directed system. Such an analysis, which applies univocally to function statements about artifacts and organisms, is defended against the following criticisms: that its choice of goals cannot be objective; that it cannot accommodate functional explanation; that it gives objects external to an organism biological functions; that it relativizes biological functions to an organism's environment; that it eliminates both maladaptive and unperformed functions; that it cannot distinguish functions from beneficial accidents; and that it conflates the functions of protective coloration in Batesian mimic and model. Applied to organisms, a GGC analysis, if supplemented by the statistical idea of the species-typical, also explains biomedical normality. Such a view is defended against writers who use typically unperformed functions, pandemic diseases, or Plantinga's cases to attack any statistical concept of biomedical normality.

This essay has a narrow aim: to answer all major criticisms, except two,[1] of a kind of analysis of function statements previously offered by several writers. Let me call it a general goal-contribution (GGC) analysis, since it defines functions generally as causal contributions to goals. This analysis seeks to cover a wide range of functions, including ordinary ones in artifacts, biological ones in living organisms, and perhaps others as well, as in social science. To achieve this breadth, the GGC analysis that I and others defend views goal-directedness as a property manifest in the behavior of, at least, all living organisms. One might, of course, argue that, since intentionality entails goal-directedness, the intentionalistic accounts of teleology by writers such as Woodfield (1976) and Nissen (1997) are one kind of GGC account. What such views lack, however, is any naturalistic way to accommodate a full range of biological functions, including those in plants and in lower animals with too little psychology for intentions. By contrast, the kind of GGC analysis I defend—for which I hereafter reserve the GGC label itself—takes goal-directedness to be an objective, non-mental

I thank Fred Adams, Robin Andreasen, Karen Neander, Joel Pust, and Michael Rea for their aid.

[1] Unanswered here are objections to the cybernetic analysis of goal-directedness and Amundson's attack (2000) on normal function. See Sect. 4.

property of all living organisms. It thus lets function statements be literally true throughout the whole biological domain, not merely metaphorically true, heuristically useful, or the like.

Since evolution in fact seems to yield organisms with the supreme goals of individual survival and reproduction (loosely, 'fitness'), within biology the GGC analysis gives the same results as one defining biological functions, specifically, as causal contributions to fitness. That approach, which I will call an S&R view,[2] has many defenders in philosophy of biology. They include Canfield (1964) and Ruse (1971), and likewise Bigelow and Pargetter (1987), who simply replace contributions with dispositions to contribute. The specificity of such an account is a defect. It does not cover artifactual functions at all, except via dubious artifactual analogues of fitness. On the other hand, it escapes one major kind of objection to the GGC approach: criticisms of the 'cybernetic' analysis of goal-directedness that GGC writers typically employ. Anyway, many function writers really care only or mainly about biological teleology, as a problem in philosophy of science. Hence, to defend the GGC account against the objections that I answer here is important even if no general analysis of goal-directedness ultimately succeeds. For GGC and S&R analyses have common biological implications, which seem superior, for reasons at which I can only gesture here, to those of rival theories.

To orient ourselves, let us begin with a quick survey of major approaches to defining function and their characteristic difficulties.[3]

1. Other Types of Function Analysis

1.1. Causal-Role (CR) Analysis

First is Cummins's theory (1975), now often called the 'causal-role' (CR) analysis of function. Cummins took an item's function simply to be its disposition causally to contribute to the output capacity of a complex containing system. To identify each part's contribution to the output of the whole is to give a 'functional analysis' of a system. That is the only kind of functional explanation Cummins accepts, rejecting any role for an item's functions in etiological

[2] One might think a natural term would be SGC, since this view defines functions as causal contributions to specific goals, survival and reproduction, and merely lacks any general goal-concept. But one might equally say it makes functions contributions to specific goods, while lacking any general concept of the good, and therefore call it a specific-value-centered (SVC) view. The neutral name S&R is most accurate.

[3] I omit details of other writers' function analyses, as well as differences in their intended scope and methodology. My taxonomy of analyses and their problems owes a large debt to Melander's monograph (1997). This work and Nissen's book (1997) are superb surveys of the function literature.

explanations of its presence. Since, in practice, virtually any effect of any item can be so viewed, this approach admits the largest class of functions. Indeed, the CR analysis is now generally seen to have a serious problem of overbreadth. For one thing, it generates functions in non-biological sciences where teleological language is absent. It implies that the function of mists is to make rainbows (Bigelow and Pargetter 1987: 184), the function of rocks in a river is to widen the river delta (Kitcher 1993: 390), 'the function of clouds [is] to make rain with which to fill the streams and rivers' (Millikan 1989*a*: 294), and the function of a piece of dirt stuck in a pipe is to regulate the water flow (Griffiths 1993: 411). Moreover, it creates false functions within biology too. Relative to our capacity to die of fluke infestation, our liver's capacity to house liver flukes is its function (Griffiths 1993: 411); relative to our ability to grow gigantic malignant tumors, oncogenes have many functions (Kitcher 1993: 390; Melander 1997: 53–4). To cure such overbreadth, rival accounts add a second condition to distinguish functions clearly from mere effects, thus essentially choosing a subset of CR functions as true functions.[4]

1.2. Etiological Analysis and Selected-Effect (SE) Variants

Several years before Cummins's paper, Wright (1973) had based his 'etiological' analysis on the idea that to cite an item's function is to explain why it is there. Seeking a univocal analysis equally applicable to artifacts and organisms, Wright defined the functions of any item as those of its effects which, via causal history, explain its presence. This general, undifferentiated etiological analysis has awkward theoretical features that I (Boorse 1976*a*) and others, including Wright himself, identified. It also suffers damaging counter-examples, such as my gas leak, which persists by asphyxiating its repairman (Boorse 1976*a*: 72), or van Gulick's stick in a stream, pinned in place by its own backwash (Bedau 1991: 648). To escape both counter-examples and theoretical defects, many later writers offer analyses requiring both a specific type of etiology, selection, and specific selection mechanisms for different domains—for example, designer's intention for artifacts, Darwinian selection of a trait's genotype for organisms. This approach, now often called a selected-effect (SE) analysis, is illustrated by Neander (1983, 1991*a*, *b*, forthcoming), Millikan (1984, 1989*a*, 1993), Griffiths (1993), and Godfrey-Smith (1994). Such theorists debate the time period over which natural selection must favor a trait's effect for it to be its function. Possibilities include distant past, recent past, present, and logical

[4] Admittedly, many theorists claim that 'malfunctioning' tokens have functions they cannot perform, whereas on Cummins's view an item must be capable of performing its function to have it. But the CR approach can handle this point just as I do. It can distinguish a token's actual function either from its own intended function, or from the normal or intended function of its type. See Sect. 3.7.

combinations thereof (Kitcher 1993: 384–7). More important than this ambiguity in selection period, however, is a different point. None of the listed writers finds an SE theory adequate for all function statements. The usual view is that, even within biology, the selectional analysis applies only to 'proper functions', to use a term introduced by Neander and Millikan. For function in general, all these writers seem to be CR–SE 'pluralists', a recently popular view to be discussed momentarily.

As is now widely recognized, counter-examples make even a specifically Darwinian SE analysis look both too broad and too narrow. It looks too broad for at least two reasons. First, objects' traits can be selected by a process isomorphic to Darwinian selection without scientists' calling them functional, such as the surface features of Schaffner's ball bearings (1993: 383–4) or the heritable variations in clay crystals that favor their survival and reproduction (Cairns-Smith 1965; Dawkins 1986; Bedau 1991: 651–4). Secondly, even in organisms, various sorts of traits maintained by natural selection are not assigned functions by biologists: now-familiar examples are junk DNA, selfish DNA, and segregation-distorter genes.[5] Conversely, the SE analysis also looks too narrow for biology, since it is increasingly clear that biologists are willing to assign a trait a function for which it is either not known to have been selected or known not to have been selected. In the former category—function attribution without evolutionary evidence—are my example of Harvey's discovery of the heart's function (Boorse 1976*a*: 74; cf. Enç 1979: 345–50); Schaffner's example (1993: 386–7) of Miller's discovery of the immunological function of the thymus; Rosenberg's example (1985: 38–40) of the function of the chemical difference between DNA and RNA; and Amundson and Lauder's many examples (1994) from functional anatomy. Schaffner (1993: 388) also offers an imaginary case of a clearly unselected function, the immunologically functional bursa of Fabricius in a mutant sterile human being born near Three Mile Island. As for real unselected functions, there is the large category of 'exaptations' offered by Gould and Vrba (1982) and others: feathers arising as insulators and co-opted for flight, bone arising as a phosphate reservoir and co-opted for rigidity, heron wings designed for flight but used to shadow aquatic prey. We are now constantly reminded that traits can persist for a long time for reasons other than selection—genetic drift, a lack of mutations, pleiotropy, genetic linkage, and so on. Nonetheless, if all sea turtles dig eggholes with their flippers (Lewontin 1978: 218), for thousands of years, there seems no doubt that such use is one current function of these organs, regardless of whether the behavior is under genetic control or results from any selective process.

[5] Godfrey-Smith (1994: 347–8); Manning (1997: 74ff). Manning argues at length that, without a goal-concept, SE writers' attempts to avoid such biological counter-examples fail.

1.3. *CR–SE pluralism*

A recently popular reply to such cases grants that biology admits both selected and unselected functions. The new view, dubbed by Godfrey-Smith (1993) 'consensus without unity', holds that there are different types of biological functions requiring different analyses. Usually, it is said that evolutionary functions follow an SE analysis, while other (for example, purely anatomical or physiological) functions follow a CR analysis.[6] An obvious defect of such a 'pluralist' analysis is its disunity. As Godfrey-Smith admits, it makes the term 'function' 'highly ambiguous' (1994: 344), and he can find 'no strong reason for using the same word' for both CR and SE functions (1993: 206). Worse yet, this view solves only one of the three scope problems we have seen. It escapes the SE view's excessive narrowness by shunting onto a CR alternative any unselected functions that arise in anatomy, physiology, exaptation, mutant individuals, and so on. But it does nothing to cure the overbreadth of either the SE analysis (clay crystals, segregation-distorter genes) or the CR analysis (rain clouds, oncogenes), which are effectively its disjuncts. The latter problem vanishes in an S&R–SE pluralism, which is nearly Melander's analysis (1997) and a much more plausible view of biological functions alone. But such a view remains disunified. It is also of narrow reach, unable literally to cover nonbiological functions in other sciences or in artifacts, since it lacks any general concept like that of a goal by which to do so.

1.4. *Value-Centered (VC) Analysis*

Finally, 'value-centered' (VC) views instead add to the CR kernel a value component, analyzing a function as an effect useful or good for some beneficiary. On such a view, every function statement is normative in the truest sense: it is or entails a value judgment. The leading current VC analysis is Mark Bedau's well-defended view.[7] His key idea is that every living organism has a 'good of its own' or 'intrinsic' good (1992*a*: 45), while such inanimate objects as clay crystals, magnetic pendulums, and sticks in a stream do not. While Bedau's distinction among three 'grades of teleology' (1992*b*) has technical problems that Melander notes (1997: 70–2), the main difficulties with Bedau's position seem to be two. First, I am unconvinced by the efforts of Bedau and his sources,

[6] As just noted, this seems to be the actual position of Neander and Griffiths, although these two writers say almost nothing about non-'proper' (i.e., for them, CR) functions in biology. The position is fully developed by Amundson and Lauder (1994), Godfrey-Smith (1993), and Millikan (1989*b*). Preston (1998) argues that Millikan is not consistent in her pluralism; Millikan (1999) replies.

[7] Bedau (1990, 1991, 1992*a*, *b*, 1993). An influential earlier VC writer is Sorabji (1964); see Bedau (1992*b*: 781 n. 1) for other theorists he finds similar. I here ignore affinities between the VC view and the intentionalistic position—e.g. Woodfield (1976).

Taylor (1986) and Callicott (1989), to show that all living organisms have intrinsic value of a kind that artifacts such as watches or pianos do not. Writers who attribute intrinsic interests to plants or bacteria, or a good or welfare of their own, are, I would argue, either anthropomorphizing, or advocating incomprehensible values, or confusedly referring to some descriptively definable property such as life or goal-directedness, in which case the VC view metamorphoses into a GGC view. Secondly, Bedau's actual analysis seems to be an SE–VC conjunction, since he seems to reject watering plants as the function of rain only because this effect plays no role in explaining rain. But, if functions are etiologically significant contributions to value, then Bedau suffers all the same counter-examples of unselected function as a pure SE view.

2. Goal-Contribution Accounts

To motivate a goal-contribution analysis, one may first observe that the functions of artifacts seem obviously determined by human goals. Cutting is the function of a knife or saw because that is the purpose for which humans use it. Objects identical to our knives and saws would have no function at all if produced by some random process on a planet devoid of life (cf. Bedau 1993: 36). Conversely, they can have many alternate functions here, if people use them differently: a knife can become a paperweight, a saw a musical instrument, an orange crate a bookcase, a jelly jar a drinking glass (Preston 1998: 220). Moreover, objects not artifacts at all can still acquire functions via people's purposive use of them, as with Cummins's bowl-shaped depression in a rock, used to hold holy water in a primitive religious ritual (1975: 747). Such pure use-functions suggest that neither design nor selection is required for artifact function, as so many writers assume, but only an object's role in a human goal-directed activity.

Goal-directed behavior also seems ubiquitous among living things, not just human beings. As Nissen (1997: p. vii) emphasizes, most of our standard vocabulary for describing organisms' behavior implies goals: 'Teleological language is not limited to the explicit use of such expressions as "purpose" and "function". Common expressions such as "flee", "protect", "hide", and "migrate" are also teleological. If an animal flees, for example, it is running in order to escape.' But, insofar as all organisms—even insects, worms, protozoa, and plants—display purposive or goal-directed behavior, it is implausible to base this concept on complex psychological notions such as concepts or intentions. To describe a fly as trying to avoid a bird's beak, or a slug as entering a cellar in search of food, is surely not to ascribe intentions to such lower organisms. Biologists tell us that 'plants exhibit heliotropism, that is, turn in varying directions *in order to* maximize their exposure to the sun. But no biologist is prepared to find the causes for this behavior in a plant's beliefs about the role of photons in photosynthesis and in its desire to maximize the number of photons landing on its leaves' (Rosenberg 1985: 44). Thus, we expect the purposiveness common to living organisms' behavior to be analyzable in some non-mental, but naturalistic, way.

To do so, writers offering a GGC analysis of functions have typically relied on some version of a 'cybernetic' account of goal-directedness using ideas from systems theory. That account derives mainly from Rosenblueth, Wiener, and Bigelow (1943), but was elaborated by Sommerhoff (1950, 1959), Braithwaite (1953), and Beckner (1959) before its first use in a GGC analysis of functions by Nagel (1961).[8] According to the Sommerhoff–Nagel view, a system S is 'directively organized', or 'goal-directed', toward a result G when, through some range of environmental variation, the system is disposed to vary its behavior in whatever way is required to maintain G as a result. Such a system, it is said, shows 'plasticity' and 'persistence' in reaching G: when one path to G is blocked, another is available and is employed. An important further requirement imposed by Sommerhoff is independence ('orthogonality') of system variables. Each must be nomologically able to vary independently of the others, a test which rules out some equilibrium-seeking systems that are intuitively not goal-directed, such as an ordinary pendulum or a marble rolling around in a bowl. Various cybernetic writers also take two further ideas as basic elements in goal-directedness: negative feedback and the internal representation of a goal-state (Nissen 1997: ch. 2). The latter helps, for example, to deal with Scheffler's two problems (1959). One is the problem of the missing goal-object, as when a roach searches for water that is not there. In this case the goal G cannot be attained no matter what the system does. The other problem is multiple goals: it seems that a cat waiting by an empty kitchen mousehole may have the goal of catching a mouse, even though its behavior is equally appropriate to catching a bowl of cream, and would be enthusiastically adjusted if a bowl of cream should suddenly appear.

Such a cybernetic account seems a promising analysis of the goal-directedness apparently shared not only by all organisms, but also by some artifacts, such as cruise missiles, stereo tuners, and even the classic simple servomechanism, a thermostatically controlled home furnace. Two general differences between organisms and artifacts are notable. First, all the gross behavior of organisms seems directed to two ultimate goals: individual survival and reproduction. Specific activities such as flying, hunting, nest-building, copulation, and parental care all seem to be part of a way of life promoting Darwinian fitness. Conversely, even in the simplest organisms, goal-directedness appears at several

[8] A brief summary of this history is in Schaffner (1993: 365–70), while Nissen's first two chapters (1997: 1–69) are a fully detailed critical account.

lower levels of internal organization. As for the physiology of vertebrates, it is a very complex hierarchy of systems and subsystems, ascending from the level of genes through organelles, cells, tissues, organs, and organ systems to gross behavior. Each level's output is directed to goals that serve as input to the next. At the organ-system level, familiar examples of internal goal-directedness are temperature regulation in homeotherms, a necessary condition of all other processes, and Nagel's example of control of blood osmolarity via the kidneys, muscles, and skin (1977: 272–3). At a more chemical level are Rosenberg's detailed flow charts for the liver's regulation of blood glucose within fixed limits via two pancreatic hormones, insulin and glucagon, and the adrenal hormone epinephrine (1985: 55–7). On the level of organelles, Rosenberg (1985: 61) notes that

> the function of the mitochondrion is to produce ATP molecules, and it does this by employing two different means: Some ATP is produced in the citric-acid-cycle process, and more is produced by oxidative phosphorylation, a different process with the same goal; moreover, this latter process generates ATP molecules by the operation of any or all of at least four different independent internal subsystems.

Countless similar examples, with appropriate flow charts, fill our textbooks of physiology and biochemistry. In short, organisms are 'vast assemblages of systems and subsystems which, in most members of a species, work together harmoniously in such a way as to achieve a hierarchy of goals', with individual fitness at the apex of the hierarchy (Boorse 1975: 57). To describe this hierarchy is to do Cummins's 'functional analysis' on a grand scale.

Given this picture, the best account of functions,[9] I believe, is to define functions simply as causal contributions to goals. To develop this view, begin with what one might call a *weak function statement*. Given system *S* directed to goal *G* at time *t*, one can say that

> *X performs the function Z* in the *G*-ing of *S* at *t* if and only if at *t*, the *Z*-ing of *X* is a causal contribution to *G*.[10]

[9] Henceforth, what I mean to defend is my own account (Boorse 1976*a*, 1977). A great deal of the defense, however, supports other GGC theorists equally.

The first GGC theorist I know of is Nagel (1961), who clearly stated twice in his text (408, 422)—though, oddly, not in his official analysis—the thesis that functions are contributions to the goals of a goal-directed system. In turn, Nagel's chapter inspired me (Boorse 1976*a*) and Adams (1979) independently to our views. In theory, Adams differs from me in holding that a goal-contribution, to be a function, must help explain the item's presence (1979: 510). Thus, he holds an etiological, but not a selectionist, GGC view. In practice, it appears that Adams' view and mine will agree on all biological cases except a very exotic one, the function of an inert part in the first generation. But the relation between our two positions is too complicated to discuss here.

[10] Cf. Boorse (1976*a*: 80). Equivalent variants are '*X serves the function Z*' and '*X functions as a Z-er*'.

On this view, functions can be performed only once and by accident, as when my telephone summons me to the dining room just before a truck crashes into my parlor, or a beesting on my nose brings me to a doctor who spots a curable melanoma on my neck. One would hardly declare defense against trucks to be the function, or even a function, of telephones, or cancer diagnosis to be a function of beestings. But there is no reason to deny that for me, on this occasion, the beesting served that function. Indeed, even etiological writers such as Wright and Neander accept such a weak function statement as true, although its truth has nothing to do with etiology and so escapes their analysis. On my view, weak function statements reveal what a function is—a contribution to a goal—and it remains only to find further conditions for the *strong function statements* that function writers usually analyze:

The function of X is Z.
A function of X is Z.
X has the function Z.

As I have said (Boorse 1976*a*: 80–1), however, the difference between weak and strong function statements seems to be simply a loose matter of how often or consistently the function gets performed. If Schaffner's mutant bursa of Fabricius (1993: 388) blocked viral infections throughout its owner's life, it would be quite natural to call antiviral defense the bursa's function in this man. But how often, or how regularly, a function *Z* must be performed by *X* in order to become a function of *X*, or the function of *X*, or *X*'s function, or the, or a, function that *X* has, cannot be precisely specified. There are no rules for converting functions performed into functions possessed. Vagueness is inevitable; but it also seems to afflict all other analyses, contrary to the impression their proponents leave. On Cummins's view, how often must a soldier's Bible stop bullets, or within what range of combat environments must it be disposed to stop them, to 'have' the function of stopping bullets? On Bedau's view, how often must the Bible thus contribute to the good of its soldier beneficiary? Cummins offers no answer, while Bedau will apparently appeal to etiological explanation. But even on an evolutionary etiological view like Neander's, it seems vague how often effect *Z* of trait *X* caused by allele *A*, to be a 'proper' function of *X*, must have made an *A*-bearer better able to reproduce than bearers of an alternative allele *A′*. A genuinely selective effect can, after all, happen only once. The organism bearing the last copy of *A′* might die, instead of a rival *A*-bearer, because the rival had a lucky accident due to effect *Z* of *X*, while in every one of the millions of other cases of *A*-superiority, *A*-bearers outcompeted *A′*-bearers by *X*'s effect *Z′* instead. The last white peppered moth in northern England could die, not because it was more visible on industrially darkened trees than the black variety, but because, as it sat equally conspicuously next to a black one on a yellow porch, a psychotic teenage satanist stomped it to death to honor the forces of darkness. Nothing in Neander's or Millikan's analyses tells us how often selection of *A* and *X* for effect *Z* must occur for *Z*, besides *Z′*, to be an additional genuine, or 'proper', function of *X*. The only principled choice, it would seem, is once. But this answer, though precise, forces one to admit many fortuitous benefits of traits that these authors surely wish to call accidents, not functions. In my scenario, I doubt any etiological writer wants to get stuck calling allure for crazed satanists a function of dark coloration in *Biston betularia*—especially if the teenager, the last of his own breed, hangs himself the next day.

A trait *X*'s contribution to a goal, then, if made sufficiently often, becomes the function of *X*, or *X*'s function, if it is *X*'s only regular contribution; and is a function of *X*, or among *X*'s functions, otherwise. So we say of Schaffner's mutant—call him Fowler—that

The function of Fowler's bursa of Fabricius is to prevent viral infection.

Conceivably, we might even be tempted to call antiviral defense the *normal* function of Fowler's bursa, especially if it blocked viral infection except, say, for short periods when it itself suffered a bacterial infection curable by antibiotics. One does see occasional medical references to what is normal for an individual. But the standard medical concept of normal function, surely, is implicitly species-relative:

The [or *a*] *normal function of X is Z in species S*,

where *S* is *Homo sapiens* for ordinary medicine or an animal species for veterinary medicine. I have argued at length that medically normal function of any token item (for example, a single human heart) is analyzable as an output within a statistically typical range of contributions to survival and reproduction by tokens of that type in an age group of a sex of a species (Boorse 1977: 554–63; 1987: 370–2). So qualified, my view is that theoretical medical normality of the organism means statistically species-non-subnormal biological function of all its parts and processes.

Finally, it is important to see that to distinguish *the function x is performing*, *the function of x*, *the normal function of x in y*, and *the normal function of X in S*[11] is only to give a few instances of a widely useful general device, the qualification of 'function' by adjectival phrases. One would think this tool too obvious to need a pretentious name, but in fact it is so commonly neglected in the function literature[12] that I will give it one: the *adjectival strategy*. In general, a good philosophical approach to analyzing a common noun N—'function',

[11] Since the token/type distinction is crucial to avoiding conceptual confusion about functions, henceforth I will employ Walsh and Ariew's clear notation (1996: 497 n. 10) that uses '*x*' for a token and '*X*' for its type.

[12] Exceptions are Boorse (1976*a*: 81–6) and Achinstein (1977: 351).

'disease', 'pornography', 'discrimination', 'art'—is as follows. First, find the broadest plausible analysis *A*: an analysis that is clearly not too narrow, but is not clearly too broad. Secondly, try to handle all debatable cases of overbreadth by adjectival qualification, that is, by conjoining properties—'normal function', 'serious disease', 'misogynistic pornography', 'unjust discrimination', 'good art'. Given general agreement that all *N* are at least *A*, one suggests to critics who deny that *x*, an *A*, is an *N* that they really mean to deny only that *x* is a particular kind of *N*: an *MN*.

One strength of the GGC analysis of function is that it fits this pattern. It is not, like CR function, hopelessly overbroad. In typical counterexamples to Cummins's analysis, either the system in question is not goal-directed (river delta, waterpipe) or the effect in question damages its goals (liver flukes, oncogenes). At the same time, the GGC analysis can say everything its other rivals can say by simple adjectival qualification. Among artifacts, one can distinguish *the function X was designed to perform* from *the function X is being used to perform*. For biological traits, one can distinguish *the past function of X* from *the present function of X*. One can admit that egghole-digging is *a function of* sea turtles' flippers, though not *an evolutionary function of* them. One can even define different types of evolutionary function, those noted by Kitcher and more besides, depending on the time period of selection or any other factor of interest: *the recent evolutionary function of X, the recent and current evolutionary function of X*, and so on. Best of all, one can suggest that such adjectival qualifications are implicit in biologists', or other function theorists', uses of the phrases *the function* and *a function*. For English definite and indefinite articles are often, if not usually, context-sensitive, with relevance conditions determined by linguistic or non-linguistic context of utterance. To a woman displaying a new engagement ring, the question 'Who's the lucky man?' asks for her fiancé, not the most recent winner of the state lottery. Similarly, when biologists say 'the function of *X*', it is plausible that evolutionists implicitly mean the evolutionary function of *X*, since their job is to explain traits' origin. Yet ordinary physiologists, seeking merely to understand how a body works with no thought of explaining its history, may intend no such qualification. Surely it is best to achieve such 'pluralism' about functions by the adjectival strategy—that is, by adding further properties to a unified, yet not overbroad, analysis of the term 'function' itself.

Before tackling ten leading objections, I wish to correct an increasingly popular misconstrual of GGC and S&R views as types of SE analysis. This confusion goes back at least to Bigelow and Pargetter (1987: 191–4), who describe the biological function of a trait both as a disposition to promote the organism's survival and as a present disposition to selection. Perhaps inspired by this idea, Walsh and Ariew's 'taxonomy of functions' divides analyses into Cummins functions and evolutionary functions. 'There are two general approaches to characterising biological functions. One originates with Cummins. According to this approach, the function of a part of a system is just its causal contribution to some specified activity of the system. . . . The other approach ties the function of a trait to some aspect of its evolutionary significance' (Walsh and Ariew 1996: 493). Walsh and Ariew then subdivide 'evolutionary functions' into two varieties: 'historical', covering the usual etiological theorists, and 'current', represented by Bigelow and Pargetter. Buller, going still further, sees among 'current theories' of biological function a 'core of agreement' that is 'as great a consensus as has been achieved in philosophy': that 'a trait or organ has a function in virtue of its role in a selection process—either its role in a selection process that a lineage bearing that trait or organ actually has undergone, or in virtue of a selection process it is currently undergoing or is set to undergo' (Buller 1999: 20). Even if true, this claim seems exaggerated in light of, say, the thousand-year consensus of all major European philosophers on the provability of God's existence. But it does not seem to me true even of Bigelow and Pargetter's account, not to mention other theorists whom Buller ignores.

The reason is that a trait can causally promote its bearer's survival in the absence of any selection process favoring the trait or its bearer.[13] In the first place, as many recent authors note,[14] for many possible reasons a trait may have no current rivals in the actual population; absence of mutations, pleiotropy, and genetic linkage are a few examples. In the causal-contribution analysis, if trait *T*'s effect *E* helps cause an individual organism *O*'s survival at time *t*, the most that could be implied is some counterfactual to the effect that, without *T* or *E*, *O* would not have survived at *t*, or at least would have been less likely to survive.[15] Suppose, for example, that all muskrats' hearts currently pump blood. That is a causal contribution to their individual survival. Still, there is no selection of muskrats with pumping hearts over muskrats without if no muskrats without exist in the first place to lose the competition. As Millikan (1993: 40) emphasizes, 'selection' implies actual differential survival of *T*-bearing organisms.

> The notion of superior fitness, as actually used in evolutionary biology, is a well-defined notion only because it is *never* taken to attach to any trait in a vacuum or absolutely but rather is understood only relative to alternative traits *actually found in the population*. . . . A trait's biological function is what it actually did—did most

[13] Schwartz (1999: S214–17) is especially clear on 'the distinction between a trait's *contributing to survival and reproduction of bearers by doing F*, and a trait's *being favored by natural selection for doing F*' (S221).

[14] See Bigelow and Pargetter (1987: 191); Millikan (1989*b*: 174–5); Kitcher (1993: 388); Melander (1997: 49 n. 7); and Buller himself (1998: 512–13).

[15] Of course, overdetermination cases are evidence that such counterfactuals are not, in fact, true in all causation.

recently—that accounts for its current presence in the population, as over against *historical* alternative traits no longer present.[16]

At most, one might claim that a causal contribution to survival implies hypothetical, not actual, selection. But now the key notions are survival and causation, not selection. Conversely, as we have noted, traits that make no contribution to organisms' survival or reproduction, such as junk DNA, selfish DNA, and segregation-distorter genes, can be preserved by natural selection; and natural selection can establish traits in objects like clay crystals from which functions are wholly absent. Finally, a goal-contribution theory awards functions to traits of sterile organisms, such as mules or Schaffner's Pennsylvania mutant. A sterile organism cannot undergo any remotely 'evolutionary' present or future process. Yet the intuitions shared by so many writers—including Bigelow and Pargetter (1987: 188), Melander (1997: 44), and even Neander (1991*a*: 169)—that the hearts of instant lions (Boorse 1976*a*: 74) would have functions are surely unchanged if the instant lions are sterile as well. Thus, it is best to view the GGC account I defend as logically independent of any evolution whatever—past, present, or future. Only then can we see how it surpasses SE views in accommodating our intuitions that unevolving, unevolved, and sterile organisms all can have parts with functions.[17]

3. Ten Leading Objections

I shall now answer ten of what seem to me the dozen best challenges to a GGC account of function. Although some of these objections were posed against other types of analysis, since they affect a GGC account equally, I will reply. Most of them can apply to all GGC analyses, and often S&R analyses too. But some are specific objections to my own statistical species-relative account of biomedical normality.

3.1. Arbitrary, Evaluative, or Circular Goal-Choice

Both Melander and Schaffner charge that the GGC approach has no naturalistic way to identify the goals of organisms, but instead requires the theorist to choose them arbitrarily or evaluatively.

Melander (1997: 32–4), adapting Nagel's example, imagines that a person's kidneys and muscles reset the target they maintain for blood water content at

[16] Similarly, Neander (forthcoming: 53 n. 13) says: 'Since there is no selection without variation, a trait that has no competition will not be selected for anything in the future (until an alternative allele arises)'. Millikan's argument (1989*b*: 174) that Bigelow and Pargetter's account is incoherent is, however, unconvincing.

[17] Thanks to Michael Rea for this concise summary.

70 per cent, rather than the 90 per cent normal for a human body. Melander says that, although 'this would doubtless be highly deleterious for the organism', on Nagel's analysis 'there seems to be no rationale whatsoever' for judging this new goal 'wrong' or 'dysfunctional'. He does immediately concede that Nagel has an obvious answer: the new water-content goal is a malfunction because it no longer serves the goals of a higher-level body system. Further, Melander recognizes that cybernetic GGC theorists intend this regress to terminate in the highest-level goals of organisms, those at the apex of the natural functional hierarchy. But he objects that to 'invok[e] the notions of health and life' as apical goals has two defects. First, since this move has no counterpart in artifacts, it destroys the 'symmetry' between biological and non-biological teleology and ruins the account's unity. Secondly, it ruins the account's naturalism, since health and life cannot, 'on pain of circularity', be analyzed in cybernetic terms. Thus, Nagel's analysis 'glaringly fails' to explain how biological science can make such '"normative" judgments' as the claim that the deviant kidneys are doing something 'they are "not supposed to" be doing'.[18]

As long as artifact functions are determined by human purposes, I see no disunity in the analysis. Human purposive behavior is one example of goal-directedness. For the analysis to be unified, there is no reason why all goals must be determined in the same way. As for the objective justification for health and life as apical goals of biological function, Sommerhoff believed that the essence of life was precisely the natural hierarchy of goal-directed processes found in organisms. He wrote:

> On the phenomenal level from which all science must proceed, life is nothing if not just this manifestation of apparent purposiveness and organic order in material systems. In the last analysis, the beast is not distinguishable from its dung save by the end-serving and integrating activities which unite it into an ordered, self-regulating, and single whole, and impart to the individual whole that unique independence from the vicissitudes of the environment and that unique power to hold its own by making internal adjustments, which all living organisms possess in some degree. (Sommerhoff 1950: 6)

I cannot here discuss the problem of defining life. But, if Sommerhoff was right that life just is a kind of natural goal-directedness, then it is not arbitrary to take life's continuation, individual survival and reproduction, as the apical goal. And what a specific organism's next-highest goals are—that is, what activities its life consists in—is, on the cybernetic analysis, a fact. As for health, on my view it is the absence of pathology, which in turn is simply species-subnormal part-function. So Melander's deviant kidneys and muscles[19] function pathologically

[18] Melander himself puts the terms 'normative' and 'not supposed to' in quotation marks because he believes that the normativity of function statements is only 'apparent' (1997: 16; see pp. 95–7).

[19] Actually, blood osmolarity is regulated by the hypothalamus, so the case should be about a deviant hypothalamus.

as compared with other human kidneys and muscles, insofar as they contribute less efficiently to the person's higher-level goals. None of these statements involves any arbitrary, evaluative, or non-naturalistic choice, given the cybernetic analysis.

Schaffner's version (1993) of the goal-choice objection is harder to follow. He complains that Nagel's account 'does not . . . present any criteria for goal states . . . since it provides no noncircular means of identifying a goal state in nongoal terminology' (p. 367). Yet he concedes that Braithwaite and Beckner characterize goal states as 'termini which are achieved under a variety of conditions'—that is, they hold that any results reached with plasticity and persistence count as goals of the organism. He also notes Beckner's claim that the goal of an organism's behavior can thus in principle be empirically 'discovered', at least up to a set of alternatives. Still, he says, 'the important philosophical problem is the introduction of the goal *concept*, expressed in goal language' (p. 368). What bothers Schaffner, then, seems not to be any kind of empirical underdetermination of goals, as with Melander, but what justification there is for calling any such empirically fixed terminus a goal at all. I can understand this worry only as based upon his belief in the 'implicit evaluative character' (p. 373) of teleological statements. He apparently assumes that to call any system output a goal, rather than just an effect, is a value judgment. Therefore, he charges, to describe any outcome, even 'decreased morbidity and mortality', as a biological goal requires either 'an unsubstantiated optimality interpretation of evolutionary theory' (p. 389) or some analogue to human purpose. Since, as he explains at length, optimality interpretations of evolution are unjustified, 'evolutionary theory does not entail any goal' (p. 375). Schaffner concludes that biological function statements are actually unwarranted and, as Hempel said, should play only a 'heuristic' (p. 410) role in science.

This criticism combines several disparate ideas. First, if one assumes that function statements are value judgments, then, of course, both biological and non-biological function statements involve taking some valuable state as our goal. So if evolution does not establish values—as I agree it cannot—it is not surprising if we must find our ultimate values by 'overt or covert postulation' (p. 384). But cybernetic GGC theorists reject Schaffner's original premise that function statements are value-laden. Moreover, I do not see how evolution could establish values even were optimality theory true. Accordingly, on the GGC view, the issue of goals of evolution is irrelevant to functions. Schaffner discusses that issue at length, attacking George Williams' doctrine (1966: 8–9) that every 'biological mechanism' has a function or goal fixed by evolution. Schaffner stresses that only one of the three factors governing evolution (natural selection) is directional, while the other two (mutation and genetic drift) are random. But the GGC view needs only individual organisms to be goal-directed, not the evolution that produced them. And it need not claim with

Williams that every biological mechanism serves a goal, only that every biological mechanism with a function does so. The existence, however common, of biological mechanisms with no function is irrelevant to an account of what a function is in mechanisms that have one.

3.2. *Lack of Explanatory Power*

For two reasons, a GGC analysis has been claimed to give functions insufficient explanatory power. One reason is Wright's, later Price's, doctrine that function statements always offer etiological explanations. Independently, both Neander (1983: 98–100) and Melander (1997: 36–8, 56) also imply that, on a GGC account, a trait's function can never explain its presence at all. Both charges are easy to rebut.

First, Wright held that function statements are inherently etiological explanations: 'The ascription of a function simply *is* the answer to a "Why?" question, and one with etiological force' (1976: 80). Price agrees: 'function statements are intrinsically explanatory: to ascribe a function to a device is to offer an explanation of its presence' (1995: 153). In the first place, however, there was never any basis for assuming function statements to be inherently explanatory of anything, any more than statements about organisms, cells, habitats, populations, satellites, polymers, dipoles, or most other objects of biology and other sciences. In the second place, even if function statements have to be inherently explanatory, a satisfactory kind of non-etiological functional explanation is available: Cummins's functional analysis (1975), undeniably prominent in biological fields like physiology. In the third place, that there are unselected biological functions is part of the current 'consensus' (Godfrey-Smith 1993) shared by writers as diverse as Schaffner (1993), Amundson and Lauder (1994), Walsh and Ariew (1996), and Melander himself (1997).[20] To attribute such functions is not to offer any etiological explanation. Thus, the Wright–Price thesis is increasingly abandoned, even by SE theorists.

Once one recognizes unevolved functions, their prevalence or rarity is, of course, an empirical question. In a passage equally relevant to the GGC view, Neander describes the position of Canfield (1964) and Wimsatt (1972) on evolutionary functional explanation as follows: 'Given that a function attribution tells us that the item contributes to fitness, and we know that natural selection selects items for such effects, we have a prima-facie case for supposing that the item was selected by natural-selection for that effect which is its function'. She then complains that 'both writers suppose the prima-facie case to be much stronger than it is' (Neander 1983: 98–9). It is, however, of no importance to

[20] Melander's final position is pluralist: there are 'two different notions of function at work in contemporary biology', which he calls weak and strong. Weak functions can be unselected (1997: 89ff).

the analysis of 'function' how strong that case is—what proportion of traits' current functions explain their presence—as long as unselected functions exist. It would be important if all function statements had to be etiological explanations. In that case, this proportion would determine the inductive strength of the generic inference from GGC function to evolutionary etiology. But if we reject Wright's thesis, Neander's criticism loses its force.

Secondly, in similar passages, Neander and Melander argue that GGC functions not only do not always explain a trait's origin, but can never do so. Neander (1983: 98) writes:

> Taken on its own, a function attribution, on a goal contribution account, cannot serve as a teleological explanation. To claim that it is the function of the heart to circulate blood is to claim that the heart's circulation of blood is a species-typical contribution to survival and reproduction. This, in itself, entails nothing about the etiology of the heart.

Likewise, in criticizing 'intrasystemic role analyses', which is his category for Cummins and Prior, Melander (1997: 56) makes a complaint that would apply equally to a GGC analysis: that they

> fail to account for the fact that citing a trait's function at least in some cases is equivalent to offering an explanation of why the trait exists. (E.g., to say that the function of the long neck of giraffes is to enable the animals to reach up into the trees can—at least in some contexts—be considered tantamount to giving an explanation of why the long neck exists.) But to be told that a disposition of a certain trait contributes to the manifestation of some high-level capacity (whether that high-level capacity is the fitness of the organism or not) is not to be told anything specific of why the trait exists.

That is 'obvious', says Melander, because the causal-contribution claim is 'compatible with just about any explanation of why the trait exists', including that it was 'miraculously created five minutes ago'. Earlier, he observed that the Hempel–Nagel deductive-nomological schema of functional explanation cannot causally explain a trait's presence because it fails to cite anything about the past (Melander 1997: 36–7).

I agree with this last point: the Hempel–Nagel schema is no model of evolutionary, or any other causal, explanation. But that schema is separable from the GGC view, and neither Adams nor I offered any version of it. What I believe, with nearly all other current writers on biology, is the following: a disposition D of a trait type T causally to contribute to the goal of individual fitness can, via evolutionary theory, explain the prevalence of present tokens of T by D's manifestation in past tokens. Since such contribution is a GGC function, that immediately solves what Melander (1997: 14) calls the 'Explanation Problem' of 'understanding how ascribing a function to a biological trait can help explain the trait's existence'. Almost invariably, evolutionary explanations of traits are in terms of their past GGC functions. No author, including Melander, gives reason to think that when evolutionary explanation of a trait by its function is possible—that is, when a currently functional trait evolved by performing that function in the past—the GGC view of functions somehow blocks this explanation.

What the analysis does not do is to write even the existence of such an explanation, let alone its details, into the meaning of the function statement itself. But there is no reason why it should. Rather, it seems to be Neander and Melander who commit a fallacy about explanation.[21] 'In itself,' that Booth shot Lincoln is also 'compatible with just about any' imaginable cause of Lincoln's death. Booth could have shot Lincoln, yet Lincoln have died instead of liver cancer, or eating toadstools, or being bitten by a rabid weasel. Still, it is a true causal explanation of Lincoln's actual death that he died because Booth shot him. It may be an incomplete explanation, since the explanans does not 'entail' the explanandum. But in the first place, all explanations are similarly incomplete in practice, even in science. Nevertheless, some deaths are explained by shootings, and so a past GGC function can still explain a trait's evolution. Secondly, even if complete explanations must be deductively valid, as on the D–N view, there is no reason why a function statement must be the sole premise of a functional explanation. For evolutionary functional explanations, presumably other premises state background theory, and still others initial conditions, such as the population distribution of rival traits.

Melander is right that 'in some contexts', such as evolutionary biology, one can agree to mean by 'the function of *X*' *the evolutionary function of X*. If so, one's function statements will have essential etiological explanatory force—they will be 'equivalent to' or 'tantamount to' an evolutionary explanation. But, even then, there is no reason why all the premises of a full evolutionary explanation, including background theory and initial conditions, must be part of a function statement's meaning. Whatever the correct schema of functional explanation and its details may be in any particular case, the GGC analysis could be inconsistent with them only if natural selection sometimes favors traits that are (i) functions but (ii) not causal contributions to goals of the organism. No writer gives any reason to think that this is so. Evolved traits of SD genes or clay crystals are not, of course, contributions to a goal of a goal-directed system, but they are also not called functional by scientists, which makes the GGC view superior to its CR and SE rivals. Conversely, however one explains the origin of traits like hearts, eyes, livers, or long necks via their fitness benefits, any such evolutionary explanation is in terms of these organs' GGC functions—namely, their causal contributions to goals of the organism, survival and reproduction. So the GGC account has no defect of missing explanatory power.

[21] I owe this point to Michael Rea.

3.3. Functions of External Objects

Nissen (1997) taxes Adams's account (1979) with unpleasant results as to functions of objects external to organisms, and his criticisms affect me equally. Regarding human artifacts, both Adams (1979: 509) and I (Boorse 1976*a*: 79) took the position that their functions are contributions to the goals of a composite system of user plus artifact—for example, writer plus fountain pen. In Adams's case of Smith's rock paperweight, Nissen finds the resulting system 'strange', 'unusual', nameless, unnatural, and unpredictable.

While it is possible to carve out a system such as Smith-rock, it is not that system to which we [ordinarily] ascribe the goal of holding down papers. No one even ventures to guess what such an unusual system might do or what goals it might have. We have no name for such a system. . . .

The awkwardness of such systems becomes even more evident when one imagines trying to explain the behavior of someone using several items successively, as a carpenter using first a hammer, then a saw, and finally a square. The explanation of the functions of these common tools requires a different system for each tool, now Smith-hammer, a moment later Smith-saw, etc. The linguistic contortions increase when one considers how the analysis would handle the functions of his shoes, belt, and shirt, all of which are used simultaneously. (Nissen 1997: 37)

Nissen (1997) concludes that explanations using such 'bizarre' and 'evanescent' (p. 54) systems are 'not credible' (p. 37) for human artifacts. Rea notes a further problem. Since the goal-directedness of the Smith-rock system resides entirely within Smith, one can add further objects at will. The rock must also have a function in the systems Smith-rock-Neptune, Smith-rock-Andromeda, Smith-rock-Gorbachev, and so on, which seems absurd.

Within biology, Nissen's problem also arises for non-human artifacts such as spiderwebs, burrows, and nests, which have genuine biological functions but are parts neither of organisms nor of any natural systems (p. 206). At the same time, GGC accounts that give such artifacts functions by the composite-system device will award false functions to any beneficial environmental feature that promotes an organism's goals.

A bush, let us say, conceals a rabbit, and is, therefore, associated with a goal-directed system . . . Therefore, it becomes correct to say that it is the function of the bush to conceal the rabbit. It, then, also becomes a function of a branch to support a sparrow . . . of an updraft to elevate an eagle, and . . . of rain to water the garden . . . (Nissen 1997: 38)

So Aristotle turns out to have been right about rain after all. The analysis, it seems, 'inserts functions throughout the realm of the physical sciences . . . where literal function talk has been conspicuously absent since the seventeenth century' (Nissen 1997: 206).[22]

[22] Here I adapt criticisms that Nissen makes of Adams and of Cummins, editing out qualifications specific to each writer in order to apply the criticisms to my own view.

One can, of course, defend the existence of the composite systems that Nissen finds so objectionable. By one common metaphysical position, mereological universalism, any two objects compose a third object. But even Nissen concedes that one 'can carve out' such systems at will. His real complaints are that they do not look like natural kinds, and that explanations involving them are too complex to resemble our ordinary thoughts about tool use.

Perhaps the best way to handle all the above objections is simply to drop the requirement that an object must be part of a goal-directed system to have a function in relation to it. The non-human artifacts Nissen mentions—spiderwebs, burrows, and nests—though not parts of the organisms, are at least created by them. But there seems to be no reason why such manufacture is required for genuine biological function. A sort of intermediate case between nest and bush occurs with organisms who decorate themselves with environmental features for camouflage. Lacewing fly larvae (*Chrysopa slossonae*) cover themselves with the 'wool' of the aphids they eat, fooling ants who otherwise would expel them from the aphid colony. An African bug, *Acanthaspis petax*, decorates itself with ant corpses for similar camouflage; a Floridian marine snail (*Xenophora concyliophora*) cements empty bivalve shells all over its own; other species of animal use pebbles or dirt for the same purpose (Owen 1980: 68). These decorations certainly have a biological function, camouflage. But they cannot be entitled to one by being manufactured by the organism—they are not—nor, it would seem, by being parts of the organism, since webs and nests are not. Nor will a disjunctive approach work: the empty sea-snail shell in which a hermit crab (*Anomura*) makes its home is equally functional, surely, yet is neither part of the crab nor manufactured by it. Analogously, it is not unnatural to speak of the function of caves in the life of bears. If so, however, we may as well admit that tree branches do have a function in the life of birds. Naturally, if the beneficial pairing occurs only once, it sounds wrong to call supporting a specific sparrow a function (let alone the function) of a specific tree branch, or hiding a specific rabbit a function of a specific bush. But surely the branch that supports a sparrow's own nest, or the bush that hides the rabbit's own burrow, does have a regular function in the life of that individual organism.

I suggest, then, that external objects have functions in relation to goal-directed systems, such as organisms, when they make regular causal contributions to their goals. Accordingly, a single object can have different functions in the life of different organisms. A given tree branch may have the (regular current) function of nest support for a sparrow, safe transportation for a squirrel, office-shading for its human owner, and support of photosynthetic leaves for the tree itself. We must merely remember that most of these functions play no role in explaining why the object is there. Usually, the only function of a part of organism O that explains its presence will be its function in relation to O; its functions for other organisms will have no etiological significance. But that is not always true. Evolutionary explanation of the galls on oak trees involves their role in wasp life, not tree life (Griffiths 1993: 416). The evolutionary explanation of plant pollen involves its role in the lives of both the plant and its insect pollinators. And the tree branch outside my window may owe its existence both to the tree's need for photosynthesis and to my need for a shade tree. As for inanimate objects or processes, their biological functions—their regular roles in the lives of organisms—usually have no etiological force at all. So there is no harm in saying that rain has a function in the lives of plants, or that the sun has a function for almost all terrestrial organisms, as long as we are clear that these functions have no tendency to explain the existence, location, or behavior of the sun and the rain.[23]

3.4. *Environmental Relativity of Function*

As Munson (1972: 529–30) charged against Ruse's analysis (1972), Melander (1997: 58) accuses Prior's account (1985) of 'radical environment relativity' of function. Since this issue also affects GGC analyses, I will answer it. One function of polar bears' white fur, Melander assumes, is camouflage, which raises the bears' fitness by helping them kill prey. But this effect

> depends on whether the bear is situated against a background of snow and ice. On a Prior-style analysis, this means that whether a function of the polar bear fur is to reflect a certain kind of light (so as to give camouflage) depends on [the bear's location]. But intuitively . . . the polar bear fur has the same functions (is *for* the same things) regardless of where the bear happens to be located. (A function of the fur of a polar bear that has strayed into some snowless and iceless region still is to reflect a certain kind of light, even though reflecting that kind of light fails to enhance its capacity to kill prey.) (Melander 1997: 58)

Further, Melander says, one cannot avoid the problem by relativizing all functions to an organism's natural habitat, since natural habitat is a '*historical* notion' that will 'reintroduce the problematical appeals to causal history that the intrasystemic role analysis was intended to avoid' (1997: 58).

I will answer this criticism together with Neander's related one.

[23] Thus it is harmless to 'insert functions throughout the realm of the physical sciences', as long as they belong to biological, not physical, science. Nissen's point damages Cummins's theory because that view, unlike mine, awards functions to components of purely physical systems, like a weather front (Nissen 1997: 206; Neander 1991*a*: 181).

3.5. *Maladaptive Functions*

In her dissertation, Neander (1983: 89) complained that a GGC account 'cannot adequately make the distinction between non-functions or dysfunctions and maladaptive functions'.

> As Ronald Munson has pointed out, the function of the fur on the polar bear is to reduce heat loss, whether the polar bear is in its natural arctic surroundings or sweltering in desert conditions.... *Individual* displacements from the natural environment can be handled by [Boorse's] 'species-typical contribution' clause, but examples of maladaptive functions can be *species-typical.* Environmental changes in the territory of a whole species of animals can and do occur which render certain traits which were once adaptive, maladaptive—that is, roughly speaking, they no longer contribute to the survival and reproduction of the organism.[24]

These objections seem to derive all their apparent force from conceptual confusion. Once one draws a type/token distinction, one can say that one function of polar-bear fur—the type—is whiteness for camouflage, and another is to reduce heat loss. These are functions typically performed in members of the polar-bear species, and they may well be functions that explain the evolution of the fur's structure and color. And, even if a given bear lives sometimes amid ice and sometimes amid desert, it may still be reasonable to say that this token bear's fur has these two functions as well.[25] In the arctic, the fur does often promote the bear's survival in these ways, while in the desert, it does not promote survival at all, so the case is a bit like Millikan's sperm (see Sect. 3.9.1), which perform the function of fertilizing ova when they perform any function at all. But there is no point in saying of a wholly desert-dwelling polar bear that its fur has either camouflage or heat conservation as functions, unless one means by this simply that this bear is a token of a type in which the fur normally has these functions. To think otherwise is to confuse normal function in a type with actual function in a token, a different matter. A similar contrast occurs in artifact functions. While the typical function of most Goodstone XRS50 tires might be in transportation, the actual function of the Goodstone XRS50 tire in my backyard may be as the seat of a child's swing (Preston 1998: 220). Writers such as Munson and Melander seem to make the following assumption: if type X normally produces Z and Z is the normal function of type X, then, if any token x produces a token effect z, z must be the function of x. But there is no basis for this assumption. If z is maladaptive, as with heat conservation in the tropics, it is much better to call this z an effect but not a function.

[24] More recently, Neander (forthcoming: 36) also explicitly makes the environmental-relativity objection.

[25] More realistically, a North American or European ermine might divide its winter activities between snowy mountains and snowless plain.

Both Neander and Melander, however, think that, even if species-normality resolved the issues for individuals, there would still be species-typical maladaptive functions. This thesis has been important in the evolution of the function debate. In Cummins's pigeon hypothetical, species-typical maladaptive function was his sole reason for rejecting an analysis of biological functions as fitness-enhancers.

> If . . . flying ceased to contribute to the capacity of pigeons to maintain their species, or even undermined that capacity to some extent, we would still say that a function of the wings in pigeons is to enable them to fly. Only if the wings ceased to function as wings, as in the penguins or ostriches, would we cease to functionally analyze skeletal structure and the like with an eye to explaining flight. (Cummins 1975: 755–6)

Likewise, a similar example of island birds is the only thing that keeps Melander from reaching a fully unified S&R analysis.

> Suppose . . . that the whole population of a species of small flying birds is blown off the continent where they once evolved and ends up on an isolated, small, windy island where having wings that make flight possible is maladaptive. Suppose further that a period of time passes . . . and suppose that due to an unfortunate absence of mutations, the birds' wings remain unaltered throughout this period despite being highly maladaptive. On Griffiths's, Godfrey-Smith's, and Millikan's analyses, the wings will have lost their proper function of bringing about flight at the end of this period, even though they are type-identical with the wings of the birds' continental ancestors and even though they still are excellent at bringing about flight . . . (Melander 1997: 97–8)
>
> Intuitively, this seems wrong: to bring about flight would still be what the wings are *for*, even though using them for flying would have no beneficial consequences. A further awkward consequence of [these analyses] is that since the wings have lost their function, there is no sense in which they can malfunction, and consequently, if one of these birds were to have its wings broken so as to make it unable to fly, there would be no sense in which that bird's wings would be malfunctioning. . . . these consequences seem quite unacceptable . . . (Melander 1997: 49–50)

On the contrary, I would say both consequences are true, except that 'malfunction' is not really a biomedical term. Schaffner (1993: 399) was right: the 'pigeon-wings counterexample . . . proves just the reverse of what Cummins is attempting to argue for', because it conflates past and present, types and tokens. For either Cummins's or Melander's birds, supporting flight is the past function of wings in birds of this type. It is likewise the past function that explains the evolution of wings in birds of this type (including the present tokens)—a stronger claim. And supporting flight is a current effect of the wings of birds of this type when they (regrettably) fly. But supporting flight is not a current function of the birds' wings; in biology, fitness-reducing effects of traits cannot be functions. The best summary is that supporting flight is a past function and a current effect, but not a current function, of the wings of birds of this type. In ostriches or penguins, by contrast, flight is not even a current effect: that is the difference. The situation with 'malfunction' is more complex, but the view I take below (Sect. 3.9.3) is as follows. A broken bird wing can be called abnormal in function by comparison with a time-slice of the species[26] long enough to include the continental phase when flying was functional. Otherwise, what is normal functioning for this species has changed, and for the island birds a broken wing is no longer abnormal in function, though it remains abnormal in structure. Under no circumstances, however, need we ever speak of maladaptive function, in individual or species.

Many recent writers assert a thesis of the 'normativity' of function, comprising two features of the function concept originally noted by Wright (1973): the function/accident distinction and the non-function/malfunction distinction.[27] Although many etiological theorists claim that a GGC analysis cannot account for these two features of functions, it can easily account for everything true about them.

3.6. *Function vs Accident*

The function/accident distinction is the thesis that an item can have a useful effect that is not its function. A stray nut may adjust a valve, a belt buckle may stop a bullet, a nose may support eyeglasses, and the heart may produce diagnostically useful sounds, but none of these effects is the item's function (Wright 1973: 147–8). Neander charges that a GGC analysis cannot handle this distinction. She admits that, in biology, it sometimes amounts to a contrast between species and individual traits. I had given the example of a lucky squirrel saved from a car by catching its tail in a crack (Boorse 1977: 557). Citing this case, she says the move to 'species-typical contributions' 'does not adequately save the distinction between proper functioning and merely fortuitous effects. This is so simply because there are *species typical* fortuitous consequences of biological items' (Neander 1983: 80–1). Her two examples are the diagnostic benefit of heart sounds and the nose's role in holding up eyeglasses. Although neither of these effects is actually typical in our species, I grant that either of them could become so. By contrast, she cites Wright's later example (1976: 110–11) of a shelf-like protuberance on the left hip, which evolution fixes in

[26] Contrary to Melander, the concept of the species-typical is not overtly historical. But, even if one brings in a chunk of species history, as I do (Sect. 3.9.3) to make the species design more stable, one needs no appeal to causal history, i.e. to etiological explanation.

[27] Explicit statements of the normativity thesis include Millikan (1989*a*: 295–6), Neander (1991*b*: 454, 467), and Price (1995: 143). As Price notes, if we use the quasi-normative operators 'what *x* is supposed to do' or 'what *x* is for', the relation between the two theses is as follows: an item produces an accidental benefit, not a function, when it does something useful that is not what it is for; an item malfunctions, not merely fails to function, when it does not do something useful that it is for.

members of a tribe because the shelf makes it easier for tribal warriors to carry heavy shields in battle. This etiology, Neander thinks, would make carrying shields the 'proper biological function' (1983: 82) of the shield-shelf. In the otherwise parallel case, however, holding eyeglasses is not the function of our nose because the nose has not been selected for that use.

With either artifacts or organisms, a GGC analysis has two independent ways to draw a function/accident distinction: via the weak/strong distinction among function statements, or via the token/type distinction. As to weak/strong, suppose an individual's token trait *x* performs a function *z* only once—that is, only once does a *Z*-effect contribute to a system goal. Then it seems natural to confine ourselves to a weak function statement. We say the soldier's Bible, on this occasion, performed the function of bullet-stopping, or the squirrel's tail, on this occasion, performed the function of traffic-avoidance. We do not say that stopping bullets was the function, or even a function, of the Bible, or that the Bible had this function, nor would we call defense against cars the, or a, function of the tail even of this particular squirrel. On the other hand, if *x* performs *Z*-functions consistently in an individual, then a strong function statement could be appropriate. So it is with Fowler's mutant bursa of Fabricius, which continually blocks viral infection in one lucky human being. Schaffner (1993: 398), a qualified physician, says that a physician would 'certainly' say that, in Fowler, this bursa has 'the function of' preventing viral infection. It is merely not a normal function in human physiology: it is not species-typical.

By contrast, provided Fowler's heart operates normally, it is not only the function of his token heart to pump blood, but also the normal function of that organ type in the human species. Similarly, when we ascend from artifact tokens to artifact types, we get the same third possible claim. In a particular artifact, part x may merely perform a function *z* once, but not do *Z* consistently enough for us to call *Z* a function of *x* in this token; or it may do the latter as well, but only in this artifact token. Or it may be that this type of part *X* typically performs *Z* in this type of artifact, whether or not it does so in this token. And with artifacts, we often wish to speak, either at token or type level, of the functions intended by designer or user. Intention introduces further complications not found in organisms. A token might be intended for a different function from its type; most or all tokens of a given type might fail to perform their intended functions; and so on. With these ample resources, there is no difficulty in drawing function/accident distinctions for artifacts or organisms.

The only real problem is Neander's view that there can be species-typical fortuitous effects, as with heartbeats and nose bridges, which are not functions, but would become so, like shield-shelves, by selection. Like other authors,[28] I will

[28] Often, I admit, this view is expressed with doubt or qualifications; see e.g. Nagel (1977: 294); Bigelow and Pargetter (1987: 195).

simply accept that, when cardiac diagnosis by heartbeat has become genuinely species-typical, for long enough, heart sounds will be a normal function of the heart in the human species. When it has long been typical to wear glasses, supporting them will be one function of the human nose. To be sure, these will not be functions that explain the organs' presence. But, as soon as one rejects the SE view by admitting some unselected functions, there is no reason to reject suitably universal cardiac diagnosis and eyeglass-propping as among them. These examples barely differ from sea-turtles' using their flippers to dig eggholes. When such behavior becomes species-typical, its effects are genuine species-typical functions, not accidents. They are merely not evolutionary functions.

3.7. Unperformed Functions

The non-function/malfunction distinction is the thesis that an item can have a function that it fails to perform. Millikan (1989*a*: 295) says: 'an obvious fact about function categories is that their members can always be defective—diseased, malformed, injured, broken, disfunctional, etc.—hence unable to perform the very functions by which they get their names'. Likewise, Neander (1991*b*: 467) writes:

> Hearts are . . . morphologically diverse within a species, because of pathological deviations from the norm, due to disease, injury, or deformity. They are all, however, *organs for pumping blood.* Not that all instances of hearts are able to pump blood. Some are too disabled. However, they are all *supposed* to pump blood; by which I mean that pumping blood is what they were selected for—it is their proper function.

Both Neander and Millikan stress that the possibility of malfunction is a defining feature of a function category like 'heart': 'Function categories are *essentially* categories of things that need not fulfill their functions in order to have them' (Millikan 1989*a*: 296). When we move from organs to artifacts, normativity of function is still stronger, according to many recent authors who claim that whole artifact-types not only do not, but cannot, perform their functions.[29] Nissen, who argues tenaciously for this thesis about artifacts, is perplexed by the refusal of so many authors to grant it. 'It is dismaying,' he writes, 'how the claim that for *X* to have a function *Y*, *X* must succeed in producing *Y* comes up again and again in the face of overwhelming and conclusive objections to it' (Nissen 1997: 213).

Since this issue is so contested, I can only summarize my answer here. Apart from occasion-specific functions for which the occasion does not arise (see

[29] See, for example, Millikan's novel can-opener, which cannot open cans (1989*a*: 294); Allen and Bekoff's early airplanes, which had the function of flight but could not fly (1995: 614); and Griffiths's tapered race-car tail, which had the function of reducing drag but was based on false physics (1993: 420–1).

Sect. 3.9.1), all claims that individual items have functions which they do not perform involve conceptual confusions between token and type, function and normal function, actual and intended function, or a combination thereof. For artifacts, the confusion goes back to Wright's broken windshield-washer button, which he said has the function of activating the windshield-washer even though it has never done so and never will (1973: 146). What is true, at most, is that this type of button activates the windshield-washer, and has the function of doing so, in cars of this model generally; or that this button, or type of button, was intended by the car's designers to activate the windshield-washer, and to have the function of doing so. But activating the washer is not the actual function of this token button; it has none, until the button is fixed. Objects that cannot fly do not have the function of flight. They have only the intended function of flight, which means, precisely, that they are intended to have the function of flight but do not have it. One may classify artifacts by intended functions if one likes, but still an object that cannot open cans will not have the function of doing so, only the intended function, whether or not you call it a can-opener.

With organisms, the classification issue has been handled by Amundson and Lauder (1994: 453ff). They argue that 'human heart', indeed 'mammalian heart', is an anatomical and morphological category, not a functional one. As to function statements, it is true, of course, that pumping blood is the normal function of the human heart—that is, its species-typical function. But, if Carla's heart cannot pump blood, then pumping blood is not, in fact, the function of her heart; it has no function. Since blood-pumping is the normal function of a human heart, it would be the function of Carla's heart if Carla's heart pumped normally; but it does not, so it is not. It is no truer to say that the function of Carla's heart is to pump blood than to say that the function of Wright's button is to wash the windshield. In both cases, what we have is a function typical in the trait type which does not exist in the token. One could, of course, agree to mean by 'Carla's heart has the function of blood-pumping' exactly the same as 'Carla is human and the human heart has the function of blood pumping'. Notice, however, that the latter is equally true of Darla, a human being who never develops a heart at all. Carla has only a functional abnormality; Darla has both a structural and a functional one. But surely their functional abnormality is identical. Neither has a heart with the function of blood-pumping; both are tokens of a type where a heart with such a function is normal.[30]

[30] Naturally, I admit that another reading of 'Carla's heart has the function of blood-pumping'—as equivalent to 'Carla has a human heart and the human heart has the function of blood-pumping'—distinguishes Carla from Darla. My point is that the true difference between Carla and Darla is structural, not functional, so this reading is pointless. Although Carla has the normal structure, a heart, and Darla does not, there is no use in saying that Carla has the normal function too. Heart or none, Carla lacks normal cardiac function every bit as much as Darla does.

At any rate, Neander (1991*b*: 460) herself stresses that biological function statements are 'primarily' about types, only 'derivatively' about tokens. Consequently, a GGC analysis has all the resources of her view. It can state the type function, and it can say the token belongs to the type; for Neander, this conjunction is equivalent to stating the token function. Thus my account can draw all the distinctions Neander and Millikan draw. But it does so without obscuring the underlying function concept, goal-contribution, which blocks such counterexamples to an SE theory as clay crystals, mutant bursas, and turtle flippers.

Finally, the type/token distinction protects my account from Melander's charge against Nagel's analysis: that it keeps unperformed functions from being explanatory: 'on Nagel's analysis only traits that actually carry out their functions can have their presence functionally explained. This effectively rules out functional explanations of the presence of malfunctioning traits and of many traits whose functions are manifested only under very specific conditions . . .' (Melander 1997: 34). Thus, regarding malfunction, if we have two tokens of the same type, one healthy and one diseased, the healthy one can be explained by Nagel's schema, but not the diseased one. Regarding functions specific to rare occasions, if a given organism is never cut, it will never bleed, and so Nagel can give no function to its blood's coagulability.

Both points are, as stated, specific to Nagel's schema of functional explanation, which a GGC analysis need not adopt. They have no force, however, against etiological explanations of traits based on GGC functions. I have just discussed malfunction and will discuss occasion-specific functions in Section 3.9.1. To avoid Melander's criticism, it suffices to note that what etiologically explains the presence of the defective organ, or the coagulable uncoagulating blood, in individuals is the performance of the type function in ancestors, not in the present individual. Token functions in the individual cannot, in fact, explain a token organ's presence even if performed. And unperformed functions, like other non-existent events, explain nothing.

3.8. *Circularity of the Reference-Class*

I make normal function in physiology or medicine a statistical concept, involving generalization over a reference-class. I defined medical normality as 'the readiness of each internal part to perform all its normal functions on typical occasions with at least typical efficiency'—that is, at an efficiency level not far below the reference-class mean. And I said that, for medical purposes, this reference-class seems to be 'a natural class of organisms of uniform functional design: specifically, an age group of a sex of a species' (Boorse 1977: 555).[31]

[31] For a diagram, see figure 1 of Boorse (1987) or (1997).

Neander (1983: 92; forthcoming: 29) first notes that choice of reference-class is important, since, for example, the mental functioning of a Down's-syndrome child is considered impaired even if it is typical of Down's children. She then charges that my specification of the reference-class

> involves an intolerable circularity.... The reason we need to know how to differentiate a reference class is because it is not until we know this that we can discover what is normal functioning for any organism.... Because the analysis of normal functioning depends upon the notion of being typical within the reference class, it is less than helpful to be told that a reference class is a natural class of organisms of uniform functional design. In order to determine which reference-class an organism belongs to we have to compare its functional organization with that of other organisms, but we can only carry out such a comparison if we already know what the functions of the parts of the organisms are. (Neander 1983: 94)

Finally, she stresses that my actual choice, an age group of a sex of a species, must be based on uniform functional design, since the time span in my example of '7–9 year old girls' would not work for faster-developing organisms, like insects (Neander 1983: 95).

The circularity charge rests on a confusion. Once one knows that functions are causal contributions to goals of the organism, one can classify the functional organization of different individuals—that is, the ways their parts contribute to their survival and reproduction—as similar or dissimilar. My analysis would be circular if I had said that normal functioning was functioning typical of a reference-class of organisms with the same normal functioning. But that is not what I said. Neander misses the fact that, on my account, three or four distinct function concepts are available: *x* is performing the function *z* in person *p*, *Z* is the function (*or*, the normal function) of *x* in *p*, and *Z* is the function of *X* in the reference class—that is, the normal function of *X* in human beings. Because on her view, within biology, 'function', 'normal function', and 'proper function' are usually synonymous, she misses the fact that they are distinct for me.[32] Moreover, her own purported non-statistical treatment of the same issue does not seem to work. She says that, on the etiological account, 'Down's-syndrome people' is an inappropriate reference-class because Down's-syndrome people 'have no special evolutionary history peculiar to them, and their brains were not selected to attain a different level of functional ability' (Neander 1983: 96). But 7–9-year-old girls likewise 'share their evolutionary background with the rest of the normal human population' (Neander 1983:

[32] In fairness, I did contrast the lucky squirrel's one-time benefit (which is 'performing the function') with a physiological function statement about the function of 'the squirrel tail' (species-normal function). And I said that 'In general, function statements describe species or population characteristics, not any individual plant or animal' (Boorse 1977: 556–7). But the analysis is still not circular even if there are no idiosyncratic individual functions, like Fowler's bursa of Fabricius. The 'uniform functional design' can simply be uniformity in functions performed.

96). Their brains were not selected to be less powerful than those of, say, 13–15-year-old girls either. Still, a 14-year-old who functions only at an 8-year-old level may be considered retarded—that is, biomedically abnormal—while a typical 8-year-old is not. As we shall see in Section 3.9.3, Neander's evolutionary view, while it can determine an organ's function, seems unable to fix a normal level for that function except by appealing to statistics, just as I do.

3.9. *Statistical Analysis of Biomedical Normality*

Objections to the view of normal function as species-typical function are a staple of one area in philosophy of medicine: defining health and disease. Here I reply only to Neander, Millikan, Plantinga, and Melander, who show no awareness that this field exists[33] but make interesting criticisms anyway.

3.9.1. *Typically unperformed functions.* Both Melander and Millikan attack a statistical account of normal function via functions that are not performed by most tokens of the type. Only 'a minute fraction' of plant seeds ever germinate, since the dispersal mechanisms do not get most of them to a suitable location. Quite possibly, in a plant's habitat, it is impossible for most of its seeds to germinate. Nevertheless, the function of seeds is to germinate, and the function of the dispersal mechanisms is to get them to a suitable location to do so (Melander 1997: 47). Millikan (1993: 161) says her 'favorite example' is sperm:

> the function of a sperm's tail is to propel it to an ovum, but very few sperm find themselves under normal conditions for proper performance of the tail. For most, no ovum chances to lie directly in the path of their random swim. Thus normal conditions for proper performance of an item are often not average conditions but, rather, ideal conditions. Sometimes they are very lucky conditions.

Throughout much of her work, to distinguish normal conditions for proper performance of an item from average conditions, Millikan writes 'Normal', with a capital letter (1984: 34).

The phenomenon of typically unperformed functions is no objection to biomedical normality of function's being species-typicality. As it happens, of course, germination is the only function—causal contribution to individual survival and reproduction—ever performed by seeds. Fertilizing an ovum is the only function ever performed by a sperm, and propelling it to do so is the only function performed by a sperm tail. So those must be the normal functions of these items if they have any. But this explanation would not make germination a normal function of seeds if they also performed other

[33] Except for Neander's citation of my (1977), these four writers' published work on functions does not cite a paper or book in philosophy of medicine. The bibliography in Boorse (1997) has over sixty such references. Good book-length introductions to issues about defining health are Nordenfelt (1987), Reznek (1987), and Fulford (1989).

functions.[34] The key point is that most biological functions are performed only on specific occasions.[35] Failure of a sperm to fertilize a non-existent ovum is like failure of blood to clot in a non-existent wound, or failure of sweat glands to release sweat when core body temperature is not above a certain level. A great many functions are occasion-specific; the occasions may be rare. What biomedical normality demands is, as I have said, the readiness of every part to perform its species-typical functions on species-typical occasions if such occasions should arise. And what that means is clear: the readiness of each token part to make, on any occasion, all types of contribution to survival and reproduction that token parts of that part-type species-typically make on that type of occasion.

Actually, however, as regards everything statistical, that is precisely Millikan's own view of normal function. 'Normal' conditions, she says, are the 'historically most usual' (1984: 33) conditions under which the function has been performed on the (possibly rare) occasions when it has been. She does stress that 'the Normal conditions for the performance of a certain function by a reproductively established family are not at all the same as average conditions under which members of that family have existed. It is for this reason that I capitalize *Normal*—to distinguish it from *normal* in the sense of *average*' (1984: 34). But a Normal explanation is still 'a preponderant explanation for those historical cases where a proper function was performed', an explanation 'uniform over as large a number of historical cases as possible' (1984: 34). So Millikan is wrong to contrast her 'Normality' with statistical normality. Her concept is just one kind of average: average function on occasions when a function was performed. Millikan and I do not differ on the statistical nature of normality; we differ on the underlying concept of function.

3.9.2. *Plantinga's cases.* Plantinga offers various examples to show that a thing's 'functioning properly' is not, as he says a passage by Pollock suggests, its functioning in the way that 'a thing of that sort *ordinarily* or *most frequently* functions'. Besides Millikan's sperm, he notes:

> Most 60-year-old carpenters have lost a finger or thumb; it is not the case that those who have not have hands that are not normal and not capable of proper manual function; and the same would hold even if we were all carpenters. Perhaps most male cats have been neutered; it hardly follows that those that haven't are abnormal and can't function properly. . . . Most baby turtles never reach adulthood; those that do are not on that account dysfunctional. (Plantinga 1993: 200–1)

Another hypothetical case, in which human beings, to propitiate the gods, have for millennia broken nearly every baby's left leg, purports to exemplify

[34] I thank Neander for this observation.

[35] See Boorse (1977: 562). Many other writers make the same point.

pandemic disease, our next topic. But Plantinga also observes that the lucky few people's legs that are unbroken would not be malfunctioning. Several other cases—the perforated aorta, scabs, and fever (1993: 206–7)—are directed at Bigelow and Pargetter's concept of a 'natural habitat' of an organ, which I have not used. Finally, in Plantinga's famous Nazi case, Nazi scientists ruin the lives of non-Aryans by saddling them with a mutation that makes their vision painful, fuzzy, and green, and by killing non-mutant non-Aryans before they reach maturity. Then the mutation spreads out of control to most of the world population, including the Nazis themselves.

None of these cases offers any problem to my analysis except as a pandemic disease. If thumbless carpenters, castrated tomcats, and dead baby turtles were species-typical, then this would be a species-typical loss of function—loss of a specific function for carpenters and cats, loss of all function for turtles. What makes having a thumb more functional than lacking one follows, of course, from the GGC analysis of function: thumbs, not the lack of them, contribute to organism goals of survival and reproduction. Naturally, unusual individuals in whom a function persists are not pathological, since what is pathological in medicine is statistically subnormal, not just statistically abnormal, function (Boorse 1977: 559). There is also no problem with functions like defense mechanisms (clotting, scabs, fever), the species-typical occasion for which is some other species-atypical dysfunction. In fact, it can be part of the function of a part to be damaged on a suitable occasion. Many butterflies and moths have conspicuous eyelike spots on their wings, which they routinely display to divert the attack of any nearby predator from the insect's vital body parts. False heads carried on the tail can have a similar target function of attracting an otherwise lethal strike (Owen 1980: 78, 81). As for Plantinga's perforated aorta, if your heart blows it out by beating with normal strength, this case is like the island birds' flight on first arrival. Your heart's strong beat is species-typical, and in most species members is functional, but in you is fatal. Here, again, a token produces a typical effect, which in the type is a function but in this token is not a function.

As for the Nazi case, one point is that Plantinga stacks the deck by his non-scientific and non-medical term 'proper function'. Naturally, what his Nazis do is not (morally) proper; why would anyone think it was? Otherwise, the case is well diagnosed by Levin. Initially, Nazi-bred painful green non-Aryan vision is just a case of 'divergent artifactual and biological functions' (Levin 1997: 92). As we saw, parts of an organism can have both biological functions and artifactual ones, as with the leaves of my office shade tree. When these two sources of functionality conflict, one is pulled in different directions. In Plantinga's case, non-Aryans' visual clarity has its normal biological function, while their visual fuzziness has the function of gratifying Nazis (Levin 1997: 91). But that is no more mystifying than that the nerve poison I spray on a wasp can be both functional (for me) and dysfunctional (for her). Finally, when the mutation spreads to the whole population, including Nazis, it becomes a case of the pandemic-disease objection. And my answer is that, at least after a sufficiently long time as pandemic, mutant vision becomes medically normal in the human species. So our only remaining subtopic is pandemic disease.

3.9.3 *Pandemic disease* Like Plantinga, Melander and Neander offer examples of pandemic disease, claiming that my type of analysis cannot distinguish such pandemic 'malfunction' from pandemic 'non-function'. Melander (1997: 36, 45, 47) imagines that the stomata of all plants of a species are clogged by fungi, blocking photosynthesis; or that all cattle have cancer of the bovine reticulum, making cellulose indigestible; or that all human kidneys stop removing metabolic waste from the blood. Actually, all those functions are essential to life, so these diseases, if universal, would simply extinguish the species. More interesting, therefore, is a case where universal disease affects a less vital function, or strikes only all species members of a certain age, or only most species members, since all these scenarios still threaten my analysis.

Melander is sure that the effect in question is still the organ's function, because, for example, digesting cellulose is what the bovine reticulum 'is *for*' (1997: 45). However, if this statement means only that the organ type evolved by performing this function, then it is true on anyone's view, including mine. If there is an issue between my view and others, it is whether the effect is now only a past function, or remains a present function though unperformed in every species member. Thus, on the suggested reading, Melander's remarks about what organs are 'for' would settle nothing. If the distinction between pandemic malfunction and pandemic non-function were only a distinction, among functions now universally unperformed, between those an organ once evolved by performing and those it never performed, anyone's theory could draw it. However, Melander contrasts the cancerous-bovine-reticulum case with the human appendix. The former is pandemic malfunction, the latter pandemic non-function (1997: 45, 48). Thus, for him, the key distinction is universal disease versus vestige.

That is also true for etiologists such as Neander and Millikan who claim superiority, regarding pandemic disease, for an etiological analysis over a goal-contribution view. As Prior noted long ago, in an essay published six years before Neander's own, for such an etiologist, vestigial traits pose a problem that is simply the dual of the goal-theorist's universal-disease problem. Prior related this issue to the two clauses of Wright's analysis, the effect clause (F.i) and the etiology clause (F.ii):

(F) The function of *X* is *Z* if and only if
  (i) *Z* is a consequence (result) of *X*'s being there
  (ii) *X* is there because it does (results in) *Z*. (Wright 1976: 81)

As Wright (1976: 89) said, his account correctly denies a function to the human appendix because the effect by which it evolved, digesting cellulose, is no longer performed, so the effect clause (F.i) is unsatisfied. But it is precisely by jettisoning (F.i) that Neander accommodates pandemic disease. Consequently, she owes us an account of vestiges. Prior (1985: 319) summarizes the situation:

Neander's account allows us to make sense of the notion of malfunction and to ascribe functions to organs at times when they cannot perform those functions. It is the function of my diseased liver to convert sugars to energy even while that liver is diseased and unable to perform that function. At those times my liver is malfunctioning because it fails to perform that function. However what we here gain on the roundabout we lose on the swings for Neander's account is unable to accommodate the phenomenon of vestigial organs. On Neander's account the function of the appendix in man is (still) the breakdown of cellulose.

Neander has never explained in print how she handles vestiges. In footnotes to her new essay, however, she at last raises the issue. Like Millikan (1984: 32), she says function-conferring selection must be 'proximal' (forthcoming: 51 n. 3); but that is just an unexplained term. She adds:

There will have to be a limit to this inheritance of functions, or else there will never be vestigial traits. On a far larger scale, the same can be said of the genes for the human appendix and their homologous relations with the genes for an ancestral mammalian appendix that was selected for digesting cellulose. How long do we continue to say that homologous traits have the function that ancestral traits were selected for? I don't think there is a precise answer to this question, but I don't think it matters if the answer is vague . . . (Neander forthcoming: 52–3 n. 11)

These remarks amount to no more than what Griffiths (1993: 413) calls Millikan's minimal gesture at distinguishing a functional trait from a vestige. They make it mystifying how Neander can criticize my reference-class for being 'vague' (forthcoming: 29), or my analysis for denying such pandemic diseases as universal blindness.

One might think that Griffiths shows the way for etiologists to handle vestiges. He offers the following definitions:

An evolutionarily significant time period for a trait T is a period such that, given the mutation rate at the loci controlling T, and the population size, we would expect sufficient variants for T to have occurred to allow significant regressive evolution if the trait was making no contribution to fitness. A trait is a vestige relative to some past function F if it has not contributed to fitness by performing F for an evolutionarily significant period. A trait is a vestige simpliciter if it is a vestige relative to all its past functions. (Griffiths 1992: 128)

Griffiths (1993: 417) says this account implies that, while ancestral appendixes had the function of cellulose digestion, ours does not, since it has not been selected for that effect in the last evolutionarily significant period. At the same time, the account saves the two kinds of universal malfunction that Neander deploys against a statistical view of normality: universal loss of function due to 'large' but 'temporary' environmental changes, and pandemic diseases, such as 'the viral infections of some plants' (Griffiths 1992: 127–8).

But it is an illusion that selection helps Griffiths distinguish vestiges from pandemic disease. First, as the quotation shows, his basic concept is not selection at all, but causal contribution to fitness. As he said earlier, his account presupposes that one can 'safely assume that useless traits tend to atrophy or decline in prevalence' (1992: 128). Thus, like me, he views the appendix as currently non-functional because it is currently useless, making no contribution to our fitness. Fitness-contribution is not, as we saw, the same as selection, since selection requires actual alternatives and causation does not. Therefore, Griffiths appeals to hypothetical selection that one might 'expect'. But this 'probabilistic' clause about expected selection over expected mutants is a contortion required only because Griffiths wishes to disguise his actual fitness-contribution view as a selectionist one.[36] It makes his account like defining honey as what a local bear would like, if there were a bear in the vicinity looking for the nectar of bees. Obviously, that is a bee theory, not a bear theory, of honey. It defines honey as bee nectar, with bears irrelevant.

One might imagine some extra role for selection in certifying or decertifying fitness-contributions as genuine functions. In particular, one might think its role is to say for how long trait *X*'s effect *Z* must increase fitness for *X* to gain *Z* as a function, or stop increasing fitness for *X* to lose it—namely, long enough for *X* to be selected for or against, if other conditions, like mutation, are right. But that is not so. In reality, all the extra content in Griffiths's analysis, beyond fitness-contribution, comes from 'significance', not selection. And the qualifier 'significant' is essential in several ways. First, without some such qualifier, the account would be an implausible view of pandemic disease. If organs can be usually diseased, what would be more natural than for such an organ to undergo regressive evolution? An unqualified account would make that impossible. In fact, selection must be qualified for Griffiths to permit pandemic disease at all. Otherwise, as soon as pandemic disease strikes organ *X*, if any individuals lack *X* altogether, then, by Griffiths's atrophy premise, selection against *X* begins, *X* loses its function, and the condition ceases to be disease. As soon as a virus makes all human eyes blind, eyeless humans have an advantage, so regressive eye evolution begins and blindness is normal, not pandemic disease. So Griffiths demands significance. *X* is to lose its function only when *X*'s fitness contribution ceases long enough for significant regressive evolution of

[36] Griffiths credits Neander for noting the need for this kind of subtlety: 'A trait might be thought to be currently contributing to fitness although it is not being selected because of an improbable absence of mutations' (Griffiths 1993: 418 n. 3). Such quotations give the game away: fitness-contribution is Griffiths's basic concept; selection is otiose.

X, and similarly for functions gained. But, if one may use vague terms like 'significant', a selectionist approach has no advantage over a GGC one. I can call current pandemic non-function pathological by comparison with a significant slice of recent species history. By contrast, I can say, universal non-function becomes normal, or the organ vestigial, if the non-function persists for a significant period—for long enough. So selection has no extra value, beyond significance, in handling vestiges or pandemic disease.

Melander (1997: 49–50) rejects Griffiths's solution to vestiges on different grounds: that it cannot properly handle the island birds. Melander's own solution is to hold that it was not, in fact, the 'current human appendix' which was selected for cellulose digestion. Rather, it was an ancestral organ which, though 'homologous', is not 'type-identical' with our appendix (1997: 96–7). This view, he argues, generates all the verdicts he wants. The human appendix has no function and is vestigial, while universally incompetent bovine reticula can still have the function of digesting cellulose, and the island birds' wings still have the function of flight.

However, Melander does not say why the organs selected for cellulose digestion were not type-identical with our appendix. If the ancestral organ is not type-identical with ours only because it was 'physiologically very different' (1997: 97 n. 2), then if physiology is about function, the explanation is circular and cannot prove our appendix's incompetence not to be a pandemic disease. In particular, Melander cannot accept Levin's (independent) proposal: 'A quicker way to vestigiality is to observe that my appendix cannot digest cellulose, so is not a token of the type of its cellulose-digesting predecessors' (1997: 87 n. 9). Then Melander's pandemic diseases become impossible. Was the ancestral appendix not our type because it was structurally different—for example, because ours has shrunk? But if our appendix, though tiny, still digested cellulose, surely everyone would wish to consider that its function. Perhaps Melander's idea is that the ancestral appendix is only the homologue of ours because it was in a different species. If that is the explanation, then his view entails that no trait can lose its function within a single species' history. To make this implausible view a doctrine of biology is an awkward constraint on taxonomy, since loss of a single function would require recognition of a new species, or, by parity of reasoning, gain of a single new one. It would also block two of Neander's recent cases, the peppered moth and the great tit, since in the former case a function is lost, in the latter a function gained, within the history of what everyone calls one species.

To return now to pandemic disease, Neander's most recent attack on a statistical account of medical normality begins as follows:

> There is simply no incoherency in the idea of a disease uniformly and universally affecting every single member of a species. A virus or other source of infection can decimate

a population, or dysfunction might become widespread or even universal due to a major environmental disaster such as an oil spill, or a drought, or radiation, or a meteor strike. A statistical definition of biological norms paradoxically implies that when a trait standardly fails to perform its function, its function ceases to be its function, and the failure to perform it ceases to be dysfunction. It seems to imply that we can cure diseases by spreading them around, which (vaccines aside) is nonsense. If we all go blind, blindness is still dysfunction. (Neander forthcoming: 30)[37]

The last statement is, of course, far from obvious. Blind eyes are normal for many species dwelling in caves or other dark places; again we see the need for clear doctrine on vestiges. As for the preceding statement, it is not much of an argument. Spreading a condition around might make it no longer a disease without constituting a 'cure', since what we normally mean by curing a condition is removing it, not engineering its reclassification.

I have offered two answers to the problem of universal disease, and Neander now attacks one of them: to use an extended time-slice of the species. Obviously, some of the species' history must be included in what is species-typical. If the whole earth went dark for two days and most human beings could not see anything, it would be absurd to say that vision ceased to be a normal function of the human eye. Actually, any time-slice shorter than a lifetime or two seems too short for the very idea of a species-typical functional design, since identifying many functions in maturation and reproduction requires a longitudinal view of an individual organism and its progeny. Originally, I spoke of including 'millennia' of the species' history (Boorse 1977: 563), and more recently I said that contemporary Western civilization was 'barely an eye-blink in the history of man' (Boorse 1997: 67).

Not only is my choice of time period vague and unprincipled, Neander complains, but it gives a wrong answer, and comes close to contradiction, in the case of the peppered moth.

During the Industrial Revolution, in some English cities the trees darkened with soot. As a result, the previously well camouflaged light-colored moths became easier prey for birds. The same species of moth also has a more heavily spotted, darker variety (one that frequently arose through mutation) and this was now better camouflaged and favored by selection. After a while the darker variety dominated the population and the lighter variety virtually disappeared, until the air cleared, the bark lightened, and the process was reversed. It would seem that at first the light pigmentation, and then the dark pigmentation, and then the light pigmentation again, had the function of camouflage. (Neander forthcoming: 31)

But if, Neander says, our time-slice is either the species' whole history or even one millennium, then my account 'tells us that the dark pigmentation never

[37] In her dissertation (1983: 86–7), Neander imagined universal blindness due to a world war. Playing with demographic factors, she then used this case to make the same argument she now makes from the peppered-moth example. So I ignore war-blindness here.

had the function of camouflage' (forthcoming: 31). If we use shorter periods, the results vary with choice of period. Suppose the total moth population is constant, and bark was light in uniform time periods T1–T3, dark in periods T4–T5, and light in T6–T8. Then in T4, dark pigmentation has the function of camouflage if we use T4–T6 as our period, but does not have it if we use T2–T4. Although this result 'is not a direct contradiction', it 'makes function attributions relative in a way that they do not seem to be' (forthcoming: 32–3).

What my account really says, however, is, at most, that dark pigment does not have a *normal* function of camouflage throughout the species' history, or for a millennium, or in T2–T4. Camouflage is certainly a function that each pigment performs at the appropriate times. One can also call camouflage the recent and current evolutionary function of dark pigment in industrial-era moths—that is, the function which explains the recent prevalence of dark coloration in that period. One simply has trouble deciding which pigment's function, if either, to call normal. But it is an illusion to think Neander's view better off. Her analysis implies that, in T4–T5, both light and dark pigment have the function of camouflage. Dark pigment has it because it is currently being selected as advantageous. But light pigment also has it, because it has previously been selected for that effect; we have precisely the polar bear in Africa, whose thick white fur is useless for camouflage and detrimental to temperature regulation. Recall the whole section of Neander's thesis charging me with excluding 'maladaptive functions' (1983: 89–92). Nor does anything in her more recent work block the conclusion that the function of the light pigment is camouflage even in dark-bark periods, when it cannot perform it.

We have not yet caught Neander in a contradiction either. But, if her theory is supposed to analyze the biomedical concept of normal function—the target of my statistical account that she is attacking—then her theory does seem to entail a contradiction. That is because, in medicine, failure of a part to perform a function that it has, or absence of the functional part itself, is pathological. So, during dark-bark periods, the dark moths' pigmentation becomes both normal (because dark pigment has the function of camouflage) and pathological (because light pigment also has a function, camouflage, but is absent). The light moths' pigmentation is also both normal and pathological. But, in medicine, 'normal' and 'pathological' are contradictories. Surely my view is better. Each pigment performs the function of camouflage when it matches the bark. It is simply unclear how long a dark-bark period is necessary for the dark pigment and its function to become 'normal', rather than being a temporarily advantageous disease, like flat feet during a draft for a murderous war.[38] But

[38] Another possibility, of course, is to call both pigments normal at all times, and view the variation as a normal polymorphism (Boorse 1977: 558). Owen (1980: 20) says that, even during the Industrial Revolution, 'black moths did not increase in frequency in rural areas which remained free of pollution. Today in Britain, 98 per cent of the peppered moths are black in areas

there can be no precise answer to this question. Inevitably, one falls back on vague terms like 'significant'.

I conclude with reasons to think that no evolutionary approach can analyze biomedical normality without appealing to statistics, as I do. As it stands, Neander's view of normality is still inchoate in two ways: it lacks any real analysis either of vestigiality or of normal levels of function. Although Melander hints at the first, he too lacks the second. Now the most obvious logical feature of medical normality is that most functions have a normal range of values. No one value of heart rate, blood pressure, blood urea nitrogen, serum glutamic-oxaloacetic transaminase, forearm strength, height, IQ, and so on is uniquely normal. Rather, there is a range of normal variation around a mean, with either one or two pathological tails. How can a purely evolutionary account determine the mean and endpoints of normal function? One obvious difficulty, of course, is that the heart rate, blood pressure, blood urea nitrogen, and so on of past human beings are unknowable to contemporary medicine. But, even theoretically, it seems impossible for a view like Neander's to avoid statistics.

First, how can evolution alone locate the mean of, say, normal human visual acuity? In her penguin case, Neander (1991*b*: 454) says that it can:

> Penguins are myopic on land and this is normal for penguins; it is a byproduct of an optical system that has the primary function of providing sharp visual focus under water where the penguins find their food. This familiar notion of 'function' (or 'proper function' as it is also called) has two very interesting features. It is normative—there is a standard of proper functioning from which actual traits can diverge. And it is teleological—the function of the penguins' eyes of providing sharp visual focus under water explains why the penguins have the eyes they have, and why they have land myopia, by explaining what their particular optical system *is for*.

This passage alone is, perhaps, ambiguous as to whether penguin land myopia is to be merely normal and a byproduct of sharp underwater vision, or normal because it is a byproduct of sharp underwater vision. But elsewhere Neander clearly says the second: 'It is normal for penguins to be myopic on land *since* the proper function of their optic lenses is sharp underwater vision' (1991*a*: 173, emphasis added). Elsewhere still (forthcoming: sect. 8, 'statistical theories'), she presents her own evolutionary analysis of normality as superior to my statistical one.

By so doing, Neander confuses two questions: what the normal level of penguin vision is, and how one explains its origin. Penguin land myopia is normal because it is typical of penguins, not because it is somehow endorsed

of heavy industry, such as Manchester, while in rural areas, such as Cornwall, all the peppered moths are of the original pale form'. A normal condition does not become pathological merely because the organism finds itself in an environment where it is disadvantageous (Boorse 1997: 41). For example, the polar bear's fur is normal even if it gives him fatal heatstroke in Africa. Perhaps, then, light moths in 1895 Manchester were normal moths unlucky in their birthplace.

by evolution as a byproduct of something else, underwater visual acuity. It is not as though a normal organism is required to have every possible ability, unless it can present a valid evolutionary excuse. In human medicine, what visual acuity is normal clearly depends on how our species actually sees. We might well benefit from distance vision as powerful as eagles', or hearing and smell as acute as dogs'. It would also be grand to be able to fly, regenerate limbs, digest cellulose, synthesize vitamin C, and breathe underwater. Still it is normal, not pathological, for us to lack such advantageous traits that other organisms have. Why? Not, surely, because we have a good evolutionary excuse. What is our excuse for lacking eagle vision—not to mention eagle wings? The whole idea of optimality is foreign to medicine, which takes the species' normal functions as it finds them. A human, to be normal, need not be able to fly simply because no human can. A normal human need not have eagle vision simply because such acuity is far above our statistical mean. That is why a normal 3-month-old royal penguin (*Eudyptes schlegeli*), but not a normal 3-month-old human baby, must be able to walk (Williams 1995: 224), while a normal 5-year-old human being, but not a normal 5-year-old royal penguin, must be able to talk—because of what the two species are actually like today. A purely evolutionary account of mean normal function seems not just implausible, but impossible. If it is as easy as Neander thinks for whole species to be diseased, why are penguins not diseased for not seeing well both on land and in the sea? It is no answer to say that their ancestral gene pool lacked the resources to evolve such a visual system. Why was that genetic deficiency not itself a pandemic penguin disease?

Finally, besides the mean, how can a purely evolutionary concept set boundaries to the normal range? If we seek to capture the biomedical idea of normality, it must be possible to have even pathologically myopic penguins. After all, we can be pathologically myopic even though normal humans lack eagle vision, and we can suffer pathological high-frequency hearing loss despite normal humans' aural inferiority to dogs. So how does penguins' evolutionary history determine a lower limit of normal penguin myopia? Pending such an explanation, I conclude that Prior was right again: even an etiological theory requires a concept of statistical normality to match basic logical features of biomedical concepts. 'Neander argues that it is one of the etiological account's great advantages that it can provide an account of normal functioning which is normal functioning *simpliciter* and not just statistically normal functioning. But this is not so. In its only defensible form the etiological account enjoys no such advantage' (Prior 1985: 320). The problems of this section affect etiological theorists as well. Perhaps, then, when suitably filled out, Neander's account of normality will not greatly differ from mine—unless the difference is that hers entails a contradiction.

Any account of normality must concede that medicine recognizes a tiny number of diseases that are typical or even universal, either in the whole species (atherosclerosis) or in an age group (osteoarthritis or prostatic cancer in men of a certain age) (Boorse 1977: 566–8). The question is how we should explain this fact. I suggested that some universal functional defects may be thought pathological as environmental effects; but this distinction between internal and external causation is much criticized, and I now doubt that its benefit was ever worth its cost. The only other explanation I can think of is that medicine assumes that, in the past, these diseases were not species-typical, so that they are atypical of a suitably broad time-slice of our species. But that is a somewhat implausible explanation of medical thinking. Also, Neander (1983: 86–7) and Plantinga (1993: 201) are right that it can get derailed by awkward demographics—for example, if most human beings are alive today. One could, of course, fix that problem by defining normal function to be what was typical of our species at most times in our chosen slice of species history. But I currently favor the view (Boorse 1977: 568) that medicine is wrong to recognize any universal diseases, since it lacks any coherent concept of pathology that can make them pathological. On this view, what is pathological is only age-excessive atherosclerosis, premature prostate cancer, and so on. After all, functional decline in 'normal aging' is a well-established medical category, although the growth of knowledge often contracts it by showing most specific causes of that decline to be atypical and so pathological—for example, Alzheimer's as a cause of senility (Boorse 1997: 91–2).

So, on balance, in Plantinga's hypothetical cases, I will embrace the conclusion of my analysis. If nearly all human left legs have been broken throughout human history, or nearly all human vision has been painful and green, then that is their normal condition regardless of its cause. Human acts causing these situations are not, of course, 'proper'; but that is morality, not biomedical science. It would be nice to have a concept of health that declared Plantinga's hypothetical cases diseases, along with atherosclerosis, a few other medical conditions, and the universal neurosis which, in psychoanalytic theory, mars ideal normality in our species (Boorse 1976*b*: 78–9; 1987: 390 n. 44). Unfortunately, no one has yet explained how such a concept can also stay faithful to the vast bulk of medical judgments of normality.

3.10. Mimicry

An interesting, wholly biological, challenge to the GGC view is Mitchell's use of Batesian mimicry to defend an SE analysis. Caterpillars of the brightly colored Monarch butterfly (*Danaus plexippus*) eat a milkweed diet rich in cardenolides, which makes them taste bad to birds both as caterpillars and as butterflies. Presumably, Monarch butterflies' bright coloration evolved as a signal of inedibility to birds. Almost identically colored are the tasty Viceroy butterflies (*Limenitus archippus*), which are less common and belong to a different family

(*Nymphalidae*). Presumably, their coloration evolved by mimicking the Monarch markings. Mitchell (1995: 47) argues that this case shows that function depends on history, not current effects.

The morphological structures have the same future consequences, i.e. avoiding predation, but have had different evolutionary histories. Do we want to say that the conspicuous coloration of the Monarch and Viceroy have the same function? No. Mimics and models are not the same. . . . The function of conspicuous coloration in the Monarch is to warn the predator of its unpalatability. The function of the Viceroy coloration is to mimic the model and deceive the predator into presuming it is unpalatable and thereby avoid predation. The same structure has two functions, one is to warn and the other is to deceive.

Mitchell is mistaken, however, that only an SE analysis distinguishes the functions of Monarch and Viceroy coloration. In the first place, given the evolutionary history she assumes, mimics and models are not 'the same' on any function analysis. Even if the common colors' present functions are the same in the two species, the species differ in past functions simply because the model's coloration evolved first. In the second place, any major type of function analysis can distinguish the colors' present functions in model and mimic, wholly in terms of effects on present fitness. That is, the essence of Batesian mimicry is analyzable as a current relationship between the two species, without appeal to evolution. As Mitchell says, in one sense it is true, on non-SE analyses, that Monarch and Viceroy colors have the same present function: avoiding predation. In the current environment, the common colors make predation less likely because many butterflies with these colors taste bad to birds. But the two functions differ when more specifically described. If functions are causal contributions to goals such as survival and reproduction, then, to specify a function fully, one must specify its full mechanism. At the species level, the function mechanism in Monarchs and Viceroys differs. In Monarchs, the function of the coloration does not depend upon traits of another species. In Viceroys, it does, namely, upon the poisonousness of Monarchs. It is this present difference between the two mechanisms, not its presumed evolutionary history, that Mitchell is describing when she says that the function of Monarch coloration is to warn, the function of Viceroy coloration to deceive.[39]

A hypothetical example may make this point easier to see, and then we will look at more real examples. Suppose southern squirrel species *S* migrates northward by evolving a specially elongated, toughened hindclaw for digging out of winter's frozen ground the acorns it has always buried in the fall. Then

[39] The two verbs 'warn' and 'deceive' also suggest a second contrast: exercise of the Monarch's, but not the Viceroy's, colors' function benefits the predator. But this contrast from the predator's viewpoint is no reason to assign Monarch and Viceroy colors different functions for the butterflies.

northern squirrel species *S′*—which buries nothing but hides its acorns in hollow trees—evolves a very similar claw by using it to dig up the acorns now buried by *S*. At a general level, the two species' elongated hindclaws have the same function: to dig up acorns from frozen ground. But the two species differ in that only *S* buries the acorns. That implies a difference in the specific details of the function's mechanism. The hindclaw's mechanism of function in *S′*, but not in *S*, depends upon the activities of another species. *S* maintains a feature of the environment necessary for their common hindclaw's function, while *S′* does not. In this way, obviously, *S* and *S′* are 'not the same' even in the present, nor do their hindclaws have the same fully specified function.[40]

To see mimicry as a present, not an evolutionary, relationship between two species, consider another major kind, Müllerian mimicry. The African Monarch *Danaus chrysippus*, also a cardenolide storer, has a mimic *Acraea encedon*, which secretes hydrogen cyanide (Owen 1980: 116). In this Müllerian pattern, two poisonous species gain extra protection by resembling each other. Clearly, two or more species in a Müllerian relationship can evolve in tandem, converging on a common coloration. The essence of such a relationship, when achieved, is the reciprocal role of each species' unpalatability in the other's fitness. But that is a simultaneous functional relation, not one of evolutionary succession. And Owen thinks 'there is probably no clear-cut distinction between Müllerian and Batesian mimicry; the two kinds are inextricably interwoven'. That is because it seems that 'the different species in a Müllerian assemblage are not all equally unpalatable, and so there are mimics, not necessarily completely palatable, but less unpalatable than the models' (Owen 1980: 116). A further complication is a phenomenon awkwardly named 'automimicry': the high variability of unpalatability even within single species, due, for example, to individual variations in caterpillars' diet, poison storage, and so on. In nature, all three types of mimicry are found interwoven in complex ways.

In Trinidad, the monarch occurs together with the related queen butterfly *Danaus gilippus* which has a similar coloration. Tests show that about 65 per cent of the monarchs but only 15 per cent of the queens are unpalatable to blue jays. This means that unpalatable monarchs and queens are Müllerian mimics of each other, that palatable queens are Batesian mimics of unpalatable monarchs, that palatable monarchs are

[40] In either example, two descriptions are possible: that the two species show different functions for the common trait, or different mechanisms for the same function. The latter description sounds more natural, and I am indebted to Joel Pust for it. However, the former seems equally true, since one can describe functions with varying generality. One can say that both squirrels' hindclaw has the same function: *to dig up acorns*, or even *to dig up acorns buried by S*. But one can also say the functions differ in that the claw's function in *S′*, but not in *S*, is *to dig up acorns buried by another species*—the kind of function parasitologists study. On either description, Mitchell's argument fails, since either way model and mimic differ in the present, without regard to evolution. Whether one calls this difference one of function or of functional mechanism is irrelevant.

Batesian mimics of unpalatable queens, and that in both species there is automimicry. (Owen 1980: 117)[41]

Since making oneself unpalatable can carry metabolic costs, costs in dietary restriction (Brower 1971), or other disadvantages, to some degree one can view all these types of mimicry as parasitism. Mimicry is, of course, a common device in other exploitative relations between species. Rove beetles may resemble ants closely enough to live in their nest and be fed by them or prey on their eggs and larvae (Owen 1980: 122). Some orchid flowers mimic female bees in order to trick amorous male bees into orchid pollination (Owen 1980: 85); cuckoos lay eggs that resemble those already in the nest of another bird species, relying entirely on the foster parents to raise the cuckoo offspring (Owen 1980: 69–71).[42] And, apart from mimicry, there is an enormous range of 'kleptobiotic' relations, in which

> one organism physically removes a resource (often food) which another organism had gathered for its survival or reproduction. In most cases the klepto interaction is interspecific although it may also be intraspecific. . . . The owner acquires . . . a resource which incurs energy, time, risk or other costs. The thief removing a resource from the owner's control invests presumably less during the theft. (Vollrath 1984: 62–3)

Into this category fall black-headed gulls' theft of earthworms from lapwings (Källander 1977), the piracy of *Argyrodes* spiders in the webs of *Nephila clavipes* (Vollrath 1984: 70), and usurpation by female digger wasps of burrows dug by other females (Brockmann *et al.* 1979).

All these forms of parasitism, interspecific or intraspecific, are present relationships among organisms. They are definable in terms of the current role of one type of organism in the life of another. Thus, as with biological functions in general, evolutionary history is not needed to define the function, but only to explain its origin. Undoubtedly, Batesian mimicry seems unlikely to arise unless model predates mimic, which is why Bates and others took his report (1862) as crucial early evidence for the theory of evolution that Darwin had published three years before. But evolution is no more necessary to define Batesian mimicry,[43] or to distinguish model from mimic, than it is to describe

[41] For a far more complicated African example involving eighteen species from six families, sex-specific coloration, and polymorphism, see Owen (1980: 128–31).

[42] Intraspecific brood parasitism, which would be the counterpart to automimicry, is also very common in birds; see Andersson (1984) for a survey.

[43] I am defining both kinds of mimicry as present relationships: Batesian mimicry as beneficial resemblance of a non-poisonous organism to a poisonous one, Müllerian mimicry as beneficial mutual resemblance among poisonous organisms. While some biological dictionaries agree, others make the presumed evolution of Batesian mimicry part of that term's definition. But the term itself is irrelevant. The present ecological relation suffices for anyone to distinguish the two species' functional mechanisms, and so to answer Mitchell's point about functions.

As usual, I also assume that, in Batesian mimicry, individual predators learn to avoid both species by experience of eating the poisonous one. If birds' distaste for Monarchs became genetically fixed,

other forms of mimicry or parasitism. Since all these relations depend on current effects of organisms' traits on other organisms' survival and reproduction, Batesian mimicry cannot show an SE analysis superior to other analyses of biological function.

4. Conclusion

So ends my rebuttal of all major objections to the GGC analysis but two: attacks on the cybernetic analysis of goal-directedness and Amundson's assault (2000) on normal function. Elsewhere I hope to answer each. For now, I will conclude by mentioning reasons for optimism despite these two critiques. Amundson's real target is the concept of medical normality itself, not my analysis of it. But surely a concept so deeply embedded in biomedicine is unlikely to be uprooted. Moreover, one might almost say the same about goal-directedness, given how many writers appeal to it. Sommerhoff is not alone in his view that directive organization typifies life. Though Rosenberg is not committed to a GGC view, he says that 'the goal-directedness of complex biological systems seems beyond doubt' (1985: 37). 'Nothing could be more obvious. Organisms have aims and purposes, which their behavior serves; their component parts serve to fulfill these purposes and have functions in meeting the needs of cells, tissues, organs, whole biological organisms, and systems like ant colonies made up of large numbers of individual organisms' (Rosenberg 1985: 43).[44]

Arguably, even leading critics of the cybernetic analysis, such as Bedau and Nissen, have trouble escaping some naturalistic concept of biological goal-directedness. Nissen (1997: 211) concludes that teleological claims about lower animals and plants are literally false without a Great Artificer. Yet in criticizing other theories, he repeatedly stresses two facts. One is how obvious the goals of animal behavior can be. 'Any successful theory of goal-directed behavior must accommodate the fact that not only humans, but animals as well, routinely make instant but highly accurate judgments concerning which actions are, and which are not, goal-directed . . . (Nissen 1997: 11). The second fact is the constant possibility of goal-failure. But, if a moth fails in its goal to escape a bird's beak, an omnipotent God's intentions are of no use in grounding that goal. Only on some non-intentional analysis, then, can it be true that moths flee

my description would change. Then all present functions of the common colors would, in fact, be identical in both butterfly species. That is the right view, since now the Monarchs' poisonousness plays no present role in avoiding predation. What differs is only the past functional mechanism of this coloration in Monarchs, on the one hand, which involved poisonousness, and its present functional mechanism in both Monarchs and Viceroys, which does not, on the other. It is a strength of the GGC view that, unlike an SE approach, it distinguishes functions in these two cases. (I thank Pust for prompting this clarification.)

[44] Rosenberg admits that this view is 'problematical', but he never rejects it.

birds.[45] As for Bedau, his function analysis assumed that every organism, unlike clay crystals and magnetic pendulums, has a 'good of its own' (1992*a*: 45). One can plausibly argue, however, that our only real concept of intrinsic good for such lower organisms as paramecia or bacteria is goal-directedness itself. Moreover, Bedau (1996) also calls clay crystals, computer programs, and even economic systems kinds of life, and says that life is definable in terms of 'supple adaptation' (1996: 338)—a notion reminiscent, to say the least, of the cybernetic analysis of goal-directedness. These observations, despite their *tu quoque* character, inspire hope that some naturalistic goal-concept evades the array of objections to the cybernetic analysis given by Bedau (1992*a*) and Nissen (1997: 1–69).

My second reason for optimism is that, even if the goal-concept collapses, both philosophy of biology and philosophy of medicine can go on much as before. One loses the feature that unifies biological with artifactual and other functions, and one is left with no explanation of why biologists use the term 'function' as they do. Otherwise, everything about biological functions can be captured by the S&R analysis of them as contributions to organism survival and reproduction, if one otherwise follows my line above. We reach a position like Bigelow and Pargetter's (1987), which would presumably also be Melander's, if he dropped the doctrine that his island birds' wings still have the function of flight. An S&R view agrees with Melander (1997: 98) that biologists can eliminate the term 'function' altogether, speaking only of 'adaptiveness' and 'adaptation'. As is now standard, one defines an adaptive trait as one contributing to its bearer's survival and reproduction, and an adaptation as a trait established by its past adaptiveness. Then, by abandoning Melander's island-bird doctrine, we reach a fully unified analysis. All biological functions are causal contributions to fitness, and the adjectival strategy distinguishes past function, present function, and evolutionary function as desired. Finally, we analyze medically normal function in the usual way, as species-typical part-contribution to individual survival and reproduction. Because a GGC view covers all types of functions without *ad hoc* definitions, it is more satisfying. But the S&R view of biological function has nearly the same consequences for both biology and medicine.

References

Achinstein, Peter (1977), 'Function Statements', *Philosophy of Science*, 44: 341–67.

Adams, Frederick R. (1979), 'A Goal-State Theory of Function Attributions', *Canadian Journal of Philosophy*, 9: 493–518.

[45] Nissen does think our confidence about teleological claims diminishes to 'less or none' (1997: 228) for lower animals and plants. So he may be willing to maintain his analysis and concede that insects never flee. Surely, however, moths can flee birds just as obviously as mice can; the difference is only in whether they know they are doing so.

Allen, Colin, and Bekoff, Marc (1995), 'Biological Function, Adaptation, and Natural Design', *Philosophy of Science*, 62: 609–22.

Amundson, Ron (2000), 'Against Normal Function', *Studies in the History and Philosophy of Biology and Biomedical Science*, 31: 33–53.

——and Lauder, George V. (1994), 'Function without Purpose: The Uses of Causal Role Function in Evolutionary Biology', *Biology and Philosophy*, 9: 443–69.

Andersson, Malte (1984), 'Brood Parasitism within Species', in Barnard (ed.), (1984), 195–228.

Barnard, C. J. (1984) (ed.), *Producers and Scroungers* (New York: Chapman & Hall).

Bates, H. W. (1862), 'Contributions to an Insect Fauna of the Amazon Valley. *Lepidoptera: Heliconidae*', *Transactions of the Linnean Society of London*, 23: 495–565.

Beckner, Morton (1959), *The Biological Way of Thought* (New York: Columbia University Press).

Bedau, Mark (1990), 'Against Mentalism in Teleology', *American Philosophical Quarterly*, 27: 61–70.

——(1991), 'Can Biological Teleology be Naturalized?', *Journal of Philosophy*, 88: 647–55.

——(1992*a*), 'Goal-Directed Systems and the Good', *Monist*, 75: 34–51.

——(1992*b*), 'Where's the Good in Teleology?', *Philosophy and Phenomenological Research*, 52: 781–806.

——(1993), 'Naturalism and Teleology', in Steven J. Wagner and Richard Warner (eds.), *Naturalism: A Critical Appraisal* (Notre Dame, Ia.: University of Notre Dame Press).

——(1996), 'The Nature of Life', in Margaret A. Boden (ed.), *The Philosophy of Artificial Life* (New York: Oxford University Press), 332–57.

Bigelow, John, and Pargetter, Robert (1987), 'Functions', *Journal of Philosophy*, 84: 181–96.

Boorse, Christopher (1975), 'On the Distinction between Disease and Illness', *Philosophy and Public Affairs*, 5: 49–68.

——(1976*a*), 'Wright on Functions', *Philosophical Review*, 85: 70–86.

——(1976*b*), 'What a Theory of Mental Health should Be', *Journal for the Theory of Social Behaviour*, 6: 61–84.

——(1977), 'Health as a Theoretical Concept', *Philosophy of Science*, 44: 542–73.

——(1987), 'Concepts of Health', in Donald VanDeVeer and Tom Regan (eds.), *Health Care Ethics: An Introduction* (Philadelphia: Temple University Press), 359–93.

——(1997), 'A Rebuttal on Health', in James M. Humber and Robert F. Almeder (eds.), *What is Disease?* (Totowa, NJ: Humana Press), 1–134.

Braithwaite, Richard (1953), *Scientific Explanation* (Cambridge: Cambridge University Press).

Brockmann, H. J., Grafen, A., and Dawkins, R. (1979), 'Evolutionarily Stable Nesting Strategy in a Digger Wasp', *Journal of Theoretical Biology*, 77: 473–96.

Brower, L. P. (1971), 'Prey Coloration and Predator Behavior', in V. Dethier (ed.), *Topics in Animal Behavior, Topics in the Study of Life* (The BIO source book, Part 6; New York: Harper & Row).

Buller, David J. (1998), 'Etiological Theories of Function: A Geographical Survey', *Biology and Philosophy*, 13: 505–27.

Buller, David J. (1999) (ed.), *Function, Selection, and Design: Philosophical Essays* (Albany, NY: SUNY Press).

Cairns-Smith, A. G. (1965), 'The Origin of Life and the Nature of the Primitive Gene', *Journal of Theoretical Biology*, 16: 53–88.

Callicott, J. Baird (1989), *In Defense of the Land Ethic: Essays in Environmental Philosophy* (Albany, NY: SUNY Press).

Canfield, John (1964), 'Teleological Explanations in Biology', *British Journal for the Philosophy of Science*, 14: 285–95.

Cummins, Robert (1975), 'Functional Analysis', *Journal of Philosophy*, 72/20: 741–65.

Dawkins, Richard (1986), *The Blind Watchmaker* (London: Longman).

Enç, Berent (1979), 'Function Attributions and Functional Explanations', *Philosophy of Science*, 46: 343–65.

Fulford, K. W. M. (1989), *Moral Theory and Medical Practice* (New York: Cambridge University Press).

Godfrey-Smith, Peter (1993), 'Functions: Consensus without Unity', *Pacific Philosophical Quarterly*, 74: 196–208.

——(1994), 'A Modern History Theory of Functions', *Noûs*, 28: 344–62.

Gould, S. J., and Vrba, E. S. (1982), 'Exaptation—a Missing Term in the Science of Form', *Paleobiology*, 8: 4–15.

Griffiths, Paul E. (1992), 'Adaptive Explanation and the Concept of a Vestige', in Paul E. Griffiths (ed.), *Trees of Life* (Dordrecht: Kluwer), 111–31.

——(1993), 'Functional Analysis and Proper Functions', *British Journal for the Philosophy of Science*, 44: 409–22.

Källander, H. (1977), 'Piracy by Black-Headed Gulls on Lapwings', *Bird Study*, 24: 186–94.

Kitcher, Philip (1993), 'Function and Design', in *Midwest Studies in Philosophy*, 18: 379–97.

Levin, Michael (1997), 'Plantinga on Functions and the Theory of Evolution', *Australasian Journal of Philosophy*, 75: 83–98.

Lewontin, R. C. (1978), 'Adaptation', *Scientific American*, 239/3: 212–30.

Manning, Richard N. (1997), 'Biological Function, Selection, and Reduction', *British Journal for the Philosophy of Science*, 48: 69–82.

Melander, Peter (1997), *Analyzing Functions: An Essay on a Fundamental Notion in Biology* (Stockholm: Almqvist & Wiksell).

Millikan, Ruth Garrett (1984), *Language, Thought, and Other Biological Categories: New Foundations for Realism* (Cambridge, Mass.: MIT Press).

——(1989*a*), 'In Defense of Proper Functions', *Philosophy of Science*, 56/2: 288–302.

——(1989*b*), 'An Ambiguity in the Notion "Function"', *Biology and Philosophy*, 4: 172–6.

——(1993), *White Queen Psychology and Other Essays for Alice* (Cambridge, Mass.: MIT Press).

——(1999), 'Wings, Spoons, Pills, and Quills: A Pluralist Theory of Function', *Journal of Philosophy*, 96: 191–206.

Mitchell, Sandra D. (1995), 'Function, Fitness and Disposition', *Biology and Philosophy*, 10: 39–54.

Munson, Ronald (1972), 'Biological Adaptation: A Reply', *Philosophy of Science*, 39: 529–32.

Nagel, Ernest (1961), *The Structure of Science: Problems in the Logic of Scientific Explanation* (New York: Harcourt, Brace & World).

——(1977), 'Teleology Revisited', *Journal of Philosophy*, 74: 261–301.

Neander, Karen (1983), 'Abnormal Psychobiology', Ph.D. dissertation, La Trobe University.

——(1991*a*), 'Functions as Selected Effects: The Conceptual Analyst's Defense', *Philosophy of Science*, 58: 168–84.

——(1991*b*), 'The Teleological Notion of Function', *Australasian Journal of Philosophy*, 69: 454–68.

——(forthcoming), 'Why History Matters: Four Theories of Functions', in M. Weingarten and Gerhard Schlosser (eds.), *Formen der Erklärung in der Biologie* (Verlag für Wissenschaft und Bildung). In German; page references are to an English manuscript supplied by the author.

Nissen, Lowell (1997), *Teleological Language in the Life Sciences* (Lanham, Md.: Rowman & Littlefield).

Nordenfelt, Lennart (1987), *On the Nature of Health: An Action-Theoretic Approach* (Dordrecht: Reidel).

Owen, Denis (1980), *Camouflage and Mimicry* (Chicago: University of Chicago Press).

Plantinga, Alvin (1993), *Warrant and Proper Function* (New York: Oxford University Press).

Preston, Beth (1998), 'Why is a Wing like a Spoon? A Pluralist Theory of Function', *Journal of Philosophy*, 95/5: 215–54.

Price, Carolyn (1995), 'Functional Explanations and Natural Norms', *Ratio*, NS 7: 143–60.

Prior, Elizabeth W. (1985), 'What is Wrong with Etiological Accounts of Biological Function?', *Pacific Philosophical Quarterly*, 66: 310–28.

Reznek, Lawrie (1987), *The Nature of Disease* (London: Routledge & Kegan Paul).

Rosenberg, Alexander (1985), *The Structure of Biological Science* (Cambridge: Cambridge University Press).

Rosenblueth, A., Wiener, N., and Bigelow, J. (1943), 'Behavior, Purpose and Teleology', *Philosophy of Science*, 10: 18–24.

Ruse, Michael (1971), 'Functional Statements in Biology', *Philosophy of Science*, 38: 87–95.

——(1972), 'Discussion: Biological Adaptation', *Philosophy of Science*, 39: 525–8.

Schaffner, Kenneth F. (1993), *Discovery and Explanation in Biology and Medicine* (Chicago: University of Chicago Press).

Scheffler, Israel (1959), 'Thoughts on Teleology', *British Journal for the Philosophy of Science*, 9: 265–84.

Schwartz, Peter H. (1999), 'Proper Function and Recent Selection', *Philosophy of Science*, 66 (Proceedings): S210–S222.

Sommerhoff, Gerd (1950), *Analytical Biology* (Oxford: Oxford University Press).

——(1959), 'The Abstract Characteristics of Living Organisms', in F. E. Emery (ed.), *Systems Thinking* (London: Harmondsworth).

Sorabji, Richard (1964), 'Function', *Philosophical Quarterly*, 14: 289–302.
Taylor, Paul W. (1986), *Respect for Nature: A Theory of Environmental Ethics* (Princeton: Princeton University Press).
Vollrath, Fritz (1984), 'Kleptobiotic Interactions in Invertebrates', in Barnard (1984), 61–94.
Walsh, Denis M., and Ariew, André (1996), 'A Taxonomy of Functions', *Canadian Journal of Philosophy*, 26/4: 493–514.
Williams, George C. (1966), *Adaptation and Natural Selection: A Critique of Some Current Evolutionary Thought* (Princeton, Princeton University Press).
Williams, Tony D. (1995), *The Penguins: Spheniscidae* (Oxford: Oxford University Press).
Wimsatt, William (1972), 'Teleology and the Logical Structure of Function Statements', *Studies in History and Philosophy of Science*, 3: 1–80.
Woodfield, Andrew (1976), *Teleology* (Cambridge: Cambridge University Press).
Wright, Larry (1973), 'Functions', *Philosophical Review*, 82: 139–68.
——(1976), *Teleological Explanations* (Berkeley and Los Angeles: University of California Press).

4. Biofunctions: Two Paradigms

Ruth Garrett Millikan

ABSTRACT

Applications to biological systems of Robert Cummins's notion of 'function' from his classic paper 'Functional Analysis' (Cummins 1975) and of my notion 'proper function' from *Language, Thought and Other Biological Categories* (*LTOBC*) (Millikan 1984) are discussed and compared. Neither notion is fully determinate in its application to life forms, always cutting decisively between 'functions' of the designated kind and 'mere effects'. Nor do we need a notion of function in biology that is fully determinate in this way. The dimensions of indeterminacy for both concepts are explored. An important tightening of the notion of a Cummins biofunction is achieved by recognizing Cummins functions that are closely analogous to 'relational', 'adapted', and 'derived' proper functions as defined in *LTOBC*. Steven J. Gould's use of the notion 'exaptation' in his argument that 'exaptations of the brain must greatly exceed adaptations by orders of magnitude' is criticized in this context, and a less amorphous notion of exaptation recommended. These are defined by reference to Cummins-style functional analyses that explain how proper biofunctions are performed.

1. Functions

Listed in *Websters 3rd International Dictionary* under 'function' are

2a. The natural and proper action of anything; special activity; office, duty, calling, operation, or the like.
2b. The natural or characteristic action of any power or faculty.
10. Physiol. The normal and special action of any organ or part of a living animal or plant.

According to *The American Heritage Dictionary*, Windows version, a function is

Parts of this paper are drawn from Millikan (1999) with kind permission of the *Journal of Philosophy*.

1. The action for which one is particularly fitted or employed.
2a. Assigned duty or activity. 2b. A specific occupation or role . . .
4. Something closely related to another thing and dependent on it for its existence, value, or significance.

I do not know of any discussion of function in philosophy or biology that does not fit under one or more of these clauses. Nor is the use of the term 'function' in contemporary biology tightly defined or univocal, either within or across subdisciplines. Argument over THE correct analysis of THE concept of function either in ordinary life or in biology is, I believe, quite pointless. On the other hand, questions about the applicability, clarity, determinateness, and usefulness to biology of various different possible notions of function, and about the relations of various notions to one another, are important.

I will explore two notions of function apparently useful in biology, along with certain subvarieties. The first notion is that of 'proper functions' as defined in my *Language, Thought and Other Biological Categories* (Millikan 1984, hereafter *LTOBC*). The second is the notion of functions described in Robert Cummins's well-known paper 'Functional Analysis' (Cummins 1975),[1] which functions I will simply call 'Cummins functions'. I will also discuss what I take to be two subspecies of Cummins functions—namely, 'exaptations' (Gould and Vrba 1982; Gould 1991) and biologically useful 'spandrels' (Gould and Lewontin 1979). When I say that I will 'explore' these various notions, nothing even remotely resembling conceptual analysis is intended. Rather, I will take the stipulative definition of 'proper function' from my *LTOBC* and I will take Cummins's description of 'functions' from 'Functional Analysis', compare the biological phenomena that fit under these two definitions, inquire how determinate each of these notions is when applied to biological phenomena, discuss possible useful adjustments to tighten them up where needed, and propose a useful tightening-up of Gould and Vrba's notion 'exaptation'. That is, I will be exploring consequences and suggesting prescriptions for usage, rather than attempting descriptions of usage, for these various terms.

[1] In Millikan (1989*a*) I remarked on Cummins's discussion of functions that its aim was entirely different from my own in *LTOBC*. I said it again (in Millikan 1989*b*) while discussing the proper functions specifically of biological kinds, and (in Millikan 1993: ch. 2), where I emphasized that the proper functions of biological kinds overlap with functions in Cummins's different but equally legitimate sense. Unfortunately, the literature on functions continues to set me over against Cummins, or to claim that I take proper functions and Cummins's functions to apply in different domains. I hope that this chapter will help to make clear why both these notions of function are useful, although it is important to keep them straight.

2. Proper Functions

Neither 'proper functions' nor Cummins's 'functions' were originally defined with specifically biological applications primarily in mind. Cummins said that his definition of function was designed to explicate 'the central explanatory use of functional language in science' (1975: 746), but what he was especially interested in was cognitive science and he wrote at a time when minds were more likely to be compared to computers than to biological systems. According to *LTOBC*, the definition of 'proper function' was needed

> in order to talk about analogies and disanalogies among things belonging to quite diverse categories—body organs, tools, purposive behaviors, language elements, inner representations, animal's signals, customs, etc. . . . [the] purpose being to make as explicit as possible analogies among categories of things, which analogies had struck me as useful to reflect on . . . the spirit in which I offer them to the reader is as a handle by which to grab hold of the analogies. (*LTOBC* 38)[2]

Again,

> 'Proper function' is intended as a technical term. It is of interest because it can be used to unravel certain problems, not because it does or doesn't accord with common notions such as 'purpose' or the ordinary notion 'function'. My program is far removed from conceptual analysis; I need a term that will do a certain job and so I must fashion one. However . . . the things that have 'proper functions' do seem to coincide with things (omitting God) that have, in ordinary parlance, 'purposes'. (*LTOBC* 2)

From these passages, and also from the definition I gave, and from the discussion that followed, much of which was about language function, I assumed it would be clear that the term 'biological' in the title of *LTOBC* was used not literally, but broadly or metaphorically. But the definition was immediately

[2] Proper functions are of great interest, I argued, because they correspond to a pattern that recurs in a large variety of forms, on many levels and in many domains, and seems to be found wherever purpose and/or intentionality are naturally ascribed. Purpose and intentionality were the phenomena (contrast 'the words', 'the concepts') I was interested in. There are lots of borderline cases of proper functions, if not so many in nature, certainly in possible worlds. What is interesting about proper functions is not their discreteness, not their dramatic difference from all other possible things, but the diversity among the actual items that have them, the great variety of their manifestations. What is surprising is that so many of the various actual things that are thought of as having purposes or meanings seem to be strung on a common thread, to be variations on a common theme. The description of 'proper functions' was an attempt to describe what is most central in these patterns, patterns that underlie, though they do not directly govern, our usage of the notions of purpose and meaning. Should they correspond to what purpose and meaning 'really are', it would be the way being HOH corresponds to what water 'really is', or the way in which loose bonding but without crystallin structure corresponds to what a liquid 'really is' (but from which it follows that glass is a liquid).

praised/blamed as a right/wrong *analysis* of the notion of *biological* function.[3] It is for this reason that I wish to be particularly careful here to make clear that the notion 'proper function' is not offered as an analysis of biologist's usage, but merely as a fairly well-defined term that is applicable and, I believe, useful in biological contexts.

The term 'proper function' was my own coinage. I intended (as suggested at *LTOBC* 2) Webster's first meaning of 'proper', which coincides with that of the Latin *proprius* meaning *one's own*.[4] Especially, 'proper' was not intended to be a prescriptive or evaluative term. I have sometimes emphasized that proper function is a 'normative' notion, but normative terms are not always evaluative.[5] Normative terms are used to indicate any kind of measure from which actual departures are possible. For example, a numerical average is one kind of norm, as is any sort of regularity: 'With that kind of sky in the west it ought to be sunny tomorrow.' (Proper functions do not correspond to averages or regularities either, of course. They define a standard of their own kind.) Certainly the point of proper functions was not to capture normativity generally. Purpose and intentionality were the original targets. The kind of error or mistake that stands over against functioning 'properly', in this technical sense, is unfulfilled purpose and its relatives (of which one form, I argued, is false representation).

I will not try to repeat here the two detailed chapters defining proper functions in *LTOBC*—chapters 1 and 2. Roughly, the idea is, first, that proper functions are associated with reproduced or, broadly speaking, copied items, certain effects of whose ancestors have helped account for the survival, by continued reproduction, of the item's lineage. This simple description, however, also fits each of the stages of any cyclical process. To have a proper function an item must also come from a lineage that has survived due to a correlation between traits that distinguish it and the effects that are 'functions' of these traits, keeping in mind that a correlation is defined by contrasting positive with negative instances. Intuitively, these traits have been selected for reproduction over actual competitors. Because the correlation must be a result of a causal

[3] See e.g. Neander (1991); Amundson and Lauder (1994); Walsh and Ariew (1996). Writing about proper functions in (Millikan 1989*a*), I said that it was fine to take the definition as stipulative, but certainly, please, not as a conceptual analysis of anything. Commentators continued to claim not just that I had given a correct/incorrect analysis of biological function but a correct/incorrect analysis of the notion of function itself... of THE notion of function! Indeed, there actually were claims that I had given an incorrect analysis of what '*proper* functions' are, though that term was entirely my own coinage.

[4] The root is found, for example, in 'property' and the verb ' to appropriate'. 'Functioning properly' is, of course, a perfectly ordinary English phrase, but there 'proper' tends instead toward an evaluative meaning.

[5] In *LTOBC* I sometimes spoke of proper functions as functions things are 'supposed to' perform. But the term 'supposed to' was defined naturalistically and, indeed, entered in the 'Glossary of Technical Terms' at the end of *LTOBC*.

effect of the trait, the trait will not merely have been 'selected' but will have been 'selected *for*' (Sober 1984). Thus a thing's proper functions are akin, intuitively, to what it does by design, or on purpose, rather than accidentally.

An important clause in the definition of proper functions allowed for certain devices that are (in a carefully defined sense) 'malformed' to have proper functions despite being unable to serve these functions. Such devices have still been 'designed' to serve their functions. For example, a mispronounced hence misunderstood word may still have the proper function(s) of the lineage from which it was copied, and a deaf ear still has enabling hearing as a proper function. Thus enters the (non-evaluative) 'normative' dimension associated with proper functions. It enters also when items do not perform their proper functions because background conditions that helped to support performance of these functions in ancestors are absent. Then they may fail to perform properly (in the defined non-evaluative sense) through no abnormality of their own.

The most familiar examples of items with proper functions are various traits of biological forms that have been selected for by Darwinian natural selection and that are reproduced genetically. Call functions of this kind 'proper biofunctions'. My focus in this chapter will be on proper biofunctions and their relation to Cummins functions found in biological contexts. Most biological traits have numerous proper biofunctions, some more proximal and others more distal, each of which helps to effect the next in the series. There is, I think, *never* such a thing as THE proper biofunction of a biological device.

Besides the most obvious examples of proper biofunctions, there are also 'relational', 'adapted', and 'derived' functions, which include, I have argued, both explicit human purposes and the functions of activities and artifacts produced in accordance with human purposes. Later I will discuss the rationale for the notions 'relational', 'adapted', and 'derived' as applied to proper functions in considerable detail. I will do so because these notions have been nearly universally overlooked or misunderstood in the literature on 'teleosemantics' and related issues.[6] Moreover, these more complex proper functions have exact parallels in the realm of Cummins functions, where it is important that they be recognized as well.

3. Cummins Functions

Cummins began his paper on functional explanation (Cummins 1975) by arguing that it is a mistake to suppose that reference to the effects of a biological

[6] See e.g. Wagner (1996); Matthen (1997, 1998); Rowlands (1997); Preston (1998); Walsh (1998).

trait helps to explain its presence in an organism. It is the genetic program, he says, that actually explains any such presence, this in turn being explained by the vagaries of mutation, rather than by the fact that the trait was selected for. This may be an error on Cummins's part. Any trait that was selected for performing its current function has historically *caused* an increase in fitness. That is what it means to say the trait was selected *for* rather than merely selected (Sober 1984). And that means the trait must actually have made a yes or no difference concerning survival and/or reproduction for at least some ancestors having the trait, hence must have helped account for the existence of current progeny having the trait. Any inherited trait that has helped to explain why an ancestor survived or reproduced *ipso facto* helps to explain why the current animal, hence the trait, is here. Nothing has a single cause. That the genetic program was a cause does not entail that nothing else was a cause (see also Neander 1995).

But if this was an error on Cummins's part, although it needs to be mentioned if the relation of Cummins functions to proper functions is to be distinctly understood, the error does not infect the central purpose of Cummins's definition of 'function'. The important point is that Cummins's definition of function is designed to clarify a completely different kind of 'functional explanation' than the 'why is it there?' kind. It is designed to clarify the sort of functional explanation that 'explains the biological capacities of the organism' (Cummins 1975: 751). It explains how the organism's system works, rather than why the traits contributing to these capacities or workings are there.

Cummins functions are such relative to a chosen 'containing system' that has, as a whole, a certain 'capacity'—that is, a certain 'disposition' or dispositions that we are interested in having explained. Cummins functions are dispositions of parts of this containing system, or simpler dispositions of the whole system that, added together, account for the complex capacity that needs explaining. Because items have Cummins functions not absolutely, but only relative to a chosen capacity or capacities of a chosen containing system, the point of ascribing a certain Cummins function to an item will depend entirely on one's explanatory interests.

> For no matter which effects of something you happen to name, there will be some activity of the containing system to which just those effects contribute, or some condition of the containing system which is maintained with the help of just those effects. (Cummins 1975: 752)

> The explanatory interest of an analytical account is roughly proportional to (i) the extent to which the analyzing capacities are less sophisticated than the analyzed capacities (ii) the extent to which [they] are different in type . . . (iii) the relative sophistication of the program appealed to, i.e., the relative complexity of the organization of component parts/processes that is attributed to the system. (Cummins 1975: 764)

In contrast to proper functions, Cummins functions, merely as such, have nothing to do with why the thing having the function is there, and nothing to do with its purposes. Cummins functions can be contrasted with 'accidental effects' only in the sense of effects that do not help to explain the capacity one has chosen to analyze. Nor are Cummins functions defined such as to contrast with malfunctions or with failures to perform functions. For example, in considering the rain-cycle system, in accordance with which the lakes and rivers are periodically refilled and so forth, the Cummins function of cumulus clouds is to produce rain, but there is no such thing as malfunction for this system nor is it the purpose of clouds to cause rain. Cummins functions are simply dispositions of a system or of parts of a system, and it is definitional of having a disposition that, if the conditions of the disposition are met, the disposition is manifested. On the other hand, just as biological traits typically have a whole series of proper functions from more proximal to more distal, more proximal or more distal Cummins functions may also be described, depending on how fine grained the functional analysis is.

4. Input, Background Conditions, and State Changes for Cummins Systems

In this section I will be describing what I take to follow from Cummins characterization of functions and of functional analysis as descriptions of *dispositions* to function rather than norms of function. It is not always easy to keep these two straight, and some of the language that Cummins uses invites a confusion that we need to avoid. When we avoid it, we will be able to see more clearly what will need to be supplied if the notion of a Cummins function is to do any work in biology.

In describing the general form that functional analyses take, Cummins mentions flow charts, circuit diagrams, and computer programs. Notice that representations of this kind generally specify ideal rather than actual systems. The circuit diagram that comes with your washing machine represents how it was designed or intended to function, not necessarily how it does function. Moreover, it was designed to function that way not unconditionally, but given quite specific background conditions and quite specific inputs. For example, it was designed to operate upright on a relatively level, stable, and rigid floor, under about one g gravitational force, surrounded by air at about one atmosphere pressure, protected from large magnetic forces, heavy blows, strong vibrations, heavily corrosive gases, and so forth. And it was designed to take as input, fed in at designated places, an electric current of about 110 volts alternating at 60 cycles, hot and cold water, certain kinds of emulsifiers, clothes or other cloth materials soiled with a reasonable amount of ordinary dirt (not completely soaked, for example, in wet tar or wet paint), and mild forces in designated directions on certain of its buttons and dials.

Specifications of this sort concerning background conditions of operation and allowable input must also be assumed for any system to be given a Cummins-style functional analysis—we can say, for any 'Cummins system'. This is because a Cummins system is analyzed as having a certain set of determinate *dispositions* of the whole and of the parts, and these dispositions will be determinate only in so far as limits on background conditions and inputs to the system are determinate. Cummins systems cannot be said to fail or malfunction as such, but they do have to be specified in relation to *delimited* possible background conditions and *delimited* possible inputs, implicitly or explicitly specified in some manner. To describe a Cummins system is to describe a set of dispositions. But a set of dispositions is a set of responses to possible, not merely actual, inputs and conditions. A set of dispositions corresponds to a set of counterfactuals. To describe a Cummins system, then, is not to describe actual historical conditions and actual historical inputs. Nor is it to describe, for some actual object, every result of every possible input under every possible condition. In this respect, Cummins systems correspond to ideal types, not to actual historical tokens or kinds. Extreme care must be taken if we are to avoid confusing, in any given case, the boundaries we must set to delimit this ideal with some kind of prior unexamined normative boundaries. For example, we must avoid implicitly assuming reference to inputs or conditions that accord with design. No reference to design is legitimate when dealing with Cummins functions pure.

It follows that no chunk of matter, such as a washing machine or an elephant, *determines* a Cummins system when considered just as such. First, as Cummins explicitly noted, we must specify which of the various output capacities of the chunk is to be analyzed. If the washing machine, for example, is to be analyzed for its capacity to turn out clean clothes rather than to rattle the dishes in the cupboard or to warm up the room, this needs to be specified. Secondly, we must specify what will count as allowable conditions of operation for the system. The washing machine may turn out nicely washed clothes if, instead of being left on after it is filled, the floor under it vibrates or rotates in just the right way. If this method of operation is not within the domain to be explained by the intended functional analysis, this must be specified. The machine taken by itself describes no such limits. Thirdly, we must specify what will count as allowable input to the system. Given a modern washer, pouring hot water in the top from a bucket, or turning the agitator by hand, are not intended inputs, although with washers of the 1920s this sort of input was part of the Cummins system intended. Possible input from a repair man also is not part of the system's intended operation. But the machine taken by itself is silent about the kind of Cummins system it prefers to exemplify.

A chunk of matter, depending on what are considered its allowable inputs and background conditions, may exemplify many different Cummins systems at once, even many different Cummins systems with the same output capacity. What counts as a Cummins function is relative to choice of an ideal type to be explained. If we are to make use of the notion of Cummins functions in biology, an important question, then, will be how to specify principled and useful ways to delimit biological Cummins systems. The essential role played by the interests and intentions of the one who analyzes a chunk of matter as exemplifying a certain Cummins system suggests that we must pay close attention to the move from a certain historical member of a species, or from the historical species as a whole, to specifying non-arbitrary and objective Cummins biofunctions of various of its traits.

Living chunks of matter do not come, just as such, with instructions about which are allowable conditions of operation and what is to count as allowable input. Similarly, they do not come with instructions telling which changes to count as state changes within the system and which instead as damage, breakdowns, or weardowns. Nor do they come with instructions about which processes occurring either within the organism or outside it are to count as occurring within and which as irrelevant or accidental to the system. What one analysis describes as the system's yielding its designated outcome according to a legitimate Cummins analysis another may describe as yielding it 'serendipitously'. What counts as 'damage' on one account may count as irrelevant change external to the system on a second, for example, because the second imposes different limitations on 'normal' input or on 'normal' surrounding conditions, adjusted to ensure that the change has no effect on the capacities of the system to be analyzed.

These problems are difficult, and I am actually not sure how satisfactorily they can be addressed.[7] They are made especially interesting by the fact that the currently controversial notions of useful 'spandrels' (Gould and Lewontin 1979) and, more generally, 'exaptations' (Gould and Vrba 1982; Gould 1991) seem to rest on a notion of function of the kind Cummins describes.[8] Thus 'exaptations' are said to be traits that have 'vital current utility based on cooptation of structures evolved in other contexts for other purposes (or perhaps for no purpose at all)' (Gould 1991: 46). That is, I believe, they have Cummins

[7] In Millikan (1999) I assumed they could not be solved. Now I am somewhat more optimistic, as will emerge below.

[8] Gould vacillates in his usage of 'spandrel', sometimes meaning merely any structure or process that is an accidental side effect of structures designed for other purposes, other times implying that the structure has current utility as well. In the latter case, I will speak of 'useful spandrels' or just 'exaptations'. Gould is also indiscriminate in his use of 'exaptation', sometimes covering with it any structure that had its origin in a prior structure originally selected for a prior purpose. But of course every adaptation has such an origin if one looks far enough back, so that no distinction at all between adaptation and exaptation is then drawn.

functions in the life system of the animal—Cummins biofunctions—that are not also proper biofunctions. But there are also indeterminacies connected with the notion of a proper biofunction—a subject to which I will turn directly.[9] I will return to the question how a biological system might be delimited and described as a Cummins system in Section 8 after more tools have been assembled in the toolbox.

But, before I proceed, one more note may help on the nature of the project I am attempting. Concerning development of the terms 'Cummins biofunction' and 'proper biofunction', it seems to me that the sensible strategy is as follows. First describe what the paradigm cases are like, the cases in which there can be no question about the application of these notions of function as already defined. Secondly, explore the various dimensions in which slippage can occur—in particular, the kinds of contingencies that can make application of these notions less than clear-cut. Thirdly, stop! Do not attempt to give these notions entirely clean boundaries. Nature has many important joints, but these joints are seldom clean. Definitions that cut sharp edges where there are none in nature are of little use in the understanding of nature. On the other hand, exploring dimensions of slippage may also reveal places where the notions themselves are inelegant or gerrymandered, or worse, open ended like a Christmas stocking with the foot cut off. Fix this if possible. (Later I will argue that the notion of an exaptation is such a Christmas stocking.) But most important, to be avoided at all costs is the attitude that there must be some 'correct' way of using these terms, some preordained way waiting to be discovered.

5. Vaguenesses in the Notion of a Proper Biofunction

In paradigm cases the genes responsible for traits with proper biofunctions have forced their alleles to extinction through their superior effects upon fitness. Even tiny differences in fitness values, if consistent over enough generations, typically lead to extinction of the less fit alleles. But differences in fitness values are not always consistent over the generations, owing, for example, to fluctuating environmental conditions. Then fluctuating changes in the gene pool may result, with certain alternative traits also surviving the generations. Theoretically possible here is a continuum of cases from traits having minimal to maximal effect on the statistics in the gene pool over a period of time. There is no benefit in inventing some definite criterion for how much influence over what period of time a trait's effect must have had on the constitution of the gene pool for this effect to be counted as a proper function. In some cases, however, systematic selection pressures have been brought to bear only on

[9] Peter Schwartz has urged this point, causing me to rethink a major chunk of this chapter.

well-circumscribed portions of a population of organisms, artificial selection by humans for certain purposes being one obvious case. It seems reasonable to say, for example, that the proper function of certain of the collie dog's behaviors is to aid in herding sheep, regardless of how local a portion of the entire dog species exhibits these behaviors. Similarly, the fitness of a trait sometimes declines as its frequency in the population increases, so that it persists but only in equilibrium with alternative traits controlled by its alleles. But its presence, when it is present, is still accounted for by its correlation with a certain effect. Perhaps we should say that it has a proper function, but only when sparsely distributed.

The genes responsible for a trait sometimes become fixed owing to more than one function that is served when perhaps none of these functions taken alone would have fixed them. There might, for example, be countervailing effects of the trait that would have led to its extinction had only one of its disjunct of positive effects been present. Just how small a contribution to increase in fitness should be allowed to count as bestowing on any one of these effects the name 'proper function'? That is exactly the kind of question I recommend we *not* ask; it seems clear that 'answering' it would not serve any theoretically interesting purpose.

A dimension of indeterminacy for proper biofunctions addressed in the literature concerns traits that once were selected for a function but that no longer serve that function (Griffiths 1993; Godfrey-Smith 1994; Schwartz, this volume). In *White Queen Psychology*, chapter 2 (Millikan 1993), I pointed out that, if a trait is now serving a different function than it was originally selected for, then it probably has more recently been under selection pressure precisely for serving that new function. I thus suggested a form of what Godfrey-Smith later termed a 'recent history' view of proper functions (1994). Schwartz correctly points out, however, that there are factors other than selection that can account for persistence of a trait that was once selected for, and suggests counting as proper functions just those effects that a trait was both selected for and still sometimes performs (this volume). This seems sensible to me. Certainly it would be confusing to say of a vestigial trait that *could* no longer serve a certain function that that was its proper function. On the other hand, one could also just allow oneself to say, in certain cases, that, although such and such is the proper biofunction of a certain trait, it seldom or never performs it any more. For example, if a human infant is dropped into very cold water, it stops breathing and instantly goes into hibernation. That is a trait that may very well have been selected for, if not in humans specifically, then in a wider clade of which we are members. This capacity very seldom serves its proper function in the modern world, but it seems reasonable just to *say* that, rather than to prune the definition of a proper biofunction artificially.

6. The Descriptive Generality Requirement

The next dimension of possible indeterminacy in the notion of a proper function applies to Cummins functions as well. It is important enough to deserve its own section and to require tightening up on both definitions. It is common for a trait to be selected for serving a function that might be defined either more broadly or more narrowly. A textbook example is offered by Elliot Sober in *The Nature of Selection* (1984). Fruit flies subjected over many generations to high-temperature stress develop thickened skins to protect themselves from the heat. Being good insulators, these skins would protect them from cold as well. Should we say that the proper function of these thickened skins is, more narrowly, protection against heat or, more broadly, insulation? Sober opts for insulation, and I think his choice is, in the end, the only reasonable one. The general principle involved is prevention of heat exchange, so that the thickened skin not only serves the same purpose—namely, keeping the fly at a uniform temperature—in both overheated and overcooled environments; it performs this function in accordance with the same explanation. A more narrow description either of the function served or of its explanation would be as inappropriate as saying that a car was wrecked because it went off an asphalt road through some grass and into an embankment rather than just because it went off the road into an embankment. Similarly, it is a proper function of our semicircular canals to help keep us upright in a gravitational field, not just in a gravitational field of one G. Moonwalkers do not employ their semicircular canals for functions they were not designed for. Nor do we employ our digestive systems for functions they were not designed for when we eat newly hybridized fruits, or even Fritos and Coke. Similarly, a proper function of my heart is to help me to wiggle my toes, but only as falling under the much more general description of supplying my organs with oxygen and nutrients so that they may do whatever their individual jobs may happen to be.

The requirement that proper biofunctions should always be described according to the most general principles available is a requirement that seems equally sensible when applied to a description of Cummins functions. To describe how my semicircular canals work by telling separately how they react under an oblique force of one G, then of 0.9 G, then of 0.8 G and so forth, naming each of these a separate Cummins function of the canals, would clearly be absurd. Similarly, a good Cummins function description will describe the digestion of carbohydrates, fats, and sugars, not the digestion of navel oranges, Fritos, Coke, and hummus. I will refer to this principle as 'the descriptive generality requirement'.

7. Relational, Adapted, and Derived Proper Functions

A central application of the descriptive generality requirement concerns what I have called 'relational', 'adapted' and 'derived' proper functions (*LTOBC*, ch. 2).[10] All adapted and derived proper functions admit also of more basic relational descriptions, and the relational descriptions are the more general. It is easy to suppose that a function is new or that it must be separately described because its adapted or derived character is novel, whereas a closer look would show that under a relational description it is the same old function over again. Let me first explain extremely carefully what these three kinds of functions are, for, as mentioned earlier, they have very often been either ignored or misunderstood. As I proceed I will explain how the descriptive generality requirement applies, and how it applies to exactly the same functions when these are viewed as Cummins functions.

Begin by considering an adding machine. Does it do the same thing every time you use it? One time it returns the number 237, the next time the number 257,000. But the important thing is that it turns out, every time, the sum of the numbers fed into it. Under this general description, it does the very same thing every time. Its effect is always production of the very same abstract relation between input and output. If one were to give a Cummins description of the adder, that is how the descriptive generality constraint would require that its basic function be described. If the adder exists as a small part of a larger machine that performs more complex functions that depend in part on the adder, the adder's function should, of course, still be described relationally. Suppose the adder needs to handle only numbers under 1,000. Then one could give a list of its 'adapted' functions. If 7 and 3 are put in, its adapted function is to output 10. If 59 and 79 are put in, its function is to output 138, and so forth. But that description would not meet the descriptive generality requirement.[11] Similarly, if we take the adding machine apart to see how it works inside, the description of each operation inside, in so far as it varies systematically with input, needs to be described relationally. Now suppose further that the adding mechanism is part of a creature designed by natural selection and that the mechanism has been selected precisely for its capacity to add something the creature needs added, say, as part of a smart foraging strategy. Then the adding mechanism's Cummins functions and the Cummins functions of

[10] The following section on relational, derived, and adapted proper functions is derived and adapted from Millikan (1999).

[11] Unless, of course, it does this by accessing a huge look-up chart where all the sums are posted in advance. In that case there is no general principle involved getting each sum right—that is, there is no correct relational description of how it adds.

its parts are proper biofunctions as well as Cummins functions and, of course, must still be described relationally.

Let me introduce some examples now that are more plausible biologically. Having relational proper functions is typical of the behaviors of organisms. These functions, in higher species controlled mostly by the perceptual and/or cognitive systems, are preformed by altering the relation between the organism and the environment as needed so that the environment will provide advantageous surrounding conditions and inputs for the organism. Some of these functions involve changing the environment to fit the organism, some involve changing the organism to fit the environment, some involve merely changing relations between organism and environment, or involve some combination of these three. The systems responsible for these changes have, first, relational proper functions of one kind or another. That is, their job is to make it the case that the organism and the environment bear some particular relation to one another. Their job is to create relational structures. As with any other proper function, a relational proper function of a mechanism corresponds to an effect that ancestors of the mechanism have historically had that helped account for their selection. In this case the effect was creation of an abstract relational structure.

The example I originally used of a mechanism with a relational proper function was the chameleon's pigment-rearranging mechanism (*LTOBC*, ch. 2). Its function is to produce the relational structure *skin-bearing-the-same-color-as-its-background* for the chameleon, a further proper function being, of course, to prevent predators from seeing the chameleon. To create this relational structure, the mechanism effects changes in the chameleon but not in the environment. Other animals effect production of a similar relational structure by moving into parts of the environment that match them—that is, by changing the spatial relation between themselves and the environment. And there are probably also animals that change the environment—say, the surroundings of their nests—in order to produce this kind of relational structure.

Notice that in order to produce the relational structure *skin-bearing-the-same-color-as-its-background* the chameleon does not need to produce both relata. It produces only one of the relata, the skin color, but in such a fashion as to bring into existence the designated relation. Nor is there any cheating here. No device can produce an effect, a result, a product, *ex nihil.* Every function is performed only by using materials of some kind, relying on certain properties of materials already at hand. Producing a relational structure by being guided by one relatum to produce the other is as legitimate a form of producing as producing a wooden spoon by choosing a piece of wood and carving it.

There are a great number of relational structures besides those involving sameness relations that behavioral mechanisms may have as proper functions to produce. Another well-worn example (I fear I have worn it out) is the mechanisms that produce the dance of the honey bee. The job of these mechanisms is to produce a relational structure having a location of nectar as one relatum and a certain aspect of the pattern of the dance as the other. The relation in question is the one given by the abstract function (mathematical sense of 'function') describing the semantic rules for the B-mese used by the particular species of bees. In most species, an angle that the dance movement marks out relative to gravity or to aspects of the hive always bears the same definite relation to the angle of the location of nectar relative to hive and sun. The properly functioning dance mechanism always produces exactly the same thing—namely, this designated relational structure. The mechanisms that produce bee dances have as a further proper function, beyond producing the dance, to send watching fellow bees off in a certain direction. The relational structure thus produced has as relata (1) the orientation of the dance relative to the hive and (2) the angle in which the watching bees fly relative to the hive and the sun. The mechanisms create this relational structure by making changes not in the organism itself but in its environment—changes not in the dancing bee itself but in the bee's fellow workers. As a result of producing, first, the proper dance/nectar-location structure, the properly functioning dance-making mechanism later effects always exactly the same thing—namely, existence of this second relational structure between angle of dance and angle of flying workers. (This requires the right environment, of course, one in which there are well-functioning fellow workers available. Their presence is what I called, in *LTOBC*, a 'historically normal condition' for managing to perform this function.) It should go without saying that a Cummins analysis of how the wider system works that effects the cyclical reproduction of honey bees needs to make reference to each of the above relational functions in exactly the same way as is required for an analysis by proper functions. Similarly, it should easily be seen that every point made below about proper biofunctions applies equally to the same functions viewed as Cummins functions. I will not keep repeating this.

Where the proper function of a trait is to produce a series of effects each effecting the next, it is also the proper function of each stage to produce the next stage. So what we have here is the production of one relational structure—dance/nectar-location—having as a proper function the production of another—dance/direction of flying. Finally, and as a logical result (given Euclidean geometry), one more relational structure is produced—namely, workers flying towards nectar. This is typical. Proper biofunctional relational structures typically do their jobs by producing further relational structures that eventually produce a relation between organism and environment yielding conditions or inputs the organism needs. Sometimes the whole process involves changing only the environment, and sometimes only the organism, but sophisticated relational proper functions typically involve both. They also typically cooperate with other structures having other relational proper functions, such as the bee dance producers' cooperation with answering mechanisms in fellow worker bees, which mechanisms could be given a similar relational analysis.

The general picture, then, is of proper functional processes that involve series of interweaving stages each of which is an abstract relational structure, some moments in these processes producing changes in the organism, others producing changes in the world, but involving always exactly the same relations, although among different relata each time they are run. They are reproduced invariant processes, always the same when described in the most general way that explains how they work, yet different in their elements each time. It is to simplify the description of these complex relational structures and processes that I introduced the terminology 'adapted proper function' and 'derived proper function' in *LTOBC*. This terminology adds nothing to the original definition of 'proper function', but affords a way of talking more easily about phenomena already captured by that notion, given that traits and mechanisms can have relational proper functions. Caution is needed in the treatment of these functions, however. Every reference to an adapted or derived proper function is really an implicit reference to one or more deeper relational functions. If we do not keep that in mind, adapted and derived functions are liable to be confused with brand-new quite separate functions—for example, as we will soon see, with exaptations. Similarly, if we do not keep in mind the descriptive generality requirement when trying to locate Cummins biofunctions.

When a mechanism has a relational proper function, it may produce one of the relevant relata while the other relatum is not affected, either (1) remaining the same or (2) undergoing its own independent course of development. A simple example of (1) is the chameleon's pigment-arranging mechanism, which changes the chameleon's color while the color of the background remains the same. An simple example of (2) I take from B. C. Smith (1996). Suppose that a species of sunflower not only tracks the sun, but continues to move when the sun goes behind a tree so as to catch up with it on the other side. Here the mechanism produces changes in the organism designed to maintain a certain organism-environment relation, and does so as the environmental relatum itself is changed, though not, of course, changed by the flower. Similarly, the relational structure consisting in the image on a male hoverfly's retina mapping the position and angle of approach of a passing female has as an eventual proper function production of another relational structure consisting in the male's path crossing the female's (for details, see Millikan 1993). As with the sun and the sunflower, the female is (as yet) unaffected by the male. Only a change in the direction of flight of the male is effected.

Now examine the relatum that is actually *produced* by a mechanism with a relational proper function. Consider the brown skin produced by the pigment-arrangers of the chameleon sitting on a brown background. The job of the pigment-arrangers is to produce the relational structure, *skin-color-matching-its-background.* But neither relatum, brown background nor brown skin, is an operative part of a historically normal set of sufficient conditions explaining the capacity of the chameleon to become camouflaged. Either relatum might have been replaced, and, so long as the other was similarly replaced, the chameleon would have been properly camouflaged. Being brown is not a part of a historically normal set of conditions for performance of any of the chameleon's proper functions. Neither the relatum produced nor the independent relatum has a proper function. Not all by itself! Only the whole relational structure has a proper function. Similarly, for a sensible Cummins functional analysis.

However, *given that brown is the color of the background*, the job of the pigment-rearrangers is certainly to make the skin brown. I call this kind of job an 'adapted proper function' of the mechanism. Turning the skin brown is not usually a proper function of the mechanism, and it will not remain a proper function of the mechanism when the chameleon no longer sits on something brown. However, right now, given that it is on a brown background, it is an adapted proper function of the mechanism to make the chameleon brown. Turning the skin brown is a proper function of the mechanism 'as adapted to' the brown color underneath. It is not, of course, a simple, but only a conditional proper function of the mechanism, an 'iffy' proper function, but the 'if' part has been asserted. The product produced by a device performing an adapted proper function I call an 'adapted device'. The chameleon's brown color is an adapted device. It is not, of course, brown itself that is an adapted device, but only the brown skin color of a chameleon produced in the right way in the right circumstances. Similarly, it is reasonable in this sort of context to speak of 'adapted Cummins functions', for notice that the use of the word 'adapted' here is not connected to its occurrence in the word 'adaptation'.

Now ask about the functions of the relata themselves in a proper functional relational structure. The function of a whole relational structure, as was said, is often to produce another relational structure. The bee-dance mechanism produces the relational structure, *dance-mapping-the-location-of-nectar*, a function of which is eventually to produce another relational structure, *worker-bees-heading-towards-nectar.* The nectar, of course, is and remains an independent relatum, so, taken by itself, it cannot have a proper function derived from that mechanism. But the other relatum, the bee dance, is produced by the mechanism in accordance with an adapted proper function: it is an adapted proper function of the mechanism, as adapted to the location of the nectar, to produce a certain bee dance, one that maps this location. Further, it is an adapted proper function of the dance-producing mechanism to produce, as a result, a certain direction of flight in fellow worker bees. *Is it also* a proper function of the dance itself to produce this direction of flight?

The answer may at first seem to be *no*, for it seems theoretically possible, at least, that the particular bee dance has no ancestors. Perhaps no bees in the bee's lineage ever danced this particular dance, because there never happened to be nectar located in this particular direction from their hives. Then this particular bee dance, having never occurred in the past, certainly could not have been selected for any effects that it had, hence could not possibly have any proper functions at all.

But notice that this overlooks the descriptive generality requirement. We must describe functions and how they are performed in the most general way possible. Because bee dances that map different directions are different from one another in specific respects does not mean they are not also the same in more general respects. Indeed, the various dances of a given subspecies of bees are very much the same—to an untutored observer, hardly discriminable. And when they function in the way that has accounted for the natural selection of their producers and of their answering mechanisms in other workers, they always do exactly the same general thing. They produce a direction of flight that is a given function (mathematical sense) of certain aspects of their form—in every case exactly the same function of that form.[12] The bee dance has been selected in part for its capacity to cause other worker bees to be guided in their direction of flight *by its form*. In this respect, all bee dances of the same bee species have exactly the same proper function—also, of course, the same Cummins function. This function is a relational function. The dance's job or, on a Cummins analysis, its disposition, is to cause the workers to fly in a direction that bears a certain relation to itself (e.g. *LTOBC* 42).

Because of this relational function, depending on the particular form of the bee dance, it has as an adapted function to cause worker bees to fly in some particular direction, say, south-south-west—just as the relational function of making the chameleon's skin match its background results in an adapted function to turn the skin brown when the chameleon is sitting on something brown. I call this kind of function a 'derived' function, derived originally from the relational function of the producing mechanism plus its context, in this case, from the adapted function of the dance-producing mechanism as operating in a certain context. All derived functions (whether derived proper functions or derived Cummins functions) are *ipso facto* adapted in this way. Things said to have derived proper functions are being named or described in accordance with their adapted aspects (they point in different directions, they are

[12] In *LTOBC* I put this rather awkwardly by saying that they have ('direct') proper functions which are adapted to their own concrete forms (*LTOBC*, ch. 2 and elsewhere). I am not sure that I have explained it less awkwardly here, but the phenomenon itself is really quite easy to grasp.

different colors), not the aspects that make them like their ancestors (they are all bee dances, they are all pigmented skins of chameleons). Their derived functions are what these adapted aspects must do in order for the whole relational structure of which they form a part to perform its further function(s). Again, derived functions can easily appear to be new functions, when they are really just fragments torn from the same old relational functions operating in new contexts.

Consider any mechanism that has as a function to produce any kind of learning on the basis of experience, for example, a mechanism that effects trial and error learning, or learning by imprinting, or learning by imitation, or learning by reasoning something out given premises derived from experience. Any such mechanism has, as such, a relational function or functions. It is designed, or (for Cummins functions) disposed, to turn out behaviors (or, say, beliefs or desires) as a certain function (mathematical sense) of certain designated kinds of input from experience. The sea otter, for example, learns what to eat from its mother, the mechanism that effects this being, in part at least, quite specific—that is, not of more general purpose or effect. Perhaps a combination of mechanisms helps to effect this learning, but a relational function of this ensemble is to create in the baby otter a complex state that produces a disposition to eat-whatever-mother-eats. Given that its mother eats sea urchins, then, it is an adapted function of the mechanism to produce a state that effects a disposition in the baby to eat sea urchins. And, once this complex state is in place in the baby, it has as a derived function to produce sea-urchin collecting and eating. Other baby otters whose mothers eat abalones acquire complex states whose derived function is to effect abalone collecting and eating.

Generalizing from this, there can also be relational functions that produce adapted devices themselves having relational functions (for example, the functions involved in effecting empirical concept formation) producing more adapted devices having further relational functions (for example, the functions involved in fixing beliefs) to any degree of nesting. Out of this sort of structure can come things (tokens) seemingly very new indeed under the sun but that still have merely derived functions when examined more closely.

8. Cummins Biofunctions and how Descriptive Generality Constrains Exaptations

According to Cummins, 'functional analysis can properly be carried on in biology quite independently of evolutionary considerations: a complex capacity of an organism (or one of its parts or systems) may be explained by appeal to functional analysis regardless of how it relates to the organism's capacity to maintain the species' (1975: 756). But a functional analysis of how the heart makes sounds, for example, is surely not what biologists are interested in, nor is ignoring evolutionary considerations the same thing as ignoring the organism's capacity to maintain the species. Rather, it seems reasonable to consider the capacity of the organism to maintain and reproduce itself to be exactly the capacity implicitly under analysis when Cummins biofunctions are in question. I take it that these are the capacities referred to by various authors as 'current uses' (e.g. Gould 1991), 'system functions' (Preston 1998), 'algorithmic functions' (Rowlands 1997), 'causal role functions' (e.g. Amundson and Lauder 1994), and, when they are not also proper functions (not 'adaptations'), as 'exaptations' (Gould and Vrba 1982) and as useful 'spandrels' (Gould and Lewontin 1979).

Earlier I noted that a difficulty in applying Cummins's notion of functions directly to biological processes is that an organism plus some capacity that it has does not by itself yield a determinate Cummins system. It does not by itself determine what are to count as allowable background conditions and allowable input for the system's operation, nor what will count as state changes within the system's operation rather than external to it, or what is to count as the system's no longer existing—that is, as its having been dissolved or broken. A first and obvious step towards making these parameters determinate is to say that the appropriate conditions, inputs, and state changes must characterize, or have characterized, various actual members of the species under analysis. But which actual members? We cannot innocently add that these must be 'normal' or 'healthy' or 'undamaged' members, for these notions have not been defined either. Perhaps best to begin by limiting the relevant members to ones that have actually maintained themselves and reproduced, and then look for commonalities among the actual historical processes that achieved those results.

Notice that this first step—surely a necessary and innocent one—moves in a direction that parallels an analysis by proper functions. First, the analysis must be given against the backdrop of the actual historic situation of the species. Secondly, because for many species survival and reproduction are not average, indeed, may be achieved by only one in hundreds or thousands, it follows that what should be considered normal conditions and normal input to the biological system might stray very far from average conditions that members of the species actually find themselves in, or average inputs received by these members. But, in searching for Cummins biofunctions, we have put aside the striking paradigm that is fixation or maintenance of a trait owing to its effects. Only statistical frequencies of the various processes propelling—or drifting—various individuals towards reproduction are left as guides for separating Cummins biofunctions from accidentally propitious effects.

A search for commonalities among these individual historical processes naturally leads us back to the requirement of descriptive generality. The only way to describe Chamileo's avoidance of a predator by turning green and Chamilea's avoidance of a predator by turning brown as exemplifying a common Cummins biofunction is to describe this function relationally. Each turns the same color as its nether environment. Similarly, the dog that brings the newspaper to its master for a reward employs its mouth in accordance with what is undoubtedly one of its *proper* functions—namely, carrying things—but carrying, specifically, *newspapers* is only an *adapted* proper function of its mouth, given the dog's previous experience. In exactly the same way, carrying newspapers should be considered only an adapted Cummins biofunction of its mouth (contrast Preston 1998). Thus it is that individual animals in a species can lead very different lives, and humans can live and have lived under widely differing cultural conditions, yet their activities may continue to exemplify very many of exactly the same Cummins biofunctions—as well, of course, as the same proper biofunctions.

Employing this principle uniformly will show, I believe, that there are many fewer exaptations, many fewer plausible Cummins biofunctions that are not also proper biofunctions, than some have supposed. Thus Gould claims, 'Surely exaptations of the brain must greatly exceed adaptations by orders of magnitude' (1991: 57). Unfortunately, many of his examples are offered without an argument that they have any kind of functions at all, proper, Cummins, or otherwise. He gives no argument, for example, that music, or religion, granted they might be spandrels, also have 'vital current utility'. Also, very many of the examples he gives turn on counting anything that originated from a structure originally designed for a different function as an 'exaptation', no matter how long there have been selection pressures on it, preserving and adapting it to newer functions. But, apart from these vaguenesses and excesses, he makes a more interesting claim: 'just make a list of the most important current uses of consciousness. Start with reading, writing and arithmetic. How many can even be plausibly rendered as adaptations?' Well, I suggest, how many can be plausibly rendered as having 'vital current utility'—that is, I suppose, Cummins biofunctions? Given the requirement of descriptive generality, they will have only adapted and derived Cummins biofunctions, and these will be exactly the same as their adapted and derived proper biofunctions. All have been learned through the adapted employment of very general learning mechanisms, such as learning from trial and error, from imitation, from instruction, by figuring things out from prior premises, and so forth. They have Cummins biofunctions only by reference, when correctly described, to very general, highly relational descriptions. Reading, writing, and doing arithmetic are like turning green and turning brown, or like collecting and eating sea urchins, or like carrying newspapers in the mouth. But then where is the argument that they are not also derived proper functions—applied, facultative adaptations?

9. Vaguenesses in the Notion of a Cummins Function: Plugging Some Leaks

9.1. The First Leak

Harder questions for a useful definition of Cummins biofunctions concern what should count, in Gould's terms, as 'having vital utility'. A Cummins biofunction is supposed to be one that 'contributes to the maintenance or reproduction' of an organism, but what, exactly, is that? Suppose we begin by trying this. We will require for paradigmatic cases of traits with Cummins biofunctions that serving this function always makes a difference between surviving and reproducing and not surviving and reproducing, or not reproducing so prolifically. Less paradigmatic Cummins biofunctions will make this difference only for some significant proportion of individuals in a species.

Suppose, for example, there are some individuals for whom presence of a trait, and others for whom absence of the same trait, would save them from dying or would aid reproduction. Consider the grey squirrel's characteristic path when fleeing from danger. Running zigzag is a very good strategy when a heavier predator is chasing you, but not when a car is approaching. Perhaps in the modern world, dropping this behavior would save more squirrels than retaining it. We will say then that only if the trait increases fitness is the mechanism or process by which it does so part of the squirrels' Cummins biosystem. That is, we will use a counterfactual analysis. For each actual squirrel we will ask whether it would have lived longer and had more babies if it had not zigzagged when fleeing from danger, and the statistics on this will determine whether zigzagging has a Cummins biofunction or not, and if so how paradigmatic or important it is.

The difficulty here is the usual one with counterfactuals. There is no such thing as taking a world and just dropping a fact from it to see what would happen without it. No determinate possible world is constructed by merely dropping a fact from the actual world. Any fact that is dropped has to be replaced with a determinate contrary fact. What will the squirrel do instead of zigzagging? Yes, it is obvious that what *we* had in mind was the squirrel running in a straight line. But perfect obviousness to *us* that this is what *we* have in mind does not make it the one objective thing to have in mind. In fact, it is entirely indeterminate what the squirrel would do instead if it did not run from danger in a zigzag line, and our agreement on what we would *like* to put in place of this zigzagging does nothing to make it less indeterminate.

A different illustration may make this easier to grasp. Consider the question what Cummins biofunctions human shoulders have, if any. For example, is one of their Cummins biofunctions to keep one's clothes from falling off, hence to keep one warm? Well, suppose you did not have shoulders, what then? Exactly! What *would* you have then? Does religion have 'vital current utility'? Suppose that people were not disposed to be religious, what then? These questions have no determinate answers in principle. It is not just that the answers are hard to discover. The notion that a trait can increase the fitness of an animal makes sense only in the context of natural selection where there are determinate traits that are selected against. Then there are determinate traits for the selected trait to be *more fit than.*

Cummins biofunctions must be abstracted then, not by using counterfactuals, but merely by reference to chains of causes and effects that culminate in reproduction and are common to historical members of the species.[13] Leave aside the question to what proportion of members of the species these causal chains should be common, but let paradigm Cummins biofunctions be performed in nearly all reproducing members of a species.

9.2. *The Second Leak*

Suppose that nearly all kittens exhibit the same sort of playing behaviors—say, they all chase their tails. Or suppose that nearly all 15-month babies tumble down and then pick themselves up again and again before moving on to a more equilibrious stage. Or suppose that all human hearts make thumping noises that occasionally are listened to by their owners at night, causing a few minutes delay in falling to sleep. Each of these processes takes place on the way to reproduction. Each is a part of the mammothly complex causal process that culminates in reproduction. Leave reference to any of these happenings out in the case of an individual animal and the result is a gap in the full explanation of the path by which the individual arrives at reproduction. Keeping in mind that there is no determinate answer to the question what would have happened in individual cases had one of these processes not occurred, which ones should be mentioned in giving a Cummins-style functional analysis of the propensity to survive and reproduce? How do we determine which processes are merely byproducts and byways and which are functional parts of the Cummins biosystem? Which loops, which physiological or behavioral detours, can be ignored in a Cummins-style analysis? Which are mere 'spandrels', accidental side effects, superfluous aspects, perhaps even detracting inefficiencies, in a perhaps far from ideal biological system?

Perhaps a sensible answer begins this way. We view the system exactly as Cummins suggested, as exemplifying something like a flow chart or circuit diagram or computer program. This involves 'modularizing' the self-sustaining

[13] Some philosophers believe that causality must be understood in terms of counterfactuals. This turns things upside down in my opinion.

and reproducing system in terms of a series of prior capacities, and modularizing many of these capacities into still prior capacities, and so forth, where the output of each subcapacity serves as input to one or more other functional modules, under stated conditions. This kind of analysis may proceed and be fully explanatory while ignoring much of the detail of actual causal chains produced within or beyond average organisms. It may ignore, for example, most of the noises produced inside and outside, most odors exuded, all those effects on the environment that do not feed back in systematic ways, in accordance with a uniform explanation so as to become input to the functioning modules. All that seems easy and obvious enough. Perhaps all we need to ask then is how regularly such a causal chain should be exemplified to be considered part of a Cummins system rather than a serendipitous occurrence, and then answer, as before, by an appeal to paradigms and a tolerance for vagueness.

9.3. The Third Leak

Many of the normal conditions and inputs for operation of particular modules of a Cummins biosystem are regularly provided by the system itself. Thus the circulating blood supplies oxygen and nutrients as input to the various organs of the body, while other body systems keep these organs at a normal temperature. Other inputs and conditions normal for the system are regularly donated by the environment—'regularly donated' in the sense that they are donated wherever individuals of the species manage to survive. Thus the oxygen in the air, and its normal pressure without which most animals would collapse or burst. Most wild seeds land on infertile ground, but those that reproduce are, quite regularly, ones that do land on sufficiently fertile ground. Few just-hatched green turtles are lucky enough to find interstices between hungry birds in their first dash for the sea, but those that survive are regularly ones that did accidentally slip through these cracks. Other normal conditions and inputs to biological systems may fall between these extremes. They are supplied by the organism itself but only with the help of the environment. Frequently they are supplied due to certain behaviors of the organism, but only under favorable environmental conditions. The hunting behaviors of most animals are met with favorable conditions (an unwary rabbit, a gazelle in poor condition) that produce edible input to the digestive systems only occasionally. Immune systems are able to control harmful bacteria or viruses they encounter only some of the time. A small running child can negotiate the ground it finds underfoot without tripping and falling only some of the time. Indeed, surely it is true of all kinds of animal behaviors that they merely raise the probability of arriving in favorable circumstances and receiving normal system inputs.

On the other hand, it seems clear enough that biological systems are also often involved *accidentally* in causing their own situations, whether these turn out to be helpful or harmful. Activities or properties of the system often result in the presence of conditions or inputs that are not produced in accordance *with* the system. They are not explained merely by a Cummins analysis of the system. If John runs after a wild turkey he is hunting, depending on accidental conditions in his vicinity it may happen that this helps cause him to find a honey tree, or that it helps cause him to fall, breaking a leg. Surely the normal (digestive) input that results from his chasing behavior in the first case is not explained on a Cummins analysis any more than his encounter with unsuitable conditions for remaining upright are in the second. But exactly what criterion distinguishes cases where, as it were, the system serendipitously helps itself to suitable input Cummins-accidentally from cases where it helps itself Cummins-systematically? When the help is not effected by traits or behaviors that have been selected for producing help in this way, this equals, of course, the question how an exaptation should be defined.

Porcupines fall out of trees surprisingly often. About 40 per cent of porcupine skeletons show broken or healed bones of a sort probably due to this infirmity. Very probably, however, porcupines are sometimes saved from breaking bones by the springiness of their quills. Is this then a Cummins biofunction of their quills? Because they are especially fond of the bark of pine trees, when porcupines fall it is often onto a bed of pine needles below, and surely this too often saves them from breaking bones. Is preventing bone fractures a Cummins function of the porcupine's especial fondness for pine bark? Given the conditions that obtain in a modern hospital, the sounds that a person's heart makes often contribute to quick diagnosis and medical attention for a life-threatening heart ailment, producing helpful inputs to the biological system via needles and pills, respirators and so forth. If these conditions and inputs are not normal inputs for a human Cummins biosystem, exactly why are they not? Compare them, for example, with what parents from all cultures typically supply for human infants. A carefully structured environment and proper input is necessary for the continued functioning of many interesting Cummins systems.

One suggestion, of a kind that should now be familiar, would be to say that in paradigm cases a Cummins function is performed in all or nearly all cases of survival and reproduction, and that, as the statistical frequency of performance decreases, the function becomes less and less of a Cummins biofunction, or a less and less important function. Another possibility would be sometimes to relativize Cummins functions to designated populations, places, periods of time, and so forth. Then the sounds of the human heart have a function but only in quite modern times and in certain places and only in certain populations, while the sickle cell gene has a function in malaria-infested areas. Similarly, the tufted-tail nosemuffs and snowshoe feet of snow leopards in zoos have no Cummins biofunctions. Is it then a function of the human disposition towards religion in certain special environments to result in Bibles in breast pockets that stop stray bullets that would otherwise kill soldiers? Is doubt about this justified only because the case happens to lie very far from the central paradigms of Cummins functions?

Notice that, on either of these latter treatments of Cummins biofunctions, the baby's disposition to hibernate when dropped into iced water falls about as far from the paradigm for having a Cummins function as does the function just suggested for the disposition to religion. Currently it probably saves a comparable number of lives. On the other hand, there is a whole host of traits that turn out to have rather bizarre yet close to paradigm functions on analyses of this kind. Mice and spiders are regularly saved from damage when they jump or fall down long distances by the fact that they weigh almost nothing. Is their slight weight an exaptation for preventing this damage? (Is the weight of elephants an exaptation for preventing them from climbing trees thus ensuring that they will not fall out?) Tape worms tend to be quite particular about their hosts and are usually adapted to resist being digested in the stomachs only of one species or small range of species. Do the stomach juices of cats then have as Cummins biofunctions digesting dog and rabbit tape worms so as to avoid infestation by them, and the other ways around? Are the poor fur and the scavenger habits of the possum exaptations for preventing humans from hunting them for their pelts or their meat? Is the beauty of butterflies an exaptation to prevent humans from swatting them as they do other insects? Is the size of the moose an exaptation to prevent its being eaten by foxes? Clearly such examples could be multiplied indefinitely.

9.4. *The Fourth Leak*

A different kind of example illustrating the open-endedness of the notion of a Cummins biofunction concerns protective reactions and healing powers of organisms. Suppose that the porcupine's quills are not regarded as exaptations for cushioning falls. Or suppose that hibernating when dropped into iced water is not regarded as a Cummins function. Since infants dropped into iced water do not survive if left there long in any event, a corollary might be that iced water is not an allowable surrounding environment for the Cummins biosystem that is a human infant. Similarly, since much or perhaps most of the time animals do not survive bone fractures either, perhaps the sorts of impacts that break bones are not allowable inputs to these Cummins biosystems. And, if these inputs are not allowable, another corollary would seem to be that bone-healing mechanisms have no Cummins biofunctions. Since invading bacteria and viruses often kill as well, perhaps the same should be said of the immune systems. And, since cold often kills, perhaps human adaptations to withstand relatively extreme cold are not part of the human

Cummins biosystem. Indeed, most people have lived in warm climates and never used these adaptations. Perhaps Cummins-style functional explanations of biological systems should treat iced water and other sources of extreme cold, bone-breaking impacts, germs, and parasites—maybe even predators—as well as respirators, zoo keepers, and antibiotics, merely as irrelevant non-allowable surrounding conditions and inputs. Similarly, as mentioned before, a Cummins-style explanation of how the washing machine gets the clothes clean might well leave out the repair man.

10. How to Fix All the Leaks at Once

The notion of a Cummins biofunction seems to be nearly as open-ended as the amorphous notion of explanation itself. Whatever happens—regularly happens, often happens, or sometimes happens—on the way to eventual reproduction is something the *how* of which can be explained. A lesson is that anyone using the notion of Cummins functions, or 'current uses', or 'system functions', or 'algorithmic functions', or 'causal role functions', or useful 'spandrels' (have I missed any?) owes us an explanation of how each of the above four leaks is to be dealt with.

Is my own conclusion that the notion of a Cummins function is pretty useless in biology? Recall once again what Cummins said about functional analyses:

> The explanatory interest of an analytical account is roughly proportional to (i) the extent to which the analyzing capacities are less sophisticated than the analyzed capacities (ii) the extent to which [they] are different in type . . . (iii) the relative sophistication of the program appealed to, i.e., the relative complexity of the organization of component parts/processes that is attributed to the system. (Cummins 1975: 764)

I think we can add, for traits of biological systems, that interest in discussing them generally depends on how improbable or unique their effects seem to be, and how improbable their functional design seems to be. That is, our interest turns on exactly the same characteristics of traits that suggest that these traits are adaptations. Whether we see the hibernation of iced babies as corresponding to an interesting Cummins function will depend, for example, on whether we suspect that it is merely an expected side effect of the mechanisms by which the metabolic systems of babies usually function, or whether there are interesting quirks and additions to these systems that produce the effect. Roughly speaking, the interesting Cummins functions associated with the mechanisms and traits of an organism will be exactly the same as the proper functions. The difference is, in the first instance, only one of emphasis. Cummins emphasized the project of finding out *how* the biological system works, not just finding out what it *does*. But, of course, finding out what it does in detail, what *all* its proper functions are and *all* the proper functions of *all* its parts, is finding out how it works.

That is true as a first approximation at least. There is, however, a very interesting kind of exception to this rule. There is one kind of Cummins function that is quite neatly delimited, cleanly definable by reference to proper functions, but that does not itself correspond to a proper function. I propose that the term 'exaptation', which we have found to be disastrously undisciplined in current usage, should be co-opted and used for this kind of function.

11. A Notion of Exaptation that might Actually Do Some Work

An adaptation is always an adaptation to or within a certain environment. Fully to explain how an adapted trait works necessarily involves reference to its interaction with its environment. Adapted traits typically have co-opted certain aspects of their environments for use. And, typically, the most important factor in the environment of an evolved trait is the rest of the surrounding organism. Many traits, of course, have co-evolved, each adapting to the other. But other times, one trait helps effect another's proper function *merely* by serving as a needed aspect of the environment in which the other has evolved. Traits already present in an organism for quite different reasons are co-opted for use along with newly fashioned traits to serve new functions. Thus they end up helping to serve functions that are proper functions, but that are not their *own* proper functions. And, of course, they must be mentioned in giving a Cummins analysis of how the new proper functions that they now help to serve are effected.

It is traits that are co-opted in this way that form the only clearly defined category of exaptations. Beyond this category, I think one should be very careful in handling the notion of exaptation, for, resting as it does on the leaky notion of a Cummins function, it easily becomes harmfully undisciplined. When understood in this way, however, it still covers all of the interesting classic examples.

Darwin's famous example, taken from Paley, was the sutures in the skulls of human fetuses that make parturition easier. These sutures exist even in reptiles and birds that only need to break out of an egg, and probably result from deep mechanisms of ontogenesis, having been merely exapted during evolution of the birth process for humans. These sutures have to be mentioned in a full Cummins analysis of the human reproductive process, but in this context they have only Cummins functions, not proper functions. Other classical likely examples of exaptations are the color of blood, which makes blushing possible (assuming that blushing has a proper function), the weight of the flying fish, which returns it to water, and the front flippers of sea turtles, designed for swimming but used for digging holes during egg laying (Lewontin 1978). Easy additions to this list are the sea otter's fat tummy, which it uses for a dinner table at sea and the warm part under the wing, where a bird instinctively tucks his beak to keep it warm.

There are many birds that literally 'feather their nests' in order to keep their young warm, using down plucked from their own breasts. It is probable that these species of birds grow breast feathers more abundantly because of this. But, if so, it is likely that the selection pressures producing this effect impinge through the birds' original need to keep their own bodies warm, rather than impinging directly on the survival of the young. Then, even if the abundant growth of breast feathers has resulted from their use to line the nest, keeping the bird warm is the only proper function of the abundant growth. This is a different variation on the theme that traits designed for another purpose (or no purpose) may be co-opted to serve new proper functions. The abundant growth is reasonably considered an exaptation for keeping the babies warm.[14]

On the other hand, there are items that have been cited in the literature as exaptations that this way of handling exaptations does not admit. Griffiths (1993) wishes to admit the snail shell that the hermit crab (soldier crab) carries on its back as having the function of protecting the crab. Since the systems that are hermit crabs do not participate in the reproduction of snail shells, this would be analogous to admitting the eggs that you eat as having the function of nourishing you (Dennett 1998) or admitting the atmosphere as having the function of helping you breathe. Yes, of course one *could* consider the egg as part of the human Cummins system and also the hen that makes the egg, and one *could* consider the oxygen as part of the human Cummins system—and also the sun that helps photosynthesis hence the production of oxygen, and so forth. But a more reasonable place to draw a line around a Cummins biosystem excludes factors that are not reproduced by the organism. Probably nonreproduced factors are better considered just normal supporting conditions or normal input to the biological system.

On this way of handling exaptations, it is also possible by appeal to the descriptive generality requirement, and the notions of relational, adapted, and derived functions, to excuse noses from being exaptations for supporting eyeglasses, to excuse shoulders from being exaptations for holding up clothes, and so forth. If we adhere to the descriptive generality requirement, our functional explanations of complex relational capacities will be relational too. For example, we will explain the sea otter's food-procuring propensities not by reference

[14] In *White Queen Psychology* (Millikan 1993: ch. 2) I suggested considering exaptations of these kinds to have 'proper functions'. I now think that would be a mistake, that it would confuse together issues that should be kept distinct. On this, see Preston (1998); Millikan (1999).

to, in one case a crab stimulus, in another an abalone stimulus, but by reference to encounter again with that-which-its-mother-has-taught-it-to-eat. Similarly, noses and shoulders will figure in biological Cummins-functional explanations of how humans sometimes equip themselves against bad eyesight and cold, not under the descriptions 'nose' and 'shoulders', but under highly relational descriptions, indeed, perhaps under the same description. The relevant relational descriptions of my nose and my shoulders will treat them abstractly as being just the same, for purposes of biological explanation, as any other objects in the world that I have learned by imitation, or trial and error, or by figuring it out, how to employ towards some purpose I have acquired from experience, in a way normal for humans. The principles involved here are highly abstract and general, certainly not peculiar to noses and shoulders.

References

Amundson, R., and Lauder, G. V. (1994), 'Function without Purpose: The Uses of Causal Role Function in Evolutionary Biology', *Biology and Philosophy*, 9: 443–69.

Cummins, R. (1975), 'Functional Analysis', *The Journal of Philosophy*, 72/20: 741–65.

Dennett, D. C. (1998), 'Preston on Exaptation: Herons, Apples and Eggs', *Journal of Philosophy*, 95/11: 576–80.

Godfrey-Smith, P. (1993), 'Functions: Consensus without Unity', *Pacific Philosophical Quarterly*, 74: 196–208.

——(1994), 'A Modern History Theory of Functions', *Noûs*, 28: 344–62.

Gould, S. J. (1991), 'Exaptation, a Crucial Tool for Evolutionary Psychology', *Journal of Social Issues*, 47/3: 43–65.

——and Lewontin, R. C. (1979), 'The Spandrels of San Marco and the Panglossian Paradigm: A Critique of the Adaptationist Programme', *Proceedings of the Royal Society of London, Series B: Biological Sciences*, 205: 581–98.

——and Vrba, E. S. (1982), 'Exaptation—a Missing Term in the Science of Form', *Paleobiology*, 8/1: 4–15.

Griffiths, P. E. (1993), 'Functional Analysis and Proper Functions', *British Journal for the Philosophy of Science*, 44: 409–22.

Lewontin, R. C. (1978), 'Adaptation', *Scientific American*, 239/3: 212–30.

Matthen, M. (1997), 'Teleology and the Product Analogy', *Australasian Journal of Philosophy*, 75: 21–37.

——(1998), 'Biological Universals and the Nature of Fear', *Journal of Philosophy*, 95/3: 105–32.

Millikan, R. G. (1984), *Language, Thought, and Other Biological Categories: New Foundations for Realism* (Cambridge, Mass.: MIT Press).

——(1989*a*), 'In Defense of Proper Functions', *Philosophy of Science*, 56/2: 288–302.

——(1989*b*), 'An Ambiguity in the Notion "Function"', *Biology and Philosophy*, 4/2: 172–6.

——(1993), *White Queen Psychology and Other Essays for Alice* (Cambridge, Mass.: MIT Press).

——(1999), 'Wings, Spoons, Pills and Quills: A Pluralist Theory of Functions', *Journal of Philosophy*, 96/4: 191–206.

Neander, K. (1991), 'The Teleological Notion of "Function"', *Australasian Journal of Philosophy*, 69: 454–68.

——(1995), 'Misrepresenting and Malfunctioning', *Philosophical Studies*, 79: 109–41.

Preston, B. (1998), 'Why is a Wing like a Spoon?: A Pluralist Theory of Function', *Journal of Philosophy*, 95/5: 215–54.

Rowlands, M. (1997), 'Teleological Semantics', *Mind*, 106: 279–303.

Smith, B. C. (1996), *The Origin of Objects* (Cambridge, Mass.: MIT Press).

Sober, E. (1984), *The Nature of Selection: Evolutionary Theory in Philosophical Focus* (Cambridge, Mass.: MIT Press).

Wagner, S. (1996), 'Teleosemantics and the Troubles of Naturalism', *Philosophical Studies*, 82: 81–110.

Walsh, D. M. (1998), 'Wide Content Indvidualism', *Mind* (July/Oct.).

——and Ariew, A. (1996), 'A Taxonomy of Functions', *Canadian Journal of Philosophy*, 26/4: 493–514.

5. On the Normativity of Functions

Valerie Gray Hardcastle

ABSTRACT

When we pick out a property's function, we are isolating one of its effects from among the many it has as the thing it is supposed to do. Proponents of an etiological view of function claim that pragmatist approaches cannot answer this question. This chapter examines this claim. I argue that, despite appearances, we can get a robust sense of normativity out of pragmatic views of function; moreover, being relative to something else does not distinguish pragmatic versions of function from etiological ones.

Function is a strange concept: normative, on the one hand, and completely naturalistic, on the other. The word both describes what a thing is supposed to do and tells us something about the way the world works. The function of my microwave oven is to heat foods at an inordinate speed. My microwave has this function even if it is brand new and sitting unused in its original packing material. It has this function even if I have never used it to cook anything but instead use it for additional countertop space. It has this function even if it is broken and can no longer emit any microwaves, even if it has always been broken and I never fix it. Regardless, its function is still to heat food.

How do we know functional facts? The functions of human artifacts depend on human intentions. We explicitly designed the microwave oven to heat food; all we have to do is talk to the microwave engineers to know what the microwave oven's function is.

But natural items have functions, too. Our legs serve a locomotive purpose; our legs are supposed to do this. Legs have this function even if they are broken or so malformed that they cannot walk.

This chapter grew out of discussions that I had with Karen Neander and Geoffrey Sayre-McCord as they tried to articulate what is wrong with my view of functions. I am doubtful that this presentation will convince them of the error of their ways, but perhaps it will serve to fuel further discussion. Conversations and other interactions with David Buller, Bill FitzPatrick, Owen Flanagan, Mark Gifford, Bob Richardson, and Matt Stewart helped focus my thoughts. I very much appreciate all their help. This chapter was completed in large measure while I was a Taft Fellow at the University of Cincinnati. My thanks go to the university for its generous support.

Despite William Paley, we discuss the function of biological 'artifacts' without appealing to a grand designer.[1] We make the normative claims about functions purely naturalistically, without appeal to anything more than our best scientific theories. Karen Neander agrees with this assessment of function: 'To attribute a natural function . . . to something is to attribute a certain kind of normative property to the thing. That is, it is to attribute an evaluative standard to it that it could fail to meet, even chronically (i.e. systematically and persistently and even under ideal circumstances)' (1999: 14).

Many philosophers of biology believe they have a lock on how to make normative and naturalistic claims of function (e.g. Williams 1966; Ruse 1971; Wright 1973, 1976; Ayala 1977; Brandon 1981, 1990; Millikan 1989; Neander 1991*a*, *b*; Godfrey-Smith 1993, 1994; Griffiths 1993; Kitcher 1993; Sober 1993). They rely on the evolutionary history of a trait to identify its function. Roughly speaking, according to this view, a trait T has the function of producing effect E in some organism O if T contributed to the fitness of O's ancestors in virtue of doing E and T is heritable.

Lots of difficulties have been identified with this approach (Bigelow and Pargetter 1987; Hardcastle 1999). One in particular is that it covers only a portion of natural properties we believe have a function, for it can explain only heritable traits that contribute to increased reproductive success. However, many natural things with functions are not heritable, such as chemical reagents, nor do they contribute to the reproductive success of anything else, such as the eyes on a worker bee. A chemist should be able to talk about the function of a substance in a chemical reaction. 'This amino acid catalyses this RNA transcription' is a functional description. Similarly, zoologists should be able to talk about the functions of the morphological features of sterile animals just as they talk about the functions of similar features in fecund animals. In both cases, eyes are for seeing.

Other philosophers have advocated a more pragmatic approach to explaining function (Wimsatt 1972; Cummins 1975; Rosenberg 1985; Bechtel 1989; Schaffner 1993; Hardcastle 1999). In this case, function is relativized to something else explicitly defined by some domain of inquiry. To take the most famous example of a pragmatic approach, Robert Cummins (1975) claims that the function of T in O is E if T is a component of O and E contributes causally to the O's capacity to do whatever it is that O does. In this case, what counts as a function depends on how we understand the larger system. Saliva functions to break down complex sugars because saliva's doing this forms one component of our digestive process.

[1] Dennett (1995) and Kitcher (1993) might be exceptions to this generalization. However, unlike Paley, they do not believe that the designer is anything more than natural selection at work.

One important advantage to this approach is that all cases of natural function fall under it. Champions of the etiological notion of function turn up their noses at pragmatic approaches, however, for the price of generality is normativity, or so the etiology advocates claim. If functions are just properties important relative to some framework, then we lose any sense of a function being *the* thing an item is supposed to do. As we shift frameworks, then the functions shift as well. From the point of view of dieticians, saliva may function as a digestive agent, but from the point of view of my 7-year-old daughter, saliva causally contributes to her capacity to make spit-wads. But surely, one wants to complain, my daughter is just wrong here. She might use saliva to make spit-wads, but that is not its function. Aiding in digestion is the thing saliva is supposed to do, regardless of how else it is used.

When we pick out a property's function, we are isolating one of its effects from among the many it has as the thing it is supposed to do. Not only do (properly functioning) human legs walk, but they also cast shadows, provide balance for sitting upright, and weigh approximately 30 pounds each. Why do we say legs are for walking and not for balancing? Those who side with the etiological view claim that the pragmatists cannot answer this question. This chapter examines this claim. I shall argue that, despite appearances, we can get a robust sense of normativity out of pragmatic views of function; moreover, being relative to something else does not distinguish pragmatic versions of function from etiological ones. Let us start by looking at my second contention first.[2]

1. The Pragmatics of Etiological Norms

Larry Wright's original formulation of the etiological view (1976: 81) is the standard against which other versions are measured. His version is very simple:

The function of *X* is *Z* means
(*a*) *X* is there because it does *Z*,
(*b*) *Z* is a consequence (or result) of *X*'s being there.

The standard example given with the standard view is that the function of the heart is to pump blood. Hearts do many things—they make noises, they contract, they consume energy—but one thing that they do is pump blood. And pumping blood is what explains why we have hearts. Organisms with hearts can circulate blood more effectively and efficiently than those without.

[2] There is actually at least a third option, the propensity view of functions. According to this version, the function of something depends upon how that thing would fare under future selection regimes in a normal environment (Tinbergen 1963; Staddon 1987; Bechtel 1989; Horan 1989). Since it is also a selectionist account of function, the criticisms and comments I make about the etiological view apply to it as well.

They can distribute precious nutrients and dispose of waste faster and better than their non-hearted conspecifics. Hence, they can out-survive and out-reproduce their non-hearted conspecifics as well. As a result, we say that pumping blood is what hearts are supposed to do; it is their function.

We determine the function by finding the effect that has been most useful for the survival and reproduction of previous generations. That effect is the function. We would say that the function of our legs is to walk because that particular effect contributed to our survival in ways that casting shadows or weighing 30 pounds did not.

There are several variants of the etiological view, ranging from a strong theory that requires selected traits to contribute positively to the larger system that contains them to a much weaker theory that requires only that there be reproduction of a series of items that have the same effects (see discussion in Buller 1998). The subtleties of the variants need not detain us here, however. What is important for our purposes is to see that no version of an etiological theory can work without first assuming some sort of framework in which we can pick out a trait and the relevant effects. That is, functions can be specified only relative to an explanatory backdrop.

This might seem a trivial point. For decades now, philosophers have been belaboring the point that all observations are theory-laden, that we can make statements of scientific fact only against a huge set of background facts, and so forth. This claim is no more than an alternative of those: functional ascriptions do not happen in a vacuum. And no one should be surprised or riled by that claim. Still, it will behoove us to pay attention to this point with respect to understanding the normative aspects of etiological theories, for it means that we cannot divorce normativity from a rich background framework either.

When we talk about selection in evolution, we can talk about it only relative to a 'common selective environment' in which conspecifics must compete for resources (Brandon 1990). The Darwinian struggle for survival takes place in particular locations against particular competitors. Therefore, one can win it only in particular locations and against particular competitors. David Buller (1998) gives a nice example of this principle. Suppose we want to compare my capacity for lifting objects with yours. Let us say that I can lift my chair off the ground, but you cannot lift yours. Am I the fitter? We do not know yet. If your chair is bolted to the ground, then your attempt at lifting your chair was thwarted by an environmental component I did not suffer. We have to compare the putative fitness of traits in a common environment before we can know. Before we can start to talk about the function of any trait, we have to isolate a population's environmental niche and examine what the trait is doing for the organism relative to the particular demands that environment makes. Lungs do not help the bottom-feeders consume oxygen, but they certainly help me. Hence, we can specify the function of T only against a particular selective regime. (See also Walsh 1996.) The function of T is to do E for O because *in context C*, E increased the fitness of O's ancestors.

However, if T increases the fitness of O's ancestors only in context C, then it would seem a mistake to claim that E is T's function *tout court*. Perhaps in context C*, having T do E does not help at all. If it turns out that all of O's ancestors lived only in C and never in C*, we can speak loosely and talk about *the* thing T is supposed to do, because in all relevant environments (namely, the one with O's ancestors), T doing E promoted O's ancestors' survival.

The wrist bone in my wrist functions to maintain the integrity of my arm, but the wrist bone in the panda functions as an opposable digit (Gould 1980). They are homologous structures, but in my environment C, it evolved to do one thing, and in the panda's environment C*, it evolved to do something else. We can talk about the thing my wrist bone did to promote the survival of my ancestors. Hence, we can also talk about its function as the thing it is supposed to be doing. Similarly, we can talk about the thing the panda's wrist bone did to promote the survival of its ancestors. Hence, we can talk about its function as the thing it is supposed to be doing. But these statements would be elliptical for saying that in context C (or C*), T's function in O is E. A champion of the etiological point of view can define the function of a trait only relative to its selective environment. There are no functions *simpliciter*.

Secondly, etiologists explicitly adopt increased relative fitness as the target state for organisms. They individuate functional effects relative to the goal of differential survival and reproduction (cf. Wimsatt 1972; Price 1995; Manning 1997). The function of T is to help attain this goal for O. There are lots of attributes E could be promoting. It could have helped O's ancestors eat better, run faster, weigh more. There are many attributes that E does in fact promote. Both hearts and legs helped my ancestors do all those things. But evolutionary biologists single out reproductive success as the important attribute in organisms. Hence, etiologists define function from the perspective of promoting reproductive success, too.

Keep in mind that natural evolution is very different from natural selection. A veritable smorgasbord of causes drives generational change, some neutral, some downright pernicious. Genes mutate and drift randomly, and, if the traits influenced have no effect on survival or reproduction, then organisms evolve all right, but they are not evolving by natural selection. Some traits hitchhike genetically on others. The effects of these traits might even interfere with reproduction or survival, but as long as the traits they are tied to promote reproduction and survival more than the hitched traits demote it, they will continue to appear across generations. Other processes, like segregation distortion, promotes the survival of genes at the expense of the organisms that carry them. In short: just because a trait is heritable and has in fact been passed down through the ancestral tree does not mean that Mother Nature selected the trait as something good or useful.

Nature herself cannot and does not sustain teleology. This is something we bring to the table when we pick out natural selection as something important we need to explain (see also Schaffner 1993; Dennett 1995). Evolution is a panoply of effects, interactions, counter-effects, and counter-interactions. From this huge causal nexus, we have isolated one sort of effect as the most important in the explanatory framework for evolutionary biology, those effects that affect survival and reproduction differentially. Functions are defined relative to this isolated effect. When we say that T is supposed to do E, we are assuming the particular explanatory interests of some evolutionary biologists. T's function is E is relative to one process in evolution we singled out as interesting or important.

Segregation distorter genes present a clear case for the relativity of natural teleology (cf. Crow 1979; Lyttle 1993). These genes tamper with meiosis by instructing their alleles to disrupt spermatogenesis in homologous chromosomes such that sperm carrying the homologous allele cannot fertilize eggs. As a result, virtually all of the viable sperm in heterozygote males carry only the allele for the distorter gene, and homozygous males either die before becoming mature or are sterile. (Females carrying the gene are reproductively normal.) However, even though, as Godfrey-Smith (1994) points out, descriptions of the segregation distorter genes satisfy etiological accounts of function, biologists do not assign a function to them. Disrupting meiosis is what these genes do, to be sure, and their doing this ensures that they out-reproduce and out-survive their competitors, the non-distorting alleles. Yet, biologists deny that distorting meiosis is their function; in fact, they deny that distorter genes have any function at all. Why? The answer has to do with how (some) biologists highlight the survival of the organism over the survival of individual alleles. Because the segregation distorter genes win their Darwinian competition at the expense of organisms, they get assigned no function. We prefer to think in organismic terms; hence, other selection processes at other levels of organization are discounted or ignored outright. We choose the level of organization at which we explain effects and assign functions; we pick what we find worthy of teleological language.

Advocates of the etiological approach are simply misguided if they believe that their sort of functional analysis reveals functions apart from any interests. Though I do not want to go as far as Daniel Dennett or Kenneth Schaffner and conclude that functional ascriptions are purely heuristic, I do think that when Millikan, Neander, and others gloss biological function merely as the effect a trait is supposed to have, they are sidestepping the foundations of evolutionary theory. In biology, a function can be an effect only relative to the selective regime and the evolutionary process we decide to privilege. Functions in biology do not fall out of the world any more than any other theoretical term in science does.

The normative aspects of function from an etiological perspective are relative through and through. But this fact should not be surprising. After all, this naturalistic sense of normativity is rather weak. Philosophers of biology merely want a way of discussing malfunction, a way of talking about the effect an organism is not exhibiting (and maybe never will exhibit) as the function. Hearts that do not beat are malfunctioning; legs that do not walk are malfunctioning. That they are not or do not is not a sign of moral turpitude or that something is wrong with the natural order of the universe; it is only that they (in some sense) should be doing these things *because* doing that in the past promoted ancestral welfare.

This sense of normativity needs to be distinguished from the sort that someone like Philippa Foot discusses, in which functional norms are tied to what promotes the welfare of the organism (Foot 1994, 1995). Here, welfare (for humans at any rate) is understood broadly to mean something like psychological flourishing, which may or may not have any interesting connection to biological heritage. These neo-Aristotlian ideas assume that there is a natural function (*ergon*) for the organism as a whole, just as each of its parts has a function. That is, there are 'characteristic activities' that belong to particular organisms, just as there are such activities for particular organs (FitzPatrick 2000). Even if we decide that there are such things, whether these activities are importantly biological is another question, and one the etiologists are not currently positioned to answer.

2. The Etiology of Pragmatic Norms

Given the considerations listed above, we can see that a pragmatic approach actually does not differ that radically from the etiological one. The real difference is in attitude: the pragmatists publicly embrace the relativity of functional attribution and then deny any sort of prejudice in assigning a goal state. We look to the various domains of science to set their own explanatory goals and we use those to make functional assignments. Biology might privilege natural selection, but psychology need not; it could emphasize information processing instead. From a psychological point of view, beliefs function in a cognitive economy as structured parameters over which minds compute. These computations may have little to do with differential reproduction. Or, even if our beliefs do result in more successful reproductive strategies, that fact may be of little interest to the psychologist concerned with mapping the beliefs in the first place.

Cummins (1975) suggests that we are actually using inference to the best explanation when we determine function. We infer the presence of T from our best explanation of O. Ascribing a function to something means that we are ascribing a capacity to it that is singled out by its role in our analysis of the system of which it is a part. Loosely speaking, the causal role T plays in O is its function. The function of the heart is to pump blood because our understanding of complex biological organisms dictates that there a blood-pumping mechanism should exist and the heart fulfills that role. Similarly, the function of our legs is to walk because our understanding of complex biological organisms dictates that a method of self-locomotion should exist and the legs fulfill that role.

The nature of the query itself defines the sorts of effects scientists are supposed to be tracking. Which parameters are chosen as important in assigning functions depends on the domain doing the abstracting. Consequently, different disciplines can analyze the same property or disposition quite differently. There is not a single function per trait. When startled, infants automatically fling their arms up and around as though to grab onto a tree limb. This is known as the Morrow reflex, and scientists have assigned it an evolutionary function. In the days of yore, when our ancestors were tree-dwellers, infants who could quickly grab onto a tree limb when in danger of falling survived better than their conspecifics who could not. The function of this startle reflex was to help infants remain treed.

We do not live in trees anymore, but we still exhibit the reflex in infancy. Nowadays, medical doctors use it to determine how mylinated an infant's brain is. As our cortex comes fully online in the first months after birth, it begins to inhibit this reflex. Determining how strong the reflex is gives doctors some information regarding how developed the cortex is. What is the function of the reflex from a medical point of view? Unlike evolutionary biologists, pediatricians are not so concerned with why the reflex is there, what led to it evolving in the past, as they are with what it indicates. In a routine physical examination, the Morrow reflex functions as a measuring rod for brain development.

Now, contrast this reflex with the palmomental reflex. Before the cortex is completely online, tickling the palm of infants causes their upper lips to curl back. Unlike the Morrow reflex, there is no good evolutionary explanation for why infants might have this reflex. There is not even a poor man's just-so story. We have no idea why the nerves in the palm are connected to the nerves controlling the lip muscles. The best guess is that this connection is a sheer accident in wiring. From an evolutionary point of view, the palmomental reflex would have no function; there is nothing it does that increases reproductive success. However, from a medical point of view, it is similar to the Morrow reflex. It too indicates how developed our brains are. The strength of the reflex, like the strength of the Morrow reflex, informs pediatricians of the underlying brain structure.

How are we supposed to understand normativity in these cases? The ways in which scientists discuss these reflexes are instructive. Because we no longer live in trees, the Morrow reflex assumes little importance from an evolutionary point of view. Though we can explain why it is we have the reflex in terms of how our ancestors evolved, that story is little more than an interesting biological side note. However, because both reflexes are very important as diagnostic tools in developmental neuroscience, scientists generally think of these reflexes in terms of what they tell us about cortical development. Further, when we talk about the malfunction of infant reflexes, we do not discuss them from an evolutionary point of view. Instead, we discuss them from a medical point of view. That is, both the Morrow and the palmomental reflexes can malfunction and they do so by misindicating the state of brain development.

If Millikan, Neander, *et al.* were correct in believing that the selective history of trait determines its function, then only the Morrow reflex could malfunction, for there is no selective story that one can tell about the palmomental reflex. But, so far as scientists are concerned, the two reflexes can each malfunction and they both malfunction in the same way. They each have things that they are supposed to do, regardless of whether they actually do them. But we determine what these reflexes are supposed to do in terms of how they function in a diagnostic context, not in terms of their selective history. In this case, we uncover normativity relative to explanatory goals of the health sciences, and not relative to the environmental niche of our ancestors.

Scientists are happy being pluralists about function. Some functions are determined by selective histories; others are not. Philosophers should be content as well. There is no conceptual reason why we should not adopt a pragmatic approach to explaining function, and there is a good conceptual reason why we should (namely, the etiological approach is already heavily pragmatic).

At the same time, being pluralist and recognizing the pragmatic aspects of assigning functions to items does not mean anything goes. To return to one of our earlier examples: my daughter is simply wrong that saliva is for spit-wads. Saliva is for digestion. How can we know this?

Some philosophers worry that a pragmatic approach allows just about anything in as a function, as long as one can describe it in terms of its contribution to something else. They conclude that a pragmatic approach does not differentiate function from mere disposition or accident; hence, it is no account of function at all.

This reaction is not quite fair, though. Cummins himself is careful to restrict functional descriptions to the subject matter of science, as are other pragmatists (see also Amundson and Lauder 1994; Hardcastle 1999). Letting the discipline set the boundaries on relevant effects is not arbitrary. Within their fields, scientists define exactly what they mean by selective advantage or health or information processing or whatever. These criteria are defined in terms of operational criteria. And, once the criteria are accepted by the relevant sciences, they help set the research agenda for the discipline. Of course, scientists can (and do) refine or completely replace their criteria for the boundaries and interests of particular fields, but that does not occur swiftly, at the behest of one or even a few, or in response to particular experimental outcomes.

Normativity is then defined in terms of the previously accepted criteria. The criteria come first, then functional assignments with concomitant norms. Relative to each explanatory framework that uses the language of functions, we find a naturalistic notion of normativity. That notion varies as the particular articulation of function does. But in each case, the function of T is to do E in O because E is necessary for answering the question of what O is doing.

We get the 'extra ingredient' that separates dispositions from functions by relying on the previously accepted explanatory goals and theories of a particular scientific discipline. The teleological goal for some trait or organism is neither arbitrary nor *ad hoc*; it depends upon the discipline generating the inquiry.

3. Ethical Norms

If I am correct and a thoroughly pragmatic version of function is the correct way to understand function in nature, then it seems that we might have room for natural ethical norms, after all. For, if what counts as a function is relative to the larger explanatory framework and that framework need not include biology or any of its relatives, then it seems we should legitimately be able to talk about human values in all their richness from the standpoint of ethical theory. In my concluding section, I shall briefly examine the tenability of this claim.

We must first keep in mind that whatever notion of normativity that one comes up with to ground morality will have to be a naturalistic notion. That is, given the constraints set out at the beginning of this chapter, only functions that stem from scientific concerns of the world are what we are considering here. Even if we want notion of human functioning that is strong enough to justify claims about how humans should behave and think in all sorts of social situations, this notion will still have to be tied to some objective study of human nature—psychology, sociology, anthropology, or, more likely, some combination of all three.

Many ethicists hold out hope that there is a conception of human nature that is objective, rigorous, and sufficiently rich that it can support any manner of ethical claims (e.g. Wallace 1978; Anscombe 1981; Foot 1994, 1995; Gaut 1997). If we know enough about how humans live their lives and what they do that promotes their well-being, then we should be able to use these facts to define particular values that enhance our tendency to flourish. These values could then form the core around which we could define most, if not all, of our ethical norms.

Let us suppose that the ethicists are correct for a moment. Let us leave aside questions of whether these norms will be specific to particular cultures or universal across our species. Indeed, let us leave aside questions of whether we can even come up with a notion of human well-being or flourishing that we can all agree is the standard against which we should measure potential values. Let us suppose that we have figured out some way for this project to get out of the philosopher's armchair and into the field. Would this be enough to ground ethical claims about what we should be doing? I am dubious that it would, and my reason for being skeptical highlights an important distinction between science and ethics. This reason also underscores why I believe that a pragmatic approach to understanding function is not all that dangerous from a normative point of view.

Over forty years ago, Peter Glassen accused Aristotle of sliding between describing the life of a good man and describing that life as something good (Glassen 1957; see also discussion in Gifford 1999). I believe that ethicists who pin their hopes on the human sciences giving them a toehold onto naturalistic views of ethical norms are making the same slide. Psychologists, sociologists, anthropologists, and so forth might be able to tell us what humans can do that will promote whatever projects the scientists deem important, but they cannot tell us that these projects are otherwise valuable for us to pursue.

Psychologists or psychiatrists might be able to tell us what we should believe or how we should act such that we can improve our mental health or degree of personal satisfaction or our level of competence in something, but they cannot tell us that doing these things is a good thing to do in and of itself. Similarly, sociologists or anthropologists might be able to define some aspects of human life that promote peace or wealth or education, but they will not be able to claim that these goals are worth pursuing *tout court.* One cannot go from claiming an activity is something humans should be doing because it helps us meet a particular goal to claiming that activity is something we should be doing because it is the right thing to do.

And here, I think, is the division between science and ethics. Science can describe what we actually do and can then list attributes that will make us better at doing what it is we do. Ethics faces a different task. Ethicists must decide what constitutes a good human or a good life and then argue that these things are worthy of pursuit. Scientists have the luxury of working within the established frameworks of their disciplines; ethicists are still struggling to fix the framework itself.

Consequently, the norms that scientists consider, the normative side to functional ascriptions, for example, are always norms relative to a well-defined and well-understood framework. In contrast, the norms of ethicists, the proscriptions on how to be a good person, are supposed to be norms appreciated and accepted apart from the framework. Scientific norms are embedded in the intellectual scaffolding that defines its various disciplines; ethical norms embed the ethical scaffolding. As a result, the norms that one gets out of science, whether from an etiological approach to biology or from a wider pragmatic view, are really only self-reflexive statements about the investigative principals that bound a particular science. They are enough to define when something is malfunctioning, when it is not performing as it should under circumstances defined by science, but they are not enough to define much more. And pragmatic approaches to explaining function, as well as etiological approaches, can do that much.

References

Amundson, R., and Lauder, G. V. (1994), 'Function without Purpose: The Uses of Causal Role Function in Evolutionary Biology', *Biology and Philosophy*, 9: 443–69.

Anscombe, G. E. M. (1981), 'Modern Moral Philosophy', repr. in *Ethics, Religion, and Politics: Collected Philosophical Papers*, iii (Minneapolis: University of Minnesota Press).

Ayala, F. J. (1977), 'Teleological Explanations', in T. Dobzhansky (ed.), *Evolution* (San Francisco: W. H. Freeman and Co.), 496–504.

Bechtel, W. (1989), 'Functional Analyses and their Justification', *Biology and Philosophy*, 4: 159–62.

Bigelow, J., and Pargetter, R. (1987), 'Functions', *Journal of Philosophy*, 84: 181–96.

Brandon, R. N. (1981), 'Biological Teleology: Questions and Explanations', in P. Asquith and T. Nicles (eds.), *PSA 1980*, vol. 2 (East Lansing, Mich.: Philosophy of Science Assocation), 315–23.

——(1990), *Adaptation and Environment* (Princeton: Princeton University Press).

Buller, D. J. (1998), 'Etiological Theories of Function: A Geographical Survey', *Biology and Philosophy*, 13: 505–27.

Crow, J. F. (1979), 'Genes that Violate Mendel's Rules', *Scientific American*, 240: 134–46.

Cummins, R. (1975), 'Functional Analysis', *Journal of Philosophy*, 72/20: 741–65.

Dennett, D. C. (1995), *Darwin's Dangerous Idea: Evolution and the Meanings of Life* (New York: Simon & Schuster).

Enç, B., and Adams, F. (1992), 'Function and Goal Directions', *Philosophy of Science*, 59: 635–54.

FitzPatrick, W. J. (2000) *Teleology and the Norms of Nature* (Boston: Garland).

Foot, P. (1994), 'Rationality and Virtue', in H. Pauer-Studer (ed.), *Norms, Values, and Society* (Dordrecht: Kluwer).

——(1995), 'Does Moral Subjectivism Rest on a Mistake?', *Oxford Journal of Legal Studies*, 15.

Gaut, B. (1997), 'The Structure of Practical Reason,' in B. Gaut and G. Cullity (eds.), *Ethics and Practical Reason* (Oxford: Oxford University Press).

Gifford, M. (1999), 'A Fallacy in Aristotle's Ethics?', presentation at the Virginia Philosophical Association, Blacksburg, Va.

Glassen, P. (1957), 'A Fallacy in Aristotle's Argument about the Good', *Philosophical Quarterly*, 66: 319–22.

Godfrey-Smith, P. (1993), 'Functions: Consensus without Unity', *Pacific Philosophical Quarterly*, 74: 196–208.

Godfrey-Smith, P. (1994), 'A Modern History Theory of Functions', *Noûs*, 28: 344–62.

Gould, S. J. (1980), *The Panda's Thumb: More Reflections in Natural History* (New York: W. W. Norton & Co.).

Griffiths, P. E. (1993), 'Functional Analysis and Proper Functions', *The British Journal for the Philosophy of Science*, 44: 409–22.

Hardcastle, V. G. (1999), 'Understanding Functions: A Pragmatic Approach', in V. G. Hardcastle (ed.), *When Biology Meets Psychology: Philosophical Essays* (Cambridge, Mass.: MIT Press), 27–46.

Horan, B. (1989), 'Functional Explanations in Sociobiology', *Biology and Philosophy*, 4: 131–58.

Kitcher, P. (1993), 'Function and Design', *Midwest Studies in Philosophy*, 18: 379–97.

Lyttle, T. W. (1993), 'Cheaters Sometimes Prosper: Distortion of Mendelian Segregation by Meiotic Drive', *Trends in Genetics*, 9: 205–10.

Manning, R. N. (1997), 'Biological Function, Selection, and Reduction', *British Journal for the Philosophy of Science*, 48: 69–82.

Millikan, R. G. (1989), 'An Ambiguity in the Notion "Function"', *Biology and Philosophy*, 4/2: 172–76.

Neander, K. (1991*a*), 'The Teleological Notion of Function', *Australasian Journal of Philosophy*, 69: 454–68.

——(1991*b*), 'Functions as Selected Effects: The Conceptual Analyst's Defense', *Philosophy of Science*, 58: 168–84.

——(1999), 'Functions and Teleology', in V. G. Hardcastle (ed.), *Where Biology Meets Psychology: Philosophical Essays* (Cambridge, Mass.: MIT Press), 3–26.

Price, C. (1995), 'Functional Explanations and Natural Norms', *Ratio*, NS 7: 143–60.

Rosenberg, A. (1985), *The Structure of Biological Science* (New York: Cambridge University Press).

Ruse, M. (1971), 'Function Statements in Biology', *Philosophy of Science*, 38: 87–95.

Schaffner, K. F. (1993), *Discovery and Explanation in Biology and Medicine* (Chicago: University of Chicago Press).

Sober, E. (1993), *Philosophy of Biology* (Boulder, Colo.: Westview Press).

Staddon, J. E. R. (1987), 'Optimality Theory and Behavior', in J. Dupre (ed.), *The Latest on the Best: Essays on Evolution and Optimality* (Cambridge, Mass.: MIT Press).

Tinbergen, N. (1963), 'On the Aims and Methods of Ethology', *Zeitschrift für Tierpsychologie*, 20: 410–33.

Wallace, J. D. (1978), *Virtue and Vices* (Ithaca, NY: Cornell University Press).

Walsh, D. M. (1996), 'Fitness and Function', *British Journal for the Philosophy of Science*, 47: 553–74.

Williams, G. C. (1966), *Adaptation and Natural Selection* (Princeton: Princeton University Press).

Wimsatt, W. (1972), 'Teleology and the Logical Structure of Function Statements', *Studies in History and Philosophy of Science*, 3: 1–80.

Wright, L. (1973), 'Functions', *Philosophical Review*, 82: 139–68.

——(1976), *Teleological Explanations* (Berkeley and Los Angeles: University of California Press).

6. Neo-Teleology

Robert Cummins

ABSTRACT

Neo-teleology is the two-part thesis that, for example, (i) we *have* hearts because of what hearts are *for*: Hearts are *for* blood circulation, not the production of a pulse, so hearts are there—animals have them—because their function is to circulate the blood, and (ii) that (i) is explained by natural selection: traits spread through populations because of their functions. This chapter attacks this popular doctrine. The presence of a biological trait or structure is not explained by appeal to its function. To suppose otherwise is to trivialize natural selection.

1. Two Species of Functional Explanation

There are two subpopulations of functional explanation roaming the earth: teleological explanation, and functional analysis. The two are in competition. In this chapter, I hope to help select the latter, and nudge the former to a well-deserved extinction.

1.1. Teleology

Teleology is the idea that some things can and should be explained by appeal to their purpose or goal or function. It is, for example, the idea that one can explain why rocks fall and fire rises by appeal to the fact that the goal of matter is to go to its natural place, and that this is down for rocks and up for fire. It is also the idea that one can explain why (though not how) an acorn grows into an oak (rather than a beech or a clam) by appealing to the fact that the goal or function of a growing acorn is to become an oak tree. More plausibly, teleological explanation seeks to account for the existence or presence of a biological trait, or structure or behavior *by appeal to its function.* It is said that

I received many hours of feedback and help with this chapter from the members of my weekly lab group: Pierre Poirier, Jim Blackman, David Byrd, and Martin Roth. I want also to thank Denise Cummins, André Ariew, and Mark Perlman for their extensive comments, as well as an anonymous reviewer for the Press.

animals that have hearts have them because of what hearts are for.[1] Hearts are for circulating the blood; they are not for generating a pulse. Therefore, circulating the blood is their function, and they are 'there'—animals have them—because they perform this function.

1.2. Functional Analysis

Teleological explanations and functional analyses have different explananda. The explanandum of a teleological explanation is the existence or presence of the object of the functional attribution: the eye has a lens because the lens has the function of focusing the image on the retina. Functional analysis instead seeks to explain the capacities of the system containing the object of functional attribution. Attribution of the function of focusing light is supposed to help us understand how the eye, and, ultimately, the visual system, works. In the context of functional analysis, a what-is-it-for question is construed as a question about the contribution 'it' makes to the capacities of some containing system.

While teleology seeks to answer a why-is-it-there question by answering a prior what-is-it-for question, functional analysis does not addres a why-is-it-there question at all, but a how-does-it-work question. These last are answered by specifying the structure (design) of the system. Rube Goldberg devices are natural candidates for this sort of explanation. In my horse pasture, I have a device that opens, at a pre-set time, a gate dividing the pasture in two. Here is how it works. There is a wind-up alarm clock. When the alarm on the wind-up alarm clock goes off, a string wound on the key-stem unwinds, releasing a ratchet on a pully. A weight on one end of a rope over the pully falls, jerking open the gate latch attached to the other end of the rope. This is, if you like, a rather abstract mechanical description. It is also a functional analysis of the capacity to open the gate as this is realized in my Rube Goldberg device. The components are identified functionally, and their interactions are described in a way that, necessarily, abstracts away from the medium-dependent details.[2] When we understand how the thing works in the way provided by a functional analysis, we understand how others might be built—how other instantiations of the same design could do the same job, and, perhaps, do it better. This is possible, because the system and its components are specified functionally, and hence in a way that allows for multiple instantiations. By substituting

[1] Paul Davies (2001) holds that something's function should not be identified with what it is for, since this builds an unacceptable sense of 'design'—one involving intentional considerations—into the concept of function. I have some sympathy with this, but am prepared, for the purposes of this chapter, to let 'What is it for?' be a way of asking the same question as 'What is its function?'

[2] Most causal analysis is like this. When we describe causal interactions between functionally characterized components, the relevant causal generalizations pretty much come for free, since a functionally characterized component is a component identified by its relevant causal powers.

functional equivalents at various points in the design, taking care to accommodate the need for adequate interfaces with other components, we can make incremental changes in the system while preserving its overall viability. This is precisely how we must understand a system to see how it could be incrementally improved, and hence how it could evolve.

Of the two forms of functional explanation, I suspect teleology is much the oldest. Teleology is a natural framework for thinking about tools, cooking and storage utensils, and shelters. These ideas extend quite naturally to the body: eyes are tools or instruments for seeing, ears for hearing, hands for grasping, teeth and jaws for chewing. Functional analysis, on the other hand, got a grip on the mind, I suspect, only with the invention of relatively complex artifacts. Carts and harnesses lend themselves to functional analysis. Machines such as catapults and water clocks are unthinkable without it. This kind of thinking extends naturally to social structures such as bureaucracies, and to complex anatomical systems: the digestive system, the circulatory system, the nervous system.

There is more to proto-teleology than attributing functions to tools and sense organs, however. What I am calling teleology is the idea that an appeal to something's function can explain 'why it is there': why there are hammers and hands. If having a function is to explain why a thing, or type of thing, exists, then there must be some background story about a mechanism or process that produces the items in question, and produces them because of their functions. It is this requirement for what I will call a *grounding process* that has proved to be the Achilles heel of teleology.

Different kinds of phenomena subject to teleological explanation have required different grounding processes. Teleological mechanics appealed to the selective attractiveness of natural places. The intentions, plans, and actions of designers, creators, and manufacturers have been rung in to support teleological explanations of quite literally everything, and remain popular as underpinnings of teleological explanations of artifacts. In the hands of Aristotle, and Hans Dreisch (1867–1941), teleological developmental biology appealed to the regulating capacites of entelochies, a sort of inner goal-directed agent. Finally, natural selection has become a popular grounding process for the teleological explanation of biological traits, and sometimes for traits of artifacts as well.

Teleological explanation of motion failed because the grounding processes were transparently insensitive to function. Even if one could make sense of natural places, any force or mechanical constraint that would get something to its natural place would get it there whether or not it was the function of the thing to go there. To take an example from Ptolemaic astronomy, if a star has its apparent motion because it is attached to a rigid moving sphere, centered at the earth, it will trace a circular orbit around the center of the sphere regardless of what its function happens to be. The same point holds of Newtonian gravitational explanations of planetary orbits. Teleological appeal to functions in mechanics therefore appears idle and misleading. Indeed, it no longer seems plausible to suppose that celestial bodies have mechanical functions at all. Like all teleology, teleological mechanics requires a grounding process. But the only grounding processes likely to satisfy render the appeal to functions utterly superfluous.[3]

Teleological explanation of growth and development fared even worse. The need for a grounding process spawned vitalism and the doctrine of entelochies. Entelochies could not be found. Moreover, appeals to them are regressive, since their own guiding and regulating behavior was itself teleologically explained, but without the hope of a corresponding grounding process. The whole misconceived enterprise rapidly became extinct when advances in cellular and molecular biology generated more adaptive theories competing for the same niche. Like post-teleological mechanics, those theories also appeal to factors that are transparently insensitive to the functions that were crucial to the teleological stories. So transparent is this, in fact, that it now seems silly to think it is the function of an acorn to develop into an oak rather than a birch, and that this explains why planting acorns never yields birch trees.

2. Neo-Teleology

Nobody much likes teleological mechanics or teleological developmental biology any more. It has been eliminated root and branch from mechanics and the other non-life sciences, recalled only by vestigial forms such as least energy principles that are without exception explained away as mere *façons de parler*. And it is similarly absent from developmental biology. But teleology survives in evolutionary biology, or anyway in the philosophy of it, as the idea that one can explain why an organism has a biological trait or structure by appeal to the function of that trait or structure. According to neo-teleologists, as I shall call them, we *have* hearts because of what hearts are *for*. Hearts are *for* blood circulation, not the production of a pulse. Hence, hearts are there—animals have them—because their function is to circulate the blood.

[3] It might seem that there is mostly a difference in attitude between saying that masses follow geodesics unless disturbed and saying that their function is to follow their natural paths. The standard contemporary reply to this sort of worry is to say that functions are normative, and, since there is no question of non-geodesic motion being a malfunction, there is no place for functions in mechanics. But this misses the point. The point is rather that the grounding process winds up accounting for the motion by appeal to factors such as forces or mechanical constraints that could not be sensitive to function in any case.

It is important to read the neo-teleologist claim transparently. Neo-teleologists hold that mammals have hearts because of something special that hearts do (or did). The heart, of course, does (did) lots of things. Among the things the heart does (did) is the thing we single out as its function, and it is *that* effect of heart presence—the one that counts as its function—that accounts for the presence of hearts. Neo-teleologists do not hold, as classical teleologists *did*, that circulation had its effect because it is or was the heart's function. They hold, rather, that the effect of heart presence that accounts for heart presence—call it e—accounts for heart presence because e has the property of being a circulating of the blood, not because e has the property of being the heart's function. Thus, neo-teleology is a mere shadow of the original (mistaken) idea, having only very limited aspirations. But, in spite of being widely influential, it is still mistaken in very much the same way that classical teleology was mistaken: the only plausible grounding processes render appeal to functions superfluous and misleading.

The idea behind neo-teleology is that evolutionary biology can provide the relevant grounding process and hence get you an answer to a why-is-it-there question from an answer to a what-is-it-for question. No doubt there is a sense of 'Why is that thing there?' that is just a way of asking what it is for. I point to a little rubber hemisphere on the carbureter of your lawnmower and ask you, 'Why is that thing there?' You reply by telling me its function—'It is for priming the engine'—and this is an appropriate and satisfactory answer. But this only means that something that looks like teleology but is not can be had cheap.[4] What I am calling neo-teleology is more than this. It is the substantive thesis that, in some important sorts of cases at least, a thing's function—the effect we identify as its function—is a clue to its existence. If it is not to degenerate into the trivial thesis that 'why is it there?' can sometimes just mean 'what is it for?', neo-teleology must be the idea that, for example, there are eyes because they enable vision, wings because they enable flight, and opposable thumbs because they enable grasping.

3. Neo-Teleology and Natural Selection

Neo-teleology as just construed has no lack of able defenders (Millikan 1984; Neander 1991; Griffiths 1993; Kitcher 1993; Godfrey-Smith 1994; Allen and Beckoff 1995).[5] Generally, these authors are associated with selectionist

[4] It might also mean that teleology used to be uncontroversial, so that the two expressions seemed to mean the same thing.

[5] See Buller (1999) for a collection of papers defending and elaborating some version or other of what I am calling neo-teleology. Notice that a defense of a selectionist account of functions is, in effect, a defense of neo-teleology, since selectionist accounts equate functional attribution with neo-teleological explanation.

etiological accounts of functions. Notice, however, that a defense of a selectionist etiological account of functions is, in effect, a defense of neo-teleology, since selectionist etiological accounts of functions equate functional attributions with what I am calling neo-teleological explanations: to say the function of the heart is to circulate the blood is, on these accounts, to offer a neo-teleological explanation of the presence of hearts. I prefer to attack the position by attacking the explanations in question rather than a thesis about what functions are, since evolutionary theory bears directly on the viability of the explanations, and only indirectly on the thesis that functional attributions are equivalent to such explanations.

Contemporary defenses of neo-teleology all share a basic selectionist strategy. The underlying idea is that traits are selected for because of the effects that count as their functions, hence exist in organisms because they have (or had) the functions they do (or did). Neo-teleology is thus packaged as what appears to be an uncontroversial part of the theory of natural (or artificial) selection. No natural places, entelechies, designer's intentions or other skyhooks (Dennett 1995) appear to taint neo-teleology; it is selectionist through and through. Since selectionist explanations are clearly legitimate scientific explanations, how could anyone object to neo-teleology? Surely we have here a subspecies of teleology that has found a legitimate grounding process.

And yet I am unpersuaded. Biological traits once explained by a teleology grounded in appeals to the intentions, plans, and actions of a creator have, in discerning minds, given way to appeals to evolution generally, and to natural selection in particular.[6] Neo-teleologists want to read this as the discovery of a legitimate grounding process for a teleological explanation of these traits. I am inclined to read the same intellectual development as analogous to what happened in mechanics and developmental biology: not a vindication but a replacement. The grounding processes of evolution, rightly understood, do not ground neo-teleology, because they are insensitive to function. Functions, I believe, have a legitimate place to play in science generally, and in biology in particular. But neo-teleology has the role of functions in selectionist explanations, and hence in biology, quite wrong: Biological traits, mechanisms, organs, etc., are not there because of their functions. They are there because of their developmental histories. Functions, I believe, enter into science legitimately as elements of functional analyses. Functional analysis is a powerful explanatory strategy that is widespread in all of the sciences. I have defended this view elsewhere (Cummins 1975, 1977, 1983), and will not comment on it

[6] Teleological appeal to designers, creators, and manufacturers to explain artifacts is still widespread. The function of an escapement in a clock is said to explain its presence in a way that is grounded in the intentions of designers, and manufacturers. However, selectionist treatments are also popular: the escapement is said to be there because it solved a problem plaguing pre-escapement clocks, leading consumers to prefer escapement clocks. The resulting pressures of the marketplace then led to the (near) extinction of pre-escapement clocks. (These, in turn, are in the process of being replaced by electronic clocks that require no escapement, of course.)

further here. My focus in this chapter is rather to expose what I think are the vices of neo-teleology. An understanding of functional analysis will be relevant only because, as mentioned above, it appears to be precisely the framework we need to understand how complex systems could evolve.

4. Against Neo-Teleology

The basic idea of my argument is quickly conveyed. Traits are acquired in a variety of ways. Some are learned. Some, like sunburn, limb loss, and the effects of disease, are the direct result of environmental influences. None of these is of interest here because they are not heritable, hence are not subject to (non-cultural[7]) selection. The traits that *are* subject to selection *develop.* For convenience, in the rest of this chapter, 'traits' will be restricted to heritable traits the expression of which is the result of development and hence highly canalized.

Development is determined by a complex interaction between genes and environment. It is utterly insensitive to the function of the trait developed. Selection, on the other hand, is sensitive to the effects that are functions, but is, in the sense relevant to neo-teleology, utterly incapable of producing traits. It can preserve them only by preserving the mechanisms that produce them. Nor can selection, in the sense relevant to neo-teleology, produce the mechanisms that underwrite a trait's development; it can preserve only whatever mechanisms it finds already there.

I say selection cannot produce traits *in the sense relevant to neo-teleology*, for there is a sense in which selection can assemble complex traits or structures. This is what gives selection its awesome explanatory power. But the creative power of selection is not the kind of process to which neo-teleologists appeal. I will return to this in a later section.

If the processes that produce traits are insensitive to their functions, how can functions account for why a trait is 'there'—that is, expressed in some specified population? The contemporary neo-teleologist answer is to concede that the processes that produce traits are insensitive to their functions, since, of course, traits do not have functions until *after* they are produced. But they argue that the processes that proliferate and preserve traits in a population are not insensitive to their functions. Certain traits spread through a population over time, and the mechanisms responsible are sensitive to function. Hence we can explain spread by appeal to function. Appeal to function thus gives us a handle on why a trait survived and proliferated, and hence a handle on why it is 'there'.

[7] I am going to ignore cultural evolution in this chapter. I think the points I make here against selectionist defenses of neo-teleology would apply to neo-teleological stories about cultural selection as well, but I have not investigated this issue.

Imagine that crab grass invades a patch of Mendel's pea plants. The short ones will soon have trouble getting enough sunlight. The tall ones will do better. (They will all have trouble competing for root space underground.) The tall ones will reproduce more than the short ones, and will soon be far more common than the short ones, though they may be less common (per square foot) then either the tall or short ones previous to the crab grass invasion, and may eventually be crowded out altogether. In the meantime, tallness will, as we say, have spread through the population, and will be maintained.

This sort of story is supposed to explain why the pea plants in Mendel's crab-grass-infested garden are tall.[8] And it does. But how does neo-teleology get into the picture? Well, the idea is that the function of tallness in plants, or at least in these pea plants, is to achieve access to sunlight. Since gaining access to sunlight is what explains, via selection, why Mendel's pea plants are tall, we have explained why Mendel's pea plants are tall by appeal to the function of tallness in those pea plants.

5. Functions and Spread

The fundamental problem with neo-teleology is that traits do not spread because of (the effects that count as) their functions.

We can distinguish strong and weak variations of neo-teleology. The strong variation holds that any biological trait that has a function was selected for because it performed that function. The weak variation holds only that some traits were selected because of their functions.

A trait can be selected because of its function only if having that function counts as an adaptive variation in the population. For wings to be selected for because they enable flight, there must be a subpopulation in which wings enable flight, while wings in the rest of the population do not. For hearts to be selected for because they circulate the blood, there must be a subpopulation in which hearts circulate the blood, while hearts in the rest of the population do not. While it is plausible to suppose that there was a first flight-enabling wing somewhere among the ancestors of today's sparrows, those ancestors were not sparrows, nor was the wing in question anything like a contemporary sparrow wing. Similarly, somewhere in our ancestral line is to be found the first appearance of centralized blood circulation. But those ancestors were not even vertebrates, and the structures in question were nothing like our

[8] It does not, in my view, explain why any particular plant is tall. See Sober (1984, 1995), and Pust (2001). Neander (1995*a*, *b*) argues for the opposing view. And it does not explain why all the pea plants in the garden are tall, since short ones will continue to occur, though they seldom reach maturity.

hearts.[9] It follows from these considerations that sparrow wings and human hearts were not selected because of their functions. Selection requires variation, and there was no variation in function in the structures in question, only variation in how well their functions were performed.[10]

Strong neo-teleology is refuted if there are legitimate targets of functional characterization that are not targets of selection. Strong neo-teleology must be rejected, since most, perhaps all, complex structures such as hearts, eyes, and wings patently have functions but were not selected because of (the effects that count as) their functions. And, since the selectionist etiological account of functions stands or falls with neo-teleology, it must be rejected as well, not because it is bad conceptual analysis (whatever that is), but because it equates functional attributions with bad evolutionary explanations.[11]

Weak neo-teleology survives this objection, but at a very considerable price. Weak neo-teleology comes out true only because of the rare though important cases in which the target of selection is also the bearer of a function that accounts for the selection of that trait. These will be cases in which genuine functional novelty is introduced; a trait present in a subpopulation that is not just better at performing some function that is also performed in competing subpopulations (though not as well), but a trait that performs a function that is not performed at all by any counterpart mechanism in competing subpopulations. This unquestionably happens, and the importance of such seeding events should not be underestimated. But complex structures such as sparrow wings and human hearts were not introduced in this way. They were selected because they were better[12] at performing some function that was also performed by the competition. It follows from the equivalence of neo-teleology and selectionist accounts of function that these accounts will limit function attribution to those traits for which neo-teleology comes out true—namely, traits in which selection was triggered by the fact that the trait in question had a function that was entirely novel in the relevant population.

[9] Even these scenarios are misleading in suggesting that flight or centralized circulators appeared suddenly on the scene. Circulation was probably centralized gradually, and early flight was no doubt a matter of short and ill-controled forays into the air.

[10] Another way of putting this point is that complex structures such as human hearts and sparrow wings are not heritable traits. What is heritable, at most, are variations in these traits. This follows from the fact that heritability is a measure of how much of the variance is accounted for by genes. Hearts in humans and wings in sparrows are not heritable because there is no variance to account for.

[11] One could deny the validity of neo-teleological explanation and still hold that functional attributions were disguised neo-teleological explanations. Presumably, someone holding this position would advocate abandoning functional attributions. Perhaps Hempel (1959) is an example. But contemporary defenders of selectionist etiological accounts of functions think of themselves as vindicating functional attribution by identifying it with what they take to be a form of viable evolutionary explanation.

[12] Even this is too strong, and will be modified shortly.

This is not merely a defense of gradualism. You do not have to be the village gradualist to be skeptical of the idea that there was variability in the presence or absence of T whenever T is rightly said to have a function. The point is rather that whether or not something has a function, and what that function happens to be, is quite independent of whether it was selected and spread. When we look for a place for selection to act on wings, say, we need to be looking for variations in wing design. All of the variant wings will have the same function—to enable flight. Thus, one cannot look to differences in the function of the wings to predict or explain selection. One must look instead to how well the various wings are functioning, and this means looking at the functions, not of the wings, but of something else: feather design, bone structure, musculature, and so on. Moreover, this argument iterates. It is the better of two muscle attachment schemes that gets selected; both the better scheme and the inferior scheme have the same function. Functions just do not track the factors driving selection. No doubt there *are* cases in which one subpopulation acquires some structure or behavior that the rest of the population just does not have, a biological analogue of adding a governor to steam engines, or an escapement to clocks. But such cases must be quite rare.[13] If they exhaust the proper domain of neo-teleology, then neo-teleology is insignificant at best. It comes out true as a kind of accident, a coincidence in the rare sort of case in which selective advantage happens to coincide with the introduction of something with a novel (in that context) function.

Selection can, to some degree at least, be explained by appeal to adaptiveness, although the connection between adaptiveness and selection is more indirect than is sometimes appreciated. What is uncontroversial is that a trait spreads because it is heritable and appears in a host that is more fit than the competition. Exactly the same thing can be said with equal truth about every trait of that host. Every trait of the winning host spreads, regardless of how adaptive it is—regardless, indeed, of whether it is adaptive (or has a function) at all. But this does not render adaptiveness irrelevant to selection, since the host in question was more fit than the competition because of some traits and in spite of others.[14] If H was a better design (in part) because of T, and all of H's traits spread because H was a better design, then T's positive contribution—its *adaptiveness*, in short—helps explain why it (and its neighbors) spread.

This suggests the possibility of saving neo-teleology by defining functions in terms of adaptiveness. This would turn neo-teleology into the idea that the proliferation and maintenence of some traits can be explained by appeal to the

[13] Mutation, for example, is much more likely to change the size, density, shape, or attachment angle of a bone than to add a new bone. The altered bone will typically have the same function as its competitors.

[14] This is Sober's distinction (1984*a*) between being selected and being selected for.

fact that they were adaptive. I certainly do not wish to take issue with that claim, though I think there are reasons for caution.[15] I do think, however, that there are good reasons to keep having a function and being adaptive distinct, and it is worth taking a brief detour to canvass these, for it will lead us back to the main point via another route.

Adaptiveness is a matter of degree; having a function is not. The more adaptive wing and its less adaptive competition both have the same function, but only the former is selected for. Functioning *better* is a matter of degree, and it is at least sometimes true that the more adaptive wing functions better. But this just makes it clear that functional analysis is prior to, and independent of, assessments of adaptiveness. When we have a system analyzed functionally, we are in a position to ask what sort of improvements could be made by substitution of functional equivalents. The substitution of a functional equivalent that is (for example) more efficient, increases adaptiveness, but, by hypothesis, does not change anything's function.

The point here is not, as selectionist etiological accounts would have it, that only the selected wing has a function. After all, the worse wing was once the better one and was itself selected for. The point is rather that having a function is not what drives selection, but rather functioning better than the competition. What the function of a wing is should be distinguished from how well it performs it. The question of what function something has is evidently prior to the question of how well it is performed in a given organismic and environmental context, and hence prior to the question of how adaptive performing that function is for a given organism in a given environment. To repeat, the better and worse wings have the same function, but only the former spreads.

It might seem that there is a link of sorts between functions and adaptiveness, and hence between functions and selection. Knowing that the function of hearts is to circulate blood might be thought to constrain what sorts of variation in heart design would be adaptive, and hence what sorts of variations might be targets of selection. Indeed, I have been saying that it is the heart design that enables better circulation that gets selected. This suggests that

[15] One needs to be careful with the idea that traits spread because they are adaptive. The underlying rationale is that adaptive traits are likely to give their hosts the kind of advantages that lead to greater reproductive success. Hence, over the long haul, the subpopulation that has the trait in question is likely to grow relative to the rest of the population. The resulting *spread* of the trait in question through the population is the essence of selection.

Two points need mentioning. First, for this story to be substantive, we require a conception of adaptiveness that makes it independent of fitness. Secondly, whether adaptive traits spread depends on the extent to which conditions approach what we might call 'full-shuffle' conditions—i.e. conditions under which there is a fixed pool of heritable traits that do not interact, and every combination of them gets tried out in the fullness of time in a fixed environment. (See Kaufman 1989, 1993, on the importance of trait interaction.) That these conditions are seldom if ever satisfied in complex organisms is evident. The wonder is that natural selection works at all, given the poor working conditions with which it is faced.

when we identify the function of a trait, we have identified the dimension of performance that is relevant to assessing the adaptiveness of that trait. Circulation, not pulse production, is the function of the heart, and so it is variations in heart design that improve circulation, not variations that improve pulse production, that matter to adaptiveness.

Attractive as this line is, I do not think it will stand scrutiny. Better wing designs need not improve flight, but simply make it more efficient, or make development less error prone, or make the structure less fragile. Hence, selected changes in wing design that accumulate to yield the current design we seek to explain need not be related to the wing's function. Indeed, they may even compromise flight in the interest of other factors. Hence, if we are trying to understand why a given trait or structure is the way we find it, we cannot simply focus on variations that affect how well that trait or structure performs its function. We need, instead, to look at the complex economy of the whole unit of selection. This is precisely what a functional analysis of the whole unit facilitates, and is neglected when we focus on the function or functions of the trait in question.

6. Paley Questions

Even if we could make sense of the idea that things like wings and eyes—salient targets of functional attribution—spread through previously wingless and eyeless populations, the serious why-is-it-there question about such things as wings and eyes would remain untouched. How did there come to be such things in the first place? To harken back to Paley's famous example (1802), when we discover a watch in the wilderness, we are likely to infer a designer, not because we wonder why watches became so popular, but because we cannot otherwise understand how such a thing could come to exist at all.[16] And this is precisely the difficulty with eyes and wings. I propose to call this sort of why-is-it-there question a Paley question.

It is pretty generally conceded, I think, that Paley questions cannot be given neo-teleological answers (Godfrey-Smith 1994). Selection presupposes something to be selected. You cannot select for creatures with eyes unless eyes already exist. So it looks like selection cannot even address Paley questions. But, of course, this is much too quick.

Selection *can* address Paley questions, but only indirectly. The selection of eyes, or sighted organisms, is the wrong place to look. Selection builds a complex structure like a human eye or a sparrow wing by successive approximation

[16] Of course, if watches are popular, you are more likely to find one. But Paley's beachcomber did not want an explanation of why a watch was found, but of why there were any watches to find.

(or what looks like it retrospectively) in relatively small steps beginning with an organism without an eye or wing and ending with what we observe today. Many of the fine details of such stories are unknown. Yet the in-principle possibility of the process is enough to provide the answer to Paley's original challenge: to explain how such things as the human eye came to exist in the first place without reference to the intentions, plans and actions of an intelligent creator and designer of eyes. Natural selection is clearly a central player in the sort of story that has successfully met this challenge. But it enters in by accounting for the spread of small modifications to precursor structures. To think of the modern human eye or sparrow wing as itself selected is, to repeat, to conjure up a scenario in which there is a population of sightless primates or wingless songbirds into which is born a sighted or winged variation whose progeny take over the land or air. No one, of course, really believes anything like this. Yet something very like this is implied by neo-teleology—by the idea that eyes are there because they enable sight and wings because they enable flight. The modern human eye or sparrow wing never spread through any population. Some small changes to earlier structures very like the modern human eye or sparrow wing may have spread. And small changes to those structures may have spread. And so on. In short, as we have already seen, targets of functional characterization and targets of selection just do not match.

To summarize: if we ask why some complex structure is 'there', in the sense in which this means how it came to exist, appeal to its function or functions, as teleology (neo and classical) requires, is only going to be misleading. Such stories either run into the fact, fatal to classical teleology, that the crucial details of evolutionary (or ontogenic) development predate anything with the function that is supposed to do the explaining, or they founder on the fact that competing traits in selection scenarios typically have the same function. Things do not evolve because of their functions any more than they develop because of their functions.

It *is* generally conceded that teleology does not address Paley questions. But we are now in a position to see that Paley questions are all the questions there are about the evolution of traits. The idea that, although eyes and wings did not come to exist because of their functions, they nevertheless *spread* because of their functions, leaves us with a distorted picture of the role of selection. It makes us think that selection can spread only what is already there. While this is true in a sense, it is seriously misleading when we focus on the kinds of traits that have salient functions. It makes us think that eyes—eyes like ours—came to be somehow (some massive mutation?), and then were selected for because they were so adaptive. When we explain how eyes like ours came to be in the first place, we have said all there is to say about spread. When we have answered Paley's question, we have answered the evolutionary question. There is nothing left over for spread to do that it has not already done.

7. Conclusion

Let us consolidate our results. Neo-teleology, the idea that traits are there because of the effects that are their functions, is a non-starter when it comes to serious why-is-it-there questions: the questions I have called Paley questions. Appeals to function fail to address Paley questions, because nothing in the relevant lineage has the function in question until the trait in question is created. When it comes to Paley questions, neo-teleology has nothing to add to classical teleology. This is quite generally acknowledged. But neo-teleology fares no better as a story about why traits spread. Substantive neo-teleology misidentifies the targets of selection with the sort of complex generically defined traits—having eyes or wings—that have salient functional specifications.

Neo-teleology, I find, dies hard. Its rejection sounds to many like rejection of evolution by natural selection. But it is not. Darwin's brilliant achievement has no more need of neo-teleology than it has for its classical predecessor. What it needs is a conception of function that makes possession of a function logically independent of selection and adaptiveness. For it is only by articulating a reasonably illuminating functional analysis of a system that we can hope to understand *what* it is that evolution has created. If we want to understand *how* it was created as well, there is no avoiding the messy historical details by the cheap trick of assuming that all we have to do to understand trait proliferation and maintenance is to attribute a function. Neo-teleology thus amounts to a license to bypass the messy and difficult details, to jump over them in a way that makes it seem that the whole process was like the progress of a heat-sensing missile, arriving more or less inevitably at its goal regardless of the vicissitudes of wind and the meanderings of the target. The idea that evolution and development are goal oriented is precisely what makes classical teleology unacceptable. Neo-teleology creates the same impression while masquerading as good Darwinian science.

There is another nexus of reasons why neo-teleology hangs on, at least in Philosophy. Twentieth-century empiricist philosophers such as Hempel (1959) were worried about function talk in science because it smacked of (classical) teleology. They set out to determine whether functions have a legitimate role in science. For reasons I am not clear about, they took this to be an issue about functional explanation, and interpreted *that* as a question about whether things could be explained by appeal to their functions. Thus, one important strand in the debate over functions simply assumed that the legitimacy of functions and the legitimacy of neo-teleology were one and the same. In Cummins (1975) I argued that this was a mistake; that functional attribution and functional analysis could be, and often are, decoupled from explaining why things are there by appeal to their functions. Still, the idea that functional attributions are equivalent to neo-teleological explanations remains widespread.

However, even if you accept that functional explanation and functional description can be decoupled from teleological explanation (some do, some do not), it might seem that the original empiricist worry remains about functions. One might continue to think that they need, in current parlance, to be *naturalized.* But most, perhaps all, of the pressure to naturalize functions is really pressure to naturalize teleology. Once functions are separated from teleology, they do not look any more likely to offend empiricist scruples than any other dispositional properties. But this point is not widely appreciated, and therefore there still is, I think, a widespread feeling that functions need naturalizing, and that this amounts to naturalizing (neo-)teleology.

There is a different sort of philosophical problem that remains, however. It is pretty generally agreed that a thing's function (or functions) is some special class of its effects. The problem of analyzing functional attributions, then, seems to require some criterion for saying which effects count as functions. Why is blood circulation a function of the heart and not production of a pulse? Selectionist etiological accounts seem to many to provide an elegant solution to this problem: the functions of an X are those effects of an X that, historically, account for Xs having been selected.[17] I have, in effect, been arguing against the selectionist etiological account of functions in biology on the grounds that the targets of functional attribution are seldom the targets of selection. If I am right, then almost nothing has a function in the sense staked out by selectionist etiological accounts of what functions are. This, I think, is what Hempel (1959) did conclude. We are better off abandoning the selectionist etiological account of functions.

References

Allen, Colin, and Beckoff, Marc (1995), 'Biological Function, Adaptation and Natural Design', *Philosophy of Science*, 62: 609–22.

Buller, David (1999) (ed.), *Function, Selection and Design: Philosophical Essays* (Albany, NY: SUNY Press).

Cummins, Robert (1975), 'Functional Analysis', *Journal of Philosophy*, 72/20: 741–65.

——(1977), 'Programs in the explanation of behavior', *Philosophy of Science*, 44: 269–87.

——(1983), *The Nature of Psychological Explanation* (Cambridge, Mass.: MIT Press).

Davies, Paul S. (2001), *Norms of Nature: Naturalism and the Nature of Functions* (Cambridge, Mass.: MIT Press).

Dennett, Daniel (1995), *Darwin's Dangerous Idea: Evolution and the Meanings of Life* (New York: Simon & Schuster).

[17] This is sometimes confused with the idea that the functions of X are those effects of X that were adaptive—i.e. that contributed to the fitness of their hosts. This could be true, even though Xs were not selected because of those effects, or even though Xs were not selected at all. Selection presupposes variability; positive contributions to fitness do not.

Godfrey-Smith, Peter (1994), 'A Modern History Theory of Functions', *Noûs*, 28: 344–62.

Griffiths, Paul (1993), 'Functional Analysis and Proper Functions', *British Journal for the Philosophy of Science*, 44: 409–22.

Hempel, C. G. (1959), 'The Logic of Functional Analysis', repr. in C. G. Hempel, *Aspects of Scientific Explanation, and Other Essays in the Philosophy of Science* (New York: Macmillan, 1965), 297–330.

Kauffman, S. A. (1989), 'Origin of Order in Evolution: Self-Organization and Selection', in B. C. Goodwin and P. Saunders (eds.), *Theoretical Biology: Epigenetic and Evolutionary Order from Complex Systems* (Edinburgh: Edinburgh University Press).

——(1993), *The Origins of Order: Self-Organization and Selection in Evolution* (New York: Oxford University Press).

Kitcher, Phillip (1993), 'Function and Design', *Midwest Studies in Philosophy*, 18: 379–97.

Millikan, Ruth Garrett (1984), *Language, Thought, and Other Biological Categories: New Foundations for Realism* (Cambridge, Mass.: MIT Press).

Neander, Karen (1991), 'The Teleological Notion of "Function"', *Australasian Journal of Philosophy*, 69: 454–68.

——(1995*a*), 'Pruning the Tree of Life', *British Journal for the Philosophy of Science*, 46: 59–80.

——(1995*b*), 'Explaining Complex Adaptations: A Reply to Sober's Reply to Neander', *British Journal for the Philosophy of Science*, 46: 583–87.

Paley, William (1802), *Natural Theology: or, Evidences of the Esistence and Attributes of the Diety, Collected from the Appearances of Nature* (London: Faulder).

Pust, Joel (2001), 'Natural Selection Explanation and Origin Essentialism', *Canadian Journal of Philosophy*, 31 (June), 201–20.

Sober, Elliot (1984*a*), 'Force and Disposition in Evolutionary Theory', in C. Hookway (ed.), *Minds, Machines and Evolution* (Cambridge: Cambridge University Press), 43–62.

——(1984*b*), *The Nature of Selection: Evolutionary Theory in Philosophical Focus* (Cambridge, Mass.: MIT Press).

——(1995), 'Natural Selection and Distributive Explanation: A Reply to Neander', *British Journal for the Philosophy of Science*, 46: 384–97.

7. Functional Organization, Analogy, and Inference

William Wimsatt

ABSTRACT

This chapter builds upon my earlier selectionist analysis (Wimsatt 1972) to characterize functional organization, to examine judgments of functional analogy, and to give conditions for evaluating functional inference. The problem of teleology was not invented for philosophers' amusement, so a good analysis should give some methodological guidance in the construction and evaluation of descriptions of functional organization that could be useful for practitioners. I thus analyze and then move beyond 'ideal' functional hierarchies to consider the kinds of pragmatic compromises necessary in the field, and to bring our analyses closer to our practice. This is not an etiological analysis, though in some ways it comes close to one, and I consider tensions arising for both human engineered designs and biological evolved ones from the gap between the origin of functional organization through a layered patchwork of kluges-aptive and exaptive fixes to earlier adaptive structures that themselves arose in the same way—and the current evaluation of functional organization into functional subsystems on the basis of current evaluated performance.

This chapter complements and builds upon my analysis of function (Wimsatt 1972),[1] whose fundamentally selectionist character was obscured to some early readers by two facts:

1. The literature then mostly sought to find acceptable 'translations' simultaneously to justify and 'eliminate' teleological language—a serious mistake. I argued that functional language was indispensable in biology (and elsewhere) as the appropriate way of applying relevant selectionist theories. Darwin's theory was obvious, but Herbert Simon (1969) and Donald Campbell (1974) also offered contemporary selectionist theories capable

[1] Written in 1970, as chapters 6 and 7 of my dissertation and published in *Evolution and Cognition*, 3/2 (1997), 2–32, this chapter was significantly rewritten for this volume, with changes in all sections and important new material in sections 1, 7, 12, and a new section 22. Old sections 22–25 are omitted to save space. They contain important material on heuristics used in the reductive analysis of systems (including functionally organized ones), which complements the practical tone of this chapter. Wimsatt (1972) was chapters 2–5 of the dissertation.

of rooting functional language for various aspects of human action, learning, practice, technology, and culture.

2. My analysis did not eliminate, but justified talk of purposes where appropriate (and gave conditions for when that was so)—as a kind of selectionist imperialism. In the context of a selectionist theory, I argued that a purpose was an ineliminable theoretical construct (roughly, an intensionally defined class of states; the highest end in a means–end hierarchy of functions serving whatever was optimized by the selection processes[2]). A natural part of the theoretical apparatus of selection theories, these did not entail the existence of intentions. Larry Wright (1976) misread this, classifying my account as one in terms of purposes *rather* than in terms of selection. No philosophers of biology misread it this way, so (unwisely!) I did not respond. (I also liked many aspects of Wright's analysis, and did not see any point in arguing with allies. His 1973 paper was in most respects a good first approximation to mine, though lacking recognition of the complexities necessary to deal with the hard cases discussed in my 1972 (and often subsequently by Millikan).)

This analysis was mostly ignored outside philosophy of biology, perhaps partly due to Wright's misrepresentation, but probably more due to the rising popularity of Cummins's approach (1975) in philosophy of psychology in the late 1970s, when evolutionary concerns were conspicuously (and erroneously!) absent. I got uncompromising resistance to missionary efforts for selectionist accounts from Ned Block, who favored a Cummins-style approach when we discussed it at length in 1976. I had other problems to work on in biology, and spent my time there rather than pushing an analysis that no one in psychology seemed to want. (I supposed that an evolutionary perspective would come in time—as it has largely due to decades of work by Dan Dennett, more recent work by Ruth Millikan, and the recent explosion of evolutionary psychology.)

An important difference remains between Wright or Millikan and me: this account is not etiological, although perhaps nearly so. Etiology is essentially inevitable for any complex adaptation—the Darwinian abhorrence of 'hopeful monsters' yields deep suspicion of claims to major adaptive progress save through cumulative accretion over *many* generations. That's 'many' on a scale where a million years may be 'geologically instantaneous'.[3] But do we need that built into the definition of function? Probably not. I took etiology seriously—at length, for much of the last section of the 1972 paper (pp. 67–78). But making it

[2] Even in 1972 this was not an unproblematic claim. I considered (1972: 62–5) problems with maximization accounts arising from satisficing, multidimensional choice and other corners, but these problems affect both accounts of purposive activity and selection in analogous ways.

[3] This has weakened only slightly with the increasing influence of developmental genetics and 'evo-devo' in recent years. See Raff (1996).

a definitional requirement would seem to forbid attribution of functions to (as yet unrealized) design elements in evolutionary optimality arguments. In the 1960s and 1970s, these were 'cutting edge' in evolutionary theory and they still are crucial to the 'engineering' analysis of function. (This is not just appealing to 'intuition' to avoid adopting the etiological account.[4])

Of course I do not need etiology as much as someone primarily motivated by teleosemantics. But even here could philosophers suffer from too much definitional rigidity? Must any good semantic theory be unitary and universal? Biology is full of piecemeal kluges, and so is culture. Why not semantics? Are there different ways to derive or to anchor meaning? Or would not teleosemantics work pretty well if there were nearly always an etiological history with a few exceptions that parasitized common practice? Then making an etiological history a contingent characteristic, but not a universal accompaniment to having a function works just fine. (Or perhaps we should use another theory of reference for cases lacking an etiology?) Perhaps the real problem is too worshipful attention to 'rigid designation'? Biology and evolution go a long way with patterns whose essential nature is that they do and must work just *tolerably often*. And does not society as well? Our semantic theories should presumably be tuned to how we actually derive meanings, and fix reference. This means that they, like our brains and our immune systems, must work in a noisy world.

Despite these reservations, the core of this chapter mostly treats problems of functional organization and functional analogy from a 'God's eye' or 'Laplacean demon' point of view, assuming that 'Darwin's demon' (Wimsatt 1980*a*) will see any functional effect, no matter how small, and incorporate it in the functional architecture: 'It may be said that natural selection is daily and hourly scrutinizing, throughout the world every variation, even the slightest, rejecting that which is bad, preserving and adding up all that is good; silently and insensibly working, whenever and wherever opportunity offers, at the improvement of each organic being in relation to its organic and inorganic conditions of life' (Darwin 1859: 84).

Assumptions of omniscience or computational omnipotence have a time-honored tradition in philosophy, and a close connection with conceptual or 'in-principle' arguments. I have attacked them throughout my professional career (Wimsatt 2003). They may be adequate for delineating the conceptual structure of functional organization (even here I have doubts, if selection is a satisficing process), but I wanted to go further—to characterize functional organization in ways useful to and recognizable by real-world practitioners. This analysis moves in that direction, but it is still frustratingly far from this goal and this led me to delay publication in 1972.

[4] The locution 'has a function' pushes harder towards etiology than 'is functional', but to turn an analysis on that would be to give 'ordinary language' more power than it deserves.

But should we not be omniscient and computationally omnipotent for constitutive questions, and leave the pragmatic pangs for methodology? Here I am not sure. In a world where quasi-deterministic selection forces and drift are on a quantitative continuum, where structured heterogeneous populations face local modulations of heterogeneous environments, and any adaptations rest on a multilayered historically extended patchwork of aptations and exaptations (Gould and Vrba 1982), it is not so clear that constitutive questions are free from such concerns. And even if a God's-eye view would do for functional architecture (but see Section 22 below!) let us not forget the (often much harder and more important) methodological questions on the way to the in-principle conclusions. To philosophers who, like myself, view their work as being as much conceptual engineering as philosophical theory (and this includes many philosophers of science), these cannot be dismissed as 'mere pragmatic complexities'.

I have not reviewed the now enormous literature on the teleological (or 'selectionist') concept of function or the competing 'role' account of Cummins (1975). These correspond to two of the senses I delimited in 1972, and I did criticize there the views of Fodor and Putnam, which are proper ancestors to any 'role' account, including Cummins. My 1972 was the first (and is perhaps still the fullest) systematic treatment of the selectionist sense of function, since known through the work of Wright (1973) and Millikan (1984). Like more recent attempts in having more moderate versions than Wright or Millikan of the etiological connection between function and past selection, it also thereby avoids most counterexamples to those analyses. It also has more structure for dealing with the diversity of uses of functional inference and provides more resources for treating issues arising particularly in the comparative uses of functional analyses, some further explored here. If any language is unfamiliar (at least some references will be, to modern eyes!), I urge you to go back to that paper to see those distinctions in context. Few if any of them are any less appropriate today.

1. Introduction

Until recently (e.g. Nagel 1961; Beckner 1969) the philosophical literature on function has been primarily concerned to determine whether functional analyses and explanations are reducible to causal analyses and explanations and whether they are scientifically acceptable. These may be interesting questions to the philosopher, but other questions are of far greater interest and import to the biologist. Two such questions are the following:

1. How is biological organization to be characterized?
2. What are the grounds for judgments of homology and functional analogy?

Biological organization is central because biological organization is functional organization. I offer an analysis of the structure of 'functional hierarchies' that bears on both questions. This analysis is addressed primarily to biology but should apply wherever one may appropriately use functional analyses as explicated here (Wimsatt 1972). This should also include ordinary explanations of human action—especially where deliberation, strategies, or planning are involved; explanations connected with learning and problem-solving behavior, and, finally, aspects of the social sciences where cultural evolution is held to be a factor.

I. The Structure of Functional Hierarchies

2. The Sense of 'Function' Intended

First, at least three distinguishable senses of 'function' are used by biologists (Wimsatt 1972 lists six):

(*a*) To say that an entity is functional is to say that its presence contributes to the self-regulation of some entity of which it is a part (Beckner 1959, 1968, 1969; Nagel 1961).

(*b*) To say that an entity is functional is to say that under at least *some* conditions it plays a (presumably causal) role in the operation of some system of which it is a part (Beckner 1959; Fodor 1965, 1968; Kauffman 1971).

(*c*) To say that an entity is functional is to say that it is being selected for or maintained by natural selection (and presumably, in the overwhelming majority of cases, that it owes its presence and form to the operation of natural selection) (Williams 1966).

Physiologists, doctors, and biologists strongly influenced by ideas of homeostasis (through the work of Cannon 1939), self-regulation, and cybernetics are prone to understand 'function' in the first sense. (Classical functionalism in the social sciences fit here also—see Hempel 1959.) Biochemists, molecular biologists, and biophysicists (when they use functional language at all) often use the second sense (though many are moving to the third sense)—as do 'intervening-variable' psychologists influenced by computer science. The third sense is favored throughout evolutionary biology, including particularly by systematists, paleontologists, geneticists, ecologists, and ethologists, but also by a sprinkling of people in the other areas.

Secondly, I use 'function' in this third sense for biology[5] and in a natural generalization of this more broadly,[6] for several reasons. (1) Biological organisms reflect their status as products of evolutionary processes at all levels of organization. So this sense of function applies at all levels and to a broader range of biological phenomena than the first sense, connected solely with self-regulatory phenomena. (2) Any theory of biological organization should reflect the fact that biological entities are evolved and evolving systems, so any sense of 'function' relevant to biological organization should be conceptually related to the operation of evolutionary processes. (3) The second sense lacks the specificity necessary to generate interesting properties for the hierarchy which result from the third sense.[7] (4) The first two senses of function can be treated in all interesting applications as special cases of the third sense, or (in non-biological contexts) of its generalization. (5) For conceptual reasons, the third sense of 'function' has the strongest reasons of any of the three to be called a 'teleological' sense of 'function'. (Wimsatt 1972).

So only the third sense has the generality (in the range of biological phenomena and entities to which it applies), relevance (in its explicit connections with evolutionary considerations), and specificity (in terms of logical constraints) to generate an interesting logical theory of biological organization.

3. The 'Function Statement' Schema

In what follows, I assume a 'normal form' for attributions of function (in the third sense). This schema (derived in Wimsatt 1972) is as follows: 'According to causal theories, *T*, *the* function of behavior *B* of item *i* in system *S* in environment

[5] This third sense is misleadingly precise. Theoretical and empirical disputes persist concerning the nature and efficacy of selection operating at various levels of organization from gene to ecosystem and species. These disputes engender arguments over the proper meaning of 'fitness' in evolutionary theory, and in definitions for 'function' in the third sense (see Thoday 1953, and Lewontin 1961 versus Dobzhansky 1968). For more recent reviews, see Wimsatt (1980*b*, 1981*b*); Brandon and Burian (1985); Lloyd (1988).

[6] I take 'function' (in any properly 'teleological' sense) to be connected with ideas of 'purpose'. My general schema for function statements (below) involves explicit reference to that concept. However, I intend no connections with consciousness or intentional action but merely as implying that certain *logical* features are present that are central in the talk of purposes in the human case. 'Purpose' may apply as a theoretical construct in areas and ways that do not postulate a conscious agent or any other mental or vitalistic properties. In this sense, evolutionary processes involve, and the third sense of 'function' relates to, purposes. Given a theoretically determined choice of evolutionary units, these purposes are, roughly, that which the net effect of selection operating at various levels promotes for these units (Wimsatt 1972).

[7] Kauffman takes the causal role of the functional entity as being relative to some 'perspective' on the system. If the 'perspective' is such that the function may be construed as promoting the attainment of some purpose, in my generalized sense, our senses converge.

E relative to purpose *P* is to do (or bring about) causal consequence *C*.' This schema will generally be represented by the logical functional equation:

F[B(i),S,E,P,T] = to do C.

A function statement fitting this schema might be: according to current physiological theories (T), the function of expansion and contractions in peripheral capillaries (B of i), in the thermoregulatory systems of a mammal (S) in environments that are appreciably hotter or colder that the normal bodily temperature of that organism (E), relative to maximizing the probability of survival to time t in the future of the appropriate evolutionary unit of which the organism is a part (P),[8] is to change the rate of heat exchange between organism and environment so as to decrease the difference between the organism's bodily temperature and its normal bodily temperature.

Most attributions of function are not this completely spelled out, and biologists frequently attribute functions to recognizable parts of organisms, as well as to their behaviors. I am *not* suggesting that all functional attributions must take this form, but only that more telescoped descriptions implicitly involve reference to all of these variables and that attributions of function to behaviors is conceptually prior to attributions of functions to physical objects.

The latter claim is roughly that we attribute functions to objects *because and only because* they exhibit certain behaviors under the appropriate conditions. This claim is supported by the concept of 'functional equivalence': two objects are functionally equivalent (or analogous) if they *do* the same (or similar) things in the same (or in similar) systems in the same (or in similar) environments, etc. The key is the emphasis on the word 'do'. No other features of the objects are relevant other than the fact that they *do* the same things under certain conditions—which is to say that it is their behavior that is important. This is why item (i) in the schema is not a separate variable—the variable is the

[8] This maximization is subject to genetic, population genetic, and environmental constraints. The appropriate evolutionary unit might be a single organism or clone (as with asexually reproducing bacteria), a breeding population or meta-population of the same or of hybridizable species, an ecosystem, or even a genus or combination of higher taxa of species. Choice of an evolutionary unit and of a time for evaluating its probability of survival are determined by factors affecting isolation of lineages and magnitudes and temporal patterns of interaction of the forces of selection operating at these and other levels. This choice of P variable for 'function' in biology is inspired by Thoday's discussion (1953) and a modification of his definition of 'fitness' suggested in part by the work of Lewontin and Levins. (Addendum: I would now describe this as a satisficing optimization rather than a maximization, and would supplement these comments in the light of subsequent work on units of selection. See e.g. Wimsatt (1980*b*, 1981*b*, Brandon and Burian 1985, Lloyd 1988, or, for the latest work on group selection, Wade 1996). Williams (1966) and Dawkins (1976) argue too quickly that the appropriate time period is a single generation. For my disagreements, see the notion of 'futurity' defined in appendix 3 of my dissertation for discussions of factors affecting the relevant time scales (Wimsatt 1971).

behavior of an item. It is one of the strengths of this analysis that it permits such a natural treatment of functional equivalence. (See Wimsatt 1972, and below.)

All of these variables are required in the function statement schema. Changes in any of these variables may result in a change in what is chosen as the functional consequence (C), and these variables are in fact independent. Thus each of them is required in the most general case.

A third feature of this schema is indicated by the use of the phrase '*the* function' in the verbal formulation and the implication that the logical equation is single valued—it is a function in the formal mathematical or logical sense. The claim (hereafter, the 'uniqueness claim') is that, when determinate values for the variables B(i), S, E, P, and T are plugged into the schema, they determine *at most one* value of the functional consequence C. (There may be cases where B(i) has only dysfunctional and/or non-functional consequences for the conditions specified.) This claim cannot be maintained in general, but it plays pragmatic and pedagogical roles in elaborating the structure of functional organization and generating the most useful representation of that structure. It increases the resolution of differentiated functional structure by determining any changes in conditions under which functional consequences are realized. Demonstrations of single or multiple functions all make heuristic use of this structure.

4. The Structure of Functional Organization: A Preliminary Attempt

Intuitively, a functional hierarchy should be a tree. I will show that this intuition is ultimately incorrect, but it is still worthwhile analyzing this structure in such a way as to make it as close to a tree as possible.

The idea that functional hierarchies have a tree structure is suggested by inclusion relations between the various systems and subsystems of a functionally organized entity. Parts of a system perform functions in that system, which may in turn be just one of several systems that perform functions in a still larger system. And each of the original parts may itself constitute a functional system, with subsidiary parts having functions in it. All of this suggests a tree structure.

The functions of the systems and parts in a given functional hierarchy are related. The function of a system determines the functions of its parts, which in turn determine the functions of the parts of the parts, and so on. Conversely, the fulfillment of function of the parts of a system contributes to (but does not entail) the fulfillment of function of that system and so on up the line to the most inclusive system in the hierarchy.

If systems and parts are represented by nodes and the functional contribution a part makes to a system is represented by a directed arrow from the part node to the system node, it appears that a simple tree structure, as represented in Fig. 7.1*a*, is generated.

But this simple picture is incorrect if systems and parts are construed as physical objects. Thus the circulatory system in mammals would include as parts the heart, arteries, veins, and peripheral capillaries. The thermoregulatory system in mammals would include parts of the endocrine and nervous systems, bodily hair, certain muscles (involved in shivering), skin pores (involved in perspiration), and the peripheral capillaries. So the peripheral capillaries are parts of two distinct functional systems if these systems are regarded as systems of physical objects. Functional systems at higher levels (for example, the whole organism) would contain both the thermoregulatory and circulatory systems as parts. Then there would be (at least) two paths from the node representing the peripheral capillaries to the node representing the organism, one via the thermoregulatory system and one via the circulatory system. (See Fig. 7.1*b*.)

But a tree structure must be 'unipathic'—such that there is *exactly one* path connecting any two points in that structure. This condition is not met here.

The result is unavoidable when the nodes—the systems and parts—are interpreted as physical *objects*. Beckner characterizes function and functional organization proceeds in terms of the functions of physical objects. Partially as a result, he speaks of the 'net-like organization' of functional systems rather than seeing anything like a tree-structure (Beckner 1969).

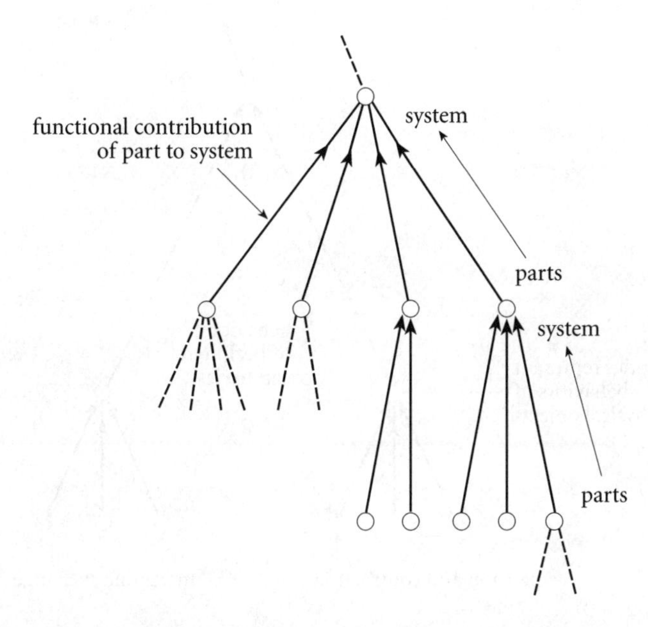

Fig. 7.1*a*. Simple tree structure (unipathic)

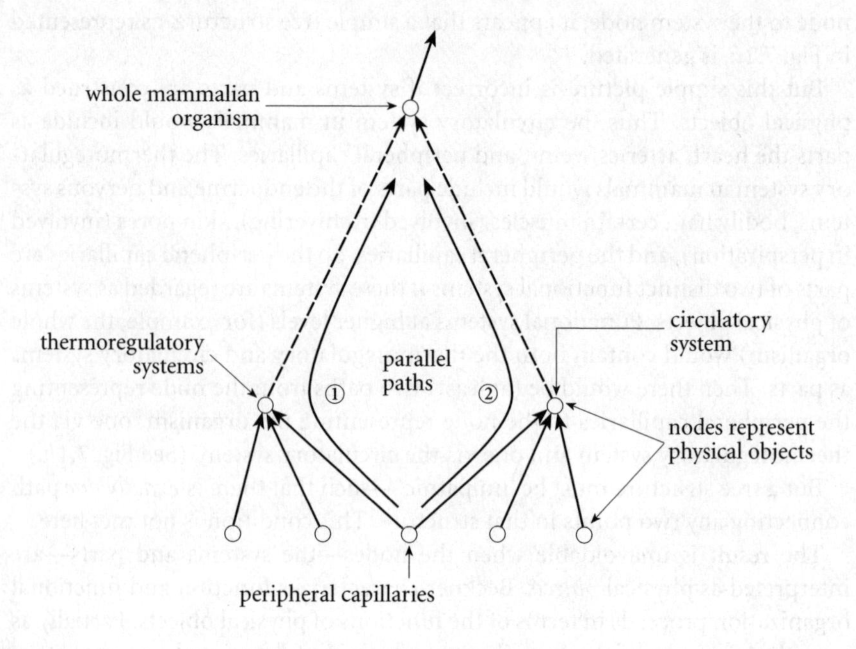

Figure 7.1*b*. Multipathicity with nodes as physical objects

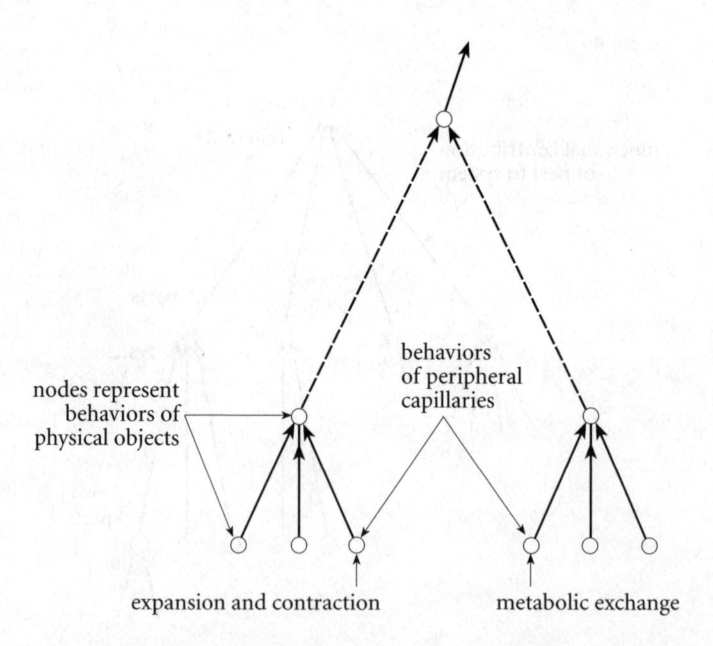

Figure 7.1*c*. Restoration of unipathicity with nodes as behaviors of physical objects

5. Nodes as Behaviors and Systems of Behavior

I argued above that attribution of functions to behaviors of objects had conceptual priority over attributions of functions to the objects or items themselves. If this is taken seriously, we should interpret the nodes as behaviors and systems of behavior. On this interpretation the case of the peripheral capillaries no longer blocks interpreting the functional hierarchy as forming a tree structure, as the distinct behaviors would individuate nodes (Fig. 7.1*c*). So there are formal as well as conceptual reasons for this move.

The peripheral capillaries have distinct behaviors in the thermoregulatory and circulatory systems. Their function in the thermoregulatory system is expansion and contraction (thereby changing the rate of heat exchange between organism and environment). Their function in the circulatory system involves (*a*) providing a closed loop for the return of blood through the veins, (*b*) allowing diffusion of nutrient materials (food and oxygen) and waste materials through their walls to and from the cells. Since these behaviors are distinct, they may be assigned separate nodes in the hierarchy, no *one* of which is contained both in the thermoregulatory behavior system and in the circulatory behavior system. So the functional hierarchy for objects (Fig. 7.1*b*) lacking a tree structure is transformed into the functional hierarchy for behaviors (Fig. 7.1*c*), which has one.

6. The 'Uniqueness Claim' and its Significance

Interpreting nodes in the hierarchy as behaviors or systems of behavior is necessary for giving the functional hierarchy a tree structure, but is it sufficient? I argue shortly that it is not—and that in general, the functional hierarchy does not have a tree structure.

The 'uniqueness claim' is both necessary and sufficient for the functional hierarchy to have a tree structure. This is the claim that there is *at most one* functional consequence C for given values of the variables B(i), S, E, P, and T.

By definition, the functional hierarchy would contain only nodes representing functional behaviors, not those that are either non-functional (selectively neutral) or dysfunctional (selected against).[9] Every node in the hierarchy will

[9] It seems sensible to require that functional hierarchies contain only functional behaviors, so 'ideal' functional hierarchies do not contain non-functional or dysfunctional nodes. But this may conflict with requirements of pragmatic usefulness, simplicity, and theoretical desirability because other desiderata apply. The tree's many nodes may be simplified by representing *sets* of different conditions by single nodes rather than individuating a node for *each* set of values of the five variables. But we may then wish to include behaviors that are non-functional or dysfunctional, under some conditions, or in some systems rather than specifying exhaustively all

be directly connected to at least one other node in the hierarchy (the functional system in which that functional behavior is included) via the arrow representing its functional consequence for that system.[10] If nodes are individuated for each set of values of the variables B(i), S, E, P, and T, for which there *is* a functional consequence, C, then the uniqueness claim says that for any given node there will be *exactly one* arrow from that node to a higher node—for there is exactly one functional consequence relative to those values of the variables. Situations like Fig. 7.1*b* (with two or more arrows from a given node) are ruled out *in principle* if the uniqueness claim is accepted.

We will see in the next section that there are cases where the hierarchy cannot have a tree structure and the uniqueness claim is false. But this does not end its usefulness. Functional analysts often *act as if* the uniqueness claim *were* true. That is, when there appear to be two functional consequences for given values of the variables B(i), S, E, P, and T, there is a tendency to redefine the values of one of the first three variables (and sometimes the fourth) to individuate two or more values for one of them so as to maintain the uniqueness claim *if possible.* Functional hierarchy is commonly elaborated in this way, producing increases in structural detail. Beckner (1959) emphasized the extent to which biological systems and subsystems are defined in functional terms. Seen in this light, acting in accordance with the uniqueness claim is one of the major forces leading to redefinition of biological systems and subsystems and the relations between them. It seems to be a *good* practice to redefine functional behaviors, system, and environments so as to maintain the uniqueness claim where possible. (It is an analogy in the functional realm to applying Mill's methods for the discovery of causal relations—see Wimsatt 1972.)

7. Closed Functional Loops and Failure of the Uniqueness Claim

With the many degrees of freedom in variables of the function statement schema and the possibility of *redefining* values of B(i), S, E, and P, why can we not always change, individuate, and redefine values of these variables to hold the uniqueness claim true come what may? There are two reasons. One of them appears to be inescapable.

combinations of conditions under which it is functional, if its presence under these conditions is a natural consequence of its presence under the conditions under which it is functional. Physical objects are a natural case for this lumping. Whenever nodes are interpreted as physical objects, these physical objects may have non-functional and dysfunctional consequences of their operation in the system as well as the functional consequences responsible for their inclusion.

[10] Even the topmost node (that node that has no arrows leaving it) is *connected*, directly or indirectly (via incoming arrows), to all other nodes in the hierarchy.

Is it always possible to redefine values for these variables in a *non-trivial* way—which does not appear to be *ad hoc*—so as to maintain the uniqueness claim? We would not want to include the same behavior in two different systems under two different names, though different aspects of the behavior could well be fairly individuated. But determining when this does or does not occur presupposes prior decision upon criteria for individuating behaviors. This is not trivial. Behaviors are generally individuated at least partially on functional criteria. If they were defined *wholly* in terms of functional criteria, no problems for the uniqueness claim could arise from this source. Nodes at different points in the functional hierarchy represent behaviors with different functional roles—behaviors that are *by definition* different behaviors if behaviors are individuated solely on functional criteria.

But behaviors are also identified and individuated on other grounds, including at least the identity of the behaving object, various kinds of similarity, and spatio-temporal location. If these criteria are used, they may conflict with the functional criteria, and defeat the uniqueness claim.

The second problem is less equivocal. A functionally organized entity may contain closed functional loops, as depicted in Fig. 7.2. Closed functional loops require closed causal loops, since on this analysis, functional consequences are causal consequences that are also functional. (Not all causal interrelations are functional, so a closed *causal* loop does not imply a closed functional loop.)

An example of a closed functional loop (due to Allan Gibbard) is as follows: the heart's pumping blood contributes through the circulatory system to the maintenance of structural integrity of all parts of the organism and thus to the maintenance of structural integrity of the ribs. But the maintenance of structural integrity of the ribs functions to protect the inner organs from injuries that would prevent or impair their functioning, and thus contributes to the heart's pumping blood.

Similarly, maintenance of homeostasis (of temperature, ionic balance, or whatever) contributes to the organism's ability to find food, and finding food contributes to the organism's ability to maintain homeostasis. More generally, any regulatory mechanism whose operation is functional should contain closed functional loops among its parts. Not every closed functional loop must be regulatory however—consider any closed functional positive feedback loop.

The presence of functional loops in the hierarchy defeats the uniqueness claim if none of the nodes in a loop is the topmost node in the hierarchy.[11] If none of the nodes in a loop is the topmost node in the hierarchy, an arrow

[11] Symmetry arguments suggest that all of the nodes in a closed functional loop be assigned the same level in that functional system. And it is hard to see how this could be the top level unless each and every one of the functional behaviors in the loop could qualify at a topmost end—at least as a part of that collectivity.

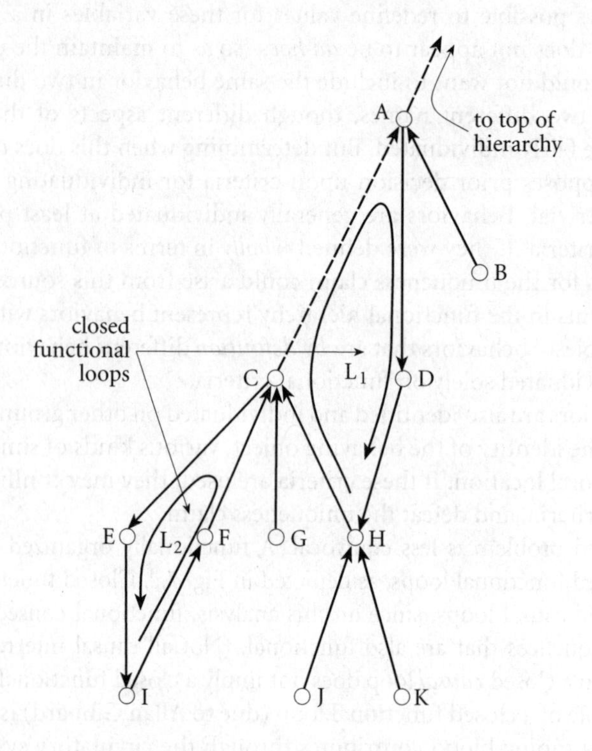

Fig. 7.2. Closed functional loops and cyclical multipathicity

must go from at least one of the nodes in the loop to a node higher in the hierarchy. But each node in the loop also has an arrow to some other node in that loop. Thus at least one node in the loop has two arrows leaving it—one to a node in the loop and one to a node outside the loop. The uniqueness claim fails for this node. It might, of course, fail for several, and even for every node in the loop. Thus it is reasonable to suggest that *each* node in a regulatory system has both internal and external arrows.

But could one save the uniqueness claim even here? Thus, one might argue that it always takes time for the functional effects at a given node to circulate around the loop and return to that node, and that the environment (E) had changed in the meantime. (It always does!) But this would trivialize the enterprise: the environment often has not changed in any relevant respects (for example, if the system is at or near a regulatory equilibrium in a constant external environment). This *ad hoc* move would compromise any usable degree of generality in talk about the functional organization of different systems, or even the same system at different times or under different conditions. And the external and internal functional consequences of a node in a loop may be produced at the same time. Individuating different nodes for the same behavior at different times removes closed loops from the hierarchy but does not do away with the occurrence of multiple arrows from the same node.

Are closed functional loops merely a logical complexity of graph theoretic structures? Why should we care? How common are they? There are at least two reasons to be concerned: we would like to know (1) how they impact on and relate to modular organization in functionally organized systems, and (2) under what conditions might they so tie things together as to render everything too complex to analyze.

Modularity is potentially important for several distinct reasons.

1. It is commonly adaptive if one can select for changes in one character without changing other characters due to pleiotropy (multiple effects) in the genes affecting the character being changed.
2. It is commonly adaptive if small changes in one character do not significantly affect the optimum values of other characters—causing a more complex adaptive problem for selection (Wimsatt 1981*b*). (Allometric growth is an important adaptive mode because it involves adaptive coordinated changes in clusters of characters so that they respond appropriately to scale changes that are important in development and in speciation.)

These first two principles are related to what Lewontin (1978) has called 'quasi-independence'. They are important for 'evolvability' or long-term evolutionary survival (Schank and Wimsatt 2000). There are two further notions relevant to modularity in evaluating how functional systems break down and respond to breakdowns:

3. It is commonly adaptive if local failures do not propagate, compromising the performance of other systems and increasing the complexity of attempts to ameliorate the problems induced by that failure.
4. It is commonly adaptive if there are multiple ways of accomplishing important functions. This is sometimes best served by having redundant independent modules serving that function.

Of these four kinds of modularity or functional independence, closed functional loops most obviously seem to compromise the third, but not to affect the other three directly. (Whether it does would depend upon details of system organization.) The fourth is independently interesting as a methodological problem when analyzing the functional organization of a system: parts playing functional roles may be deleted and that function still be served if there is redundancy for it.

I do not have further answers to how closed functional loops affect modularity, except to say that *ceteris paribus* they would seem to decrease it. So also does generative entrenchment. The evolution and degree of modularity is a hot issue now in both evolutionary and developmental biology (Schank and Wimsatt 2000) Wimsatt and Schank 2002).

Secondly, when closed functional loops involve regulative interactions (as they so often do), they are likely to affect the whole functional system and promote or degrade performance across the board, though there usually are some components more severely affected by a given decrement than others. To the extent that they all affect the whole organism, they may not so much confound interactions as create a common milieu or context against which other changes can be evaluated. Thus, while it is true in principle that extra functional connections indicate other relationships that must be studied and controlled for, if they have the effect of stabilizing the environment, they may actually simplify the task of analyzing some relationships within the system.

8. The Elimination of P and T from Node Specifications

I suggested above (Section 6) that individual nodes be included in the hierarchy for each set of values of the variables, B(i), S, E, P, and T. But this is too much: it would include every function statement made about any entity in the *same* functional hierarchy! *The values of P and T may be regarded as constant within a given functional hierarchy* without causing problems for any nodes not already included in closed functional loops.

Three kinds of cases could lead to a change in the T variable:

(1) Different theories apply to different function statements because they refer to entirely different realms of phenomena such that one realm is not held to be a description of the other at a different level of organization or theoretical level. In this case, one is probably talking about two distinct functional hierarchies. Thus, decision theory and learning theory might be held to apply to function statements about human action, and evolutionary theory to the structure and behavior of organisms. But—questions of reduction aside—the functional hierarchy for the choice and explanation of human actions is normally distinct from the functional hierarchy for the selection and explanation of the biological behavior of organisms.

(2) One theory of the operation of a system is replaced by another theory. On his physiological theories, Aristotle thought that the function of the brain was to cool the blood. On modern theories, this is not even a candidate. Functional hierarchies have been represented in different ways with the changes in physiological theories, often with one functional hierarchy gradually transforming into another, but sometimes with quite major changes if a major functional assessment is changed, because that often forces reassessment of many other functional assignments. (Such major rearrangements

indicate what I have more recently called *generative entrenchment*, which arises when failure of a functional entity has widespread consequences (Riedl 1978; Wimsatt 1986; Schank and Wimsatt 1988).) But no one thinks that the functional hierarchy of a biological entity should contain any functions attributed on the basis of past rejected theories. I suppose that, at any given state of theoretical knowledge, T is constant within a functional hierarchy, though it must sometimes be taken as a *set* of mutually compatible theories, $\{T_i\}$, rather than a single theory. This leads naturally to the next kind of case.

(3) The level of investigation and description of a system is changed, and this change in level carries with it a change in which theories are relevant. Biologists frequently apply different theories at different levels simultaneously in describing, analyzing, and explaining the organization of living things. Thus, in a chapter on hemoglobins, Jukes (1966) discusses the amino-acid substitutions in the hemoglobin molecule characteristic of sickle-cell anemia victims, their effects upon the tertiary structure of that molecule under certain biochemical conditions, speculates on the cellular effects in an attempt to analyze the characteristic breakdown of the erythrocytas (red blood cells) at low concentrations of oxygen in the bloodstream, and discusses the resultant pathological effects on the organism. Population genetics then explains why the genes which cause sickle-cell anemia are maintained in a population in spite of their dysfunctional traits.[12]

Though different theories are applied at different levels here, these are different *functional* levels of organization. These different theories will characteristically add nodes and arrows (functional behaviors and consequences) at different levels of the functional hierarchy. The uniqueness claim is not invalidated by different theories unless they add functional consequences at the *same* node—and thus, trivially, at the same level in the hierarchy. The net effect of using such theories at different levels will be to force interpretation of T as referring to the set of such theories used, rather than to regard them as competing theories having competing claims on the structure of the functional hierarchy.

In the three types of case discussed, changes in the T variable cannot disturb the structure of a given functional hierarchy. While a set of theories may be determinative of the structure of a given hierarchy, changes in the T variable need not individuate nodes *within* that hierarchy.

[12] The gene is recessive for sickle-cell anemia and over-dominant for increased resistance to malaria, resulting in higher frequencies of the gene in tropical and subtropical regions but lower frequencies in the temperate zones where malaria resistance is less important. The discovery of the altered protein sequence in 'sickling' hemoglobin trumpeted recognition of this as a 'molecular' disease. Fuller investigations suggest that mosquito transfer of the malaria parasite to humans (from birds and animal stocks) was facilitated by the enormous growth in larval sites in pools of water accompanying the advent of 'slash-and-burn' agriculture, so malaria is equally a 'social' or a 'cultural' disease. (See Durham 1991 for a rich discussion at population and cultural levels and Sarkar 1998 for more on the molecular effects.)

Changes in the 'purpose' or P variable are simpler to discuss. Purpose attainment is promoted by the operation of the topmost system in the functional hierarchy and determines functions of the nodes all of the way down through various levels of that hierarchy. A variation in P that produces another functional consequence can be ignored because the other functional consequence is *not contained* in the given hierarchy. Changes in P values change the hierarchy under consideration.[13]

Defense of the uniqueness claim requires that P be a 'simple' purpose—one whose attainment can be characterized without using two or more logically independent criteria (Wimsatt 1972). Purposes requiring two or more logically independent criteria for the description of their attainment states are complex. Many purposes are complex in this sense, and it may be worthwhile to speak of a single functional hierarchy for a given complex purpose. In this case, it would be necessary to regard changes in the value of the P variable as individuators of nodes. But, also in such a case, it is natural to regard such a complex hierarchy as a set of 'simple' hierarchies (with a constant P within each) 'strung together' at the top to give a hierarchy for the complex purpose.

So, in sum, it is sufficient to maintain the uniqueness claim within a given hierarchy (without cycles) that a separate node is individuated for each set of values of the variables B(i), S, and E. Variations of P and T can be ignored within a given hierarchy because they are constant within it.

9. Characterizing the Environment

It might seem to be a good idea to treat changes in the environment like changes in the P variable, individuating different functional hierarchies for each environment, and perhaps 'stringing them together' at the top to get the total hierarchy under different environmental conditions. But it is not possible to separate characterizations of the environment from the individuation of nodes within a given hierarchy and still organize the hierarchy sensibly—due to practical constraints on how to characterize the environment in studying such systems. Because systems are characterized and analyzed at different levels in the hierarchy, it makes sense to characterize and analyze the environment in the same way because the two are complementary. Each system or subsystem is characterized and studied in its *own* environment—complementing each other all the way down.

[13] The first case cited for changes in the T variable also fits here. Human purposes generally differ from (and often are at odds with) evolutionary ones. Different purposes are more important than the fact? that the theories apply to diverse realms in deciding that two distinct functional hierarchies are involved.

Different values of P may plausibly involve different functional hierarchies, but not different values of E. Even with nodes interpreted as behaviors, the hierarchy gives the functional organization of a complex *object* that behaves. We apply spatio-temporal criteria in identifying objects, and suppose persistence of this object in a variety of environments. Each organism faces environmental changes, both within individual life cycles and in successive generations. Selection operates upon the differential ability of different evolutionary units (organisms, groups, and so on) to survive under such environmental variations, and the functional hierarchy of an evolutionary unit is the product. It would be extremely artificial to treat this hierarchy as a composite of separate hierarchies for distinct environments.

If individual subhierarchies were used for each value of E, would E be constant within them? That depends upon how E is characterized. If each environmental subhierarchy is individuated solely on the basis of changes in the environment of the system at the topmost node of the subhierarchy, the environments of subsystems of those systems will not be totally specified. There will be conditions *internal* to any system that are *external* to any one of its subsystems (such as the states of its *other* subsystems). Changes in these conditions will not be included in the environmental specifications of higher nodes, and thus may individuate nodes at each level in the hierarchy.

If the environment of the topmost system includes not only its environment, but also its internal state description, there would be no branching at lower levels, but this move has little else to recommend it. On this alternative, each subhierarchy would require 'apocalyptic' specification of the environment at all levels. One could not investigate one subportion of a hierarchy at a time in local fashion (as is the usual practice), and the difficulties in filling in all of this information would be insurmountable.

The first alternative reflects and facilitates comparative investigations of the behavior of a system and its subsystems in different environments. The environment can be 'cut up' by levels and conceived of as varying one level at a time, and subparts of the system may be bounded and studied locally. Any node represents a system in its given environment. The nodes below it represent its subsystems in variations of their environments internal to that system and its environment. Comparative investigation at some level will tend to fill in information concerning immediately connected nodes on the first mode of representation. Investigation would fill in nodes in widely separated 'corresponding subbranches' of different environmental subhierarchies on the 'apocalyptic' second. The first proposal, but not the second, captures the intuition that a given investigation is performed at a certain level for conditions holding at that level by localizing the information discovered in a compact subregion of the hierarchy.

Finally, the first proposal has an advantage over the second if the tree is simplified by merging nodes for functionally equivalent environments. (Assume that functional equivalence of two environments relative to a set of systems and subsystems is the isomorphism of the nodes and arrows assigned to those systems and subsystems in the two environments.) But, if an environment is an entire state description of the topmost node and its environment, two environments are functionally equivalent if and only if the entire hierarchy of systems and parts is isomorphic. If the environment is characterized by levels, two environments isomorphic only at certain levels of the hierarchy can be grouped together at those levels.

To summarize the results of the last few sections, the structure of the functional hierarchy is as follows: With no closed functional loops, the hierarchy is a composite tree composed as follows (see Fig. 7.3 and note at end of paper):

1. If the hierarchy is for a complex purpose, it consists of a number of 'simple-purpose' hierarchies 'strung together' at the top.
2. Each 'simple-purpose' hierarchy in turn consists of a number of hierarchies, one for each set of functionally equivalent environments of the topmost system, strung together to get the functional hierarchy for that system in the range of environments considered.
3. The hierarchy for each set of functionally equivalent environments for the topmost system node is a composite of trees that branch solely with respect to the system–subsystem relation and the environment–subenvironment relation for those systems, resulting in a single tree that branches simultaneously with respect to both relations.
4. Operations (2) and (3) are reapplied to nodes at the next level until the bottom-most nodes are reached in all branches.

Fig. 7.3. *(opposite)* Different elements of a schematic biological functional hierarchy
Notes:
[a] If a social unit is a functional unit, it should be represented as a single node rather than as it is here. (This node would be above that for organisms and below that for breeding populations.) (Added, 2000: At the time I wrote this I had not yet developed the views on how to choose the unit of selection that I had later (Wimsatt 1980*b*), so I supposed that the breeding population, if a social group, could also be a unit of selection.)
[b] Selection acts as feedback, which changes the structure of the hierarchy. It can act at different levels with different units and can affect traits at other levels. Since selection changes genes and gene frequencies, thereby affecting the behavior of genes in their genetic, somatic and environmental milieu, changes induced by selection are shown as acting downwards from the relevant unit of selection whose functions are served to a gene or genes as represented through their behaviors, and then propagating upwards from the bottom-most nodes in the hierarchy.
[c] The functional hierarchy does not normally include non-functional and dysfunctional behaviors and interactions, though these may be responsible for the presence of many functional subsystems, which are included and influence the form and architecture of the functional hierarchy in many ways. They are represented here to indicate that they can connect different sub-branches of the hierarchy.
[d] Systems individuated on non-functional criteria (e.g. genes, anatomical organs, and physiological systems) generally map into multiple different nodes on the functional hierarchy—they are pleiotropic for function.

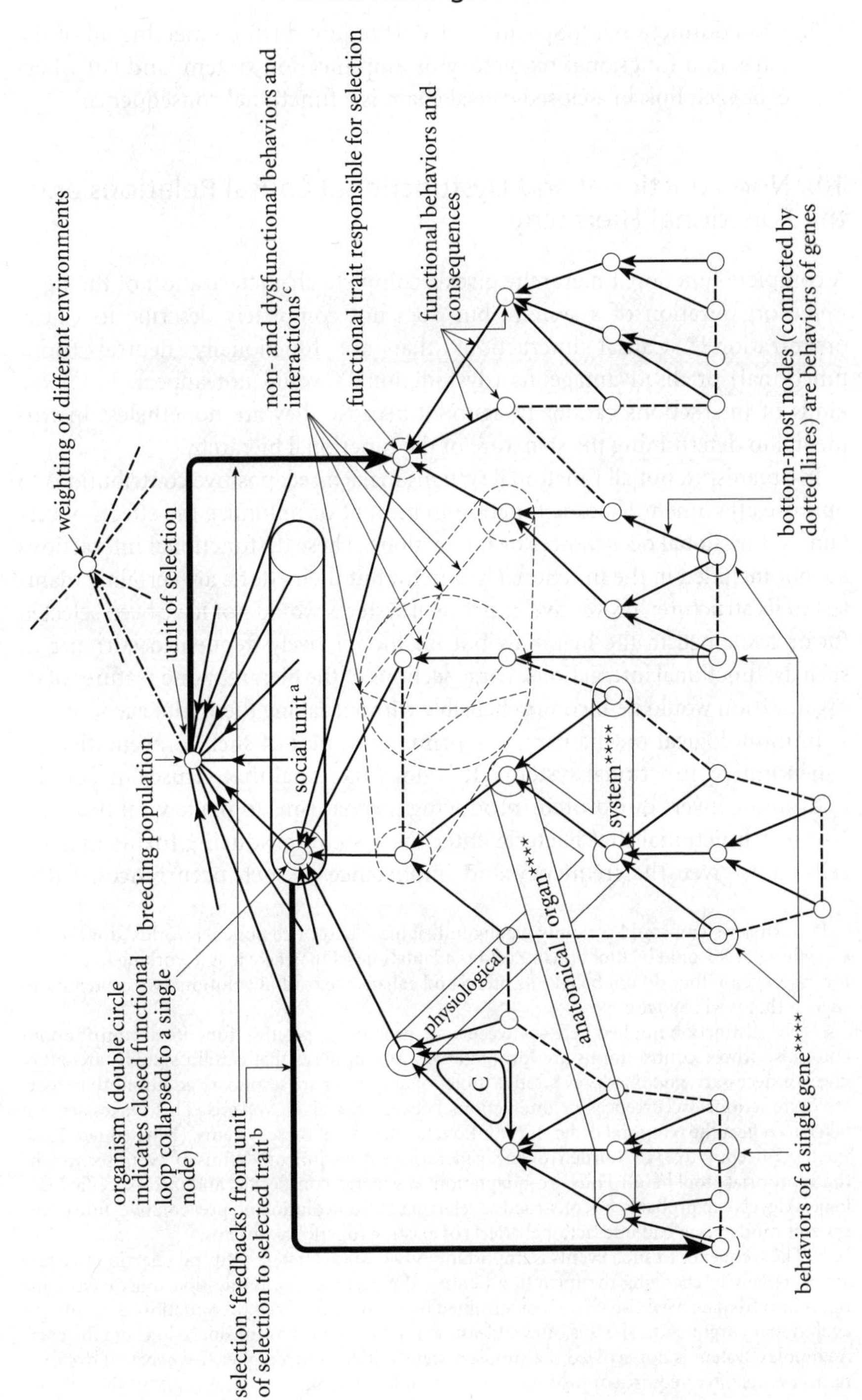
weighting of different environments
unit of selection
breeding population
social unit[a]
non- and dysfunctional behaviors and interactions[c]
functional trait responsible for selection
functional behaviors and consequences
bottom-most nodes (connected by dotted lines) are behaviors of genes
behaviors of a single gene****
system****
anatomical organ****
physiological
C
organism (double circle indicates closed functional loop collapsed to a single node)
selection 'feedbacks' from unit of selection to selected trait[b]

5. Closed functional loops are added as required (*a*) connecting all of the parts in a functional regulatory or amplification system, and (*b*) wherever each link in a closed causal chain is a functional consequence.

10. Non-Functional and Dysfunctional Causal Relations and the Functional Hierarchy

A complete functional hierarchy gives a complete characterization of the *functional* organization of a system, but does not completely describe its causal organization.[14] Causal interactions that are functionally neutral (non-functional) or disadvantageous (dysfunctional) would not appear. But these kinds of interactions cannot be ignored because they are nonetheless instrumental in determining the structure of the functional hierarchy.

In organisms, not all functional systems make direct positive contributions to purpose attainment.[15] Some function to prevent or minimize the effects of certain dysfunctional occurrences or interactions. These dysfunctional interactions are not included in the functional hierarchy, but their effects are certainly manifest in its structure: 'preventive' functional systems would not have been selected for or contained in the hierarchy but for the relatively frequent occurrence of such dysfunctional interactions. Large sections of the hierarchy and features of its organization would be incomprehensible without taking them into account.

Immunological mechanisms are prime examples of such 'preventative' or 'ameliorative' functional systems. It is not functional that viruses invade the body and convert our protein-producing mechanisms to make viral proteins. Nor is it functional that bacteria enter the system, producing toxins that kill cells. But, given the frequency and importance of such occurrences,[16] it is

[14] Such a system would certainly belong in Leibniz's 'best of all possible worlds', for literally everything in it would be functional. Zealous adaptationists may take it as a working hypothesis for traits even if they do not believe its literal and exhaustive truth. Evolutionary considerations suggest that it will be *false*.

[15] The distinction implied here (between systems making 'positive' functional contributions and those whose contributions are 'preventative') presupposes that certain systems are either given or necessary, and that 'preventative' functional systems are secondary additions to moderate dysfunctional occurrences or interactions between the given systems. This makes sense if taken to reflect the temporal order of evolutionary changes in these systems. (Note added, 1997: Riedl's concept (1978) of 'burden' or my 'generative entrenchment' (Wimsatt 1986) seem to be the appropriate tool here.) Thus, 'co-adaptation' at genetic, conspecific, and intra-specific (ecological) levels can probably be construed as referring to the evolution of 'preventative' functional systems moderating bad interactional effects of existing functional systems.

[16] The frequency of such events is important in two ways. First, for natural selection to operate effectively in changing the form of an entity, the occurrences in question must have some minimum frequency whose value is determined by the parameters of the situation. Secondly, in evolutionary arguments, there is often at least an implicit appeal to considerations of efficiency. A complex system is not evolved if a simple system will do, and a higher frequency of dysfunctional events often requires or justifies a more complex system.

functional to have a system—whose complexity reflects its importance—for neutralizing or disposing of viruses and bacteria. Similar remarks apply for various healing mechanisms.

It is not just the dysfunctional operation of foreign agents within the system that is left out of the functional hierarchy. There may be many functionally neutral traits of biological organisms.[17] Functional systems may interact in ways that are dysfunctional or antagonistic. None of these occurrences, traits, or interactions is included in the functional hierarchy.

In certain circumstances, the immunological mechanisms can become sensitized, not only to foreign proteins, but also to the body's own proteins. In this case, they can act to destroy the body rather than protecting it, and are interacting with other bodily functional systems in a dysfunctional manner.

With extreme heat loss in mammals the thermoregulatory system cannot keep the temperature of the whole body at an acceptable level, circulation to the extremities decreases radically, and their temperature decreases. The thermoregulatory system can thereby resist temperature decreases in the crucial organs. So normal functioning (under normal conditions) of the thermoregulatory system is impaired under extreme conditions to increase the ultimate chances of survival. (A man with frozen extremities does better than a man with a frozen brain.)

It would be better (other things being equal) if the thermoregulatory system had sufficient capacity to meet all emergencies without restricting circulation to the extremities, but any system has physical limitations, which are themselves products of selection. These make it unlikely that all causal interactions in a system are functional under all conditions. Interactions that are non- or dysfunctional, and the conditions under which they are non- and dysfunctional place constraints on the structure of the functional hierarchy even though they never appear there explicitly. Things that seem locally disadvantageous might in fact be globally advantageous—or the reverse. These are among the most difficult issues to resolve in complex functionally organized systems.[18]

[17] *To say that two alternative genes or genotypes are selectively neutral relative to each other is not to say that either is non-functional but rather that they are equally functional.* Substitution of one gene for another is a non-functional transition if the two are selectively neutral. Fitch and Margoliash have argued that many mutations causing amino-acid substitutions in cytochrome-c molecules appear to be non-functional transitions. Some biologists feel that the amount of polymorphism in enzymes found by Hubby and Lewontin (Hubby and Lewontin 1966; Lewontin and Hubby 1966) can be explained only if these polymorphisms are either selectively neutral or very nearly so. Even though there are now more recent treatments, the reasons behind this thinking—and why it has been difficult to resolve—are nicely reviewed in Lewontin, 1974.

[18] Added, 1997: As Buss (1987) nicely illustrates, this problem is compounded when one considers multiple possible levels and units of selection, and the possibility of intra-organismal competition among cell-lineages.

We must be methodologically attuned to the analysis of dysfunctional effects even when we are seeking only the functional organization of a system: the study of how the operation of a functional system changes 'when things go wrong' is a powerful way of determining its functional structure, but must be done with care and detailed consideration of all three types of causal interactions in the system.[19]

Any theory of the behavior of systems picks out certain key features and ignores others. For evolutionary or selectionist theories, the structure and nature of the functional hierarchy is of prime importance, and other features of the system are of theoretical importance only insofar as they affect this structure. That functional hierarchies do not capture all of the causal interactions in their systems is appropriate to the nature and modes of generalization of these theories.

II. Pragmatic Issues of Functional Hierarchies and the Logic of Functional Analogy

11. 'Real' versus 'Ideal' Functional Hierarchies

If 'function' is conceptually more closely connected with behavior than with objects (Section 3), and functional hierarchies have a logically more desirable structure when constructed in that way (Sections 4 and 5), why do we tend to attribute functions to physical objects and think of functional systems primarily as systems of physical objects?[20] This puzzling anomaly derives from practical considerations in how we elaborate and investigate functional hierarchies.

The conceptual priority of behavior over objects in the analysis of function does not give behavior observational priority. Behavior is always behavior of an object, and behavioral interactions are interactions between objects. It is objects and changes in objects that are characteristically observed—behaviors

[19] When a functional part is removed from a system and the system fails as a result to perform some task, it cannot generally be inferred that the function of the part is to accomplish that task. But this mode of inference has been widely used in attempts to analyze brain function via ablation experiments and behavioral function via deprivation experiments. Gregory (1962) has criticized ablation studies, and Konrad Lorenz (1965) has analyzed the shortcomings of a simplistic use of deprivation experiments. Gregory compares supposing that the function of a part of the brain is to accomplish something not done when that part is removed to supposing that the function of the spark plugs in an automobile engine is to prevent the engine from sputtering because the engine sputters more severely as more spark plugs are removed.

[20] Of course biological functional systems *are* systems of physical objects, but the best way of describing their *functional* organization may not cut a physical system up into physically recognizable (spatio-temporally compact) parts (Wimsatt 1974).

are reifications of these. It is because of this observational priority of objects that we attribute functions to objects and systems.

Most investigations of functional organization start by investigating the static physical structural properties of the system. The 'statics' of a system are generally easier to analyze than its dynamical interactions. Physics students are taught statics before dynamics, and Vesalius conducted detailed anatomical investigations while physiology was still an infant. Substantial knowledge of system structure is more easily acquired than, and must commonly *precede* even relatively primitive analyses of, its dynamical interactions. (This does not mean that knowledge of structure must always stay 'ahead' of knowledge of dynamical interactions, or that dynamical investigations cannot contribute anything to knowledge of structure, but only that, in the beginning of an investigation of a system, some minimum knowledge of structure must come first.) So it is not surprising that the ontology of a system's *static* structure—individual physical objects and their spatial relations—should be applied also to its *dynamic* structure, though with somewhat less success.

The structure of a functionally organized system is partially determined by its functional organization, and thus affords clues as to the form of this organization. Since static physical structure is characterized by the individuation and interrelation of physical objects, the simplest hypothesis concerning the relation of structure and function is that functions are in one–one correspondence with these objects. This hypothesis seems to be confirmed in some cases, especially at early stages in an investigation, but frequently turns out to be wrong. Yet, even after investigations have progressed to the stage that mutually exclusive functional systems are shown not to correspond to mutually exclusive systems of physical objects, the tendency persists to regard functional systems as systems of physical objects—but with 'something added'.[21]

To suggest that functional hierarchies should be regarded as functionally interrelated behaviors and systems of behavior does *not* mean that functional analysts can follow this recommendation in practice. One would expect just the opposite! While an ideal completely elucidated functional hierarchy might

[21] Vitalistic and anti-reductionistic thinking in biology and psychology was surely facilitated by the failure of functional systems to correspond to spatially compact and well-delineated physical systems. That functional organization can differ from the most readily observable physical organization—the organization of physical *objects*—does not mean that functional organization is not physical. Lashley's failures to localize specific functions in specific areas of the cerebral cortex led to the rise of Gestalt psychology, and cut ties between psychology and physiology for many years. Fodor (1965, 1968) and Putnam (1967) give equally fallacious arguments that functional and physical descriptions and analyses are intrinsically incommensurable (Wimsatt 1972). Their arguments turn on the fact that functional equivalence or isomorphism of two systems does not entail the equivalence or similarity of the physical *objects* in those systems. But there is more in the physical world than objects!

contain only behaviors of physical objects and no direct reference to the objects themselves, especially in the early stages, functional analysis must proceed in terms of systems of objects. Even in a world of processes operating on different time scales, a set of processes changing much more slowly than another set will appear as fixed and (if they have delimited and compact boundaries) as object-like relative to the more rapidly changing set of processes. As progress is made in analyzing the functional organization of a system, it is found more often that individual functions are performed jointly in a series or independently in parallel by more than one physical object in the system, and that individual objects in that system may perform a variety of functions.[22] This is to be expected, because modifications accumulated in design or selection processes should far more commonly have pleiotropic rather than monotropic effects.

These considerations or others deriving from issues of functional analogy or total or (almost invariably) various degrees of partial multiple realizability may make it increasingly conceptually profitable to construe functional systems as systems of behavior rather than as systems of objects. This will generally be done piecemeal in various branches and subbranches of the functional hierarchy. Thus, at any intermediate stage in the elaboration of a functional hierarchy, one could expect to find some nodes interpreted as behaviors and other nodes interpreted as physical objects. But, even after the analysis is finished, physical objects are not completely dispensable: it would be impossible

[22] This will not happen if the system is such that, in all environments and at all levels of analysis, there is a one–one correspondence between functions and physical objects. We rarely find this in natural systems, but it is more closely approximated in artifacts: common (but unsophisticated) design procedure separates functions of the proposed mechanism and designs individual parts or sets of parts to perform each function. This is an accident of our way of analyzing problems into sequential steps or operations. Minimization techniques have resulted in substantial increases in economy and reliability with given components by combining the functions of components. (See exposition of the Cowan–Winograd theorem on functional multiplexing in Arbib 1965.) Efficiency and reliability are presumably both selected for in evolutionary processes. It is thus not surprising that organisms do not exhibit one–one correspondences between functions and physical parts or systems of objects. (Added 1997: In the 1960s programming and design time were cheap and components expensive, so one sought to get the biggest computational bang out of the smallest number of components, and functional multiplexing of components seemed the way to go for increased efficiency in computer design. The invention of the integrated circuit, economies of scale in production, and Moore's law: the rough exponential doubling of speed computational power (and number of components on chips) every two years has made it easier to use standard CPU chips for a wide variety of tasks for which they are profoundly overqualified, and the excess power and memory capacity has made program speed and efficiency relatively unimportant in most applications. (And programmers who are paid by the line of code have no incentive to undertake additional work to construct compact code.) Thus at the chip level there may be an increase in one–one mappings between functions and components, though the components are now polyfunctional, and identical chips are programmed differently to perform different tasks. So, within different instances of a given chip type, the same component may be wildly polyfunctional. *Note that similar things can happen with polyploidy and gene duplication.*)

to verify that a given system exemplified a given functional hierarchy without analyzing the static and dynamical relations between various physical parts of that system.

12. Pragmatic Criteria for the Attribution of Functions

I have so far discussed two extremes in the elaboration of functional hierarchies. In early stages, with comparatively scanty information concerning static and dynamic characteristics of a system, one would expect nodes to be interpreted most commonly as (functionally) behaving objects. At the other extreme, individual *behaviors* in well-defined *systems of behavior* in given environments characterized down to the level of that behavior would be assigned nodes in a given hierarchy if their occurrence contributes under those conditions to the attainment of the purpose at the top of that hierarchy. The latter picture would reflect much more explicitly the conditionalizations and contextualizations of performance expected in any complex system.

But how are functions initially assigned to physical objects, and how are they assigned and individuated in the long course of successive analyses that lead to the completed 'ideal' functional hierarchy? Functional systems are kinds of machines, whose articulated parts contribute to the ends specified by a selection process—whether internal or external and whether of natural or artificial origin (Wimsatt 1972). Roughly, each functional item must produce or contribute to its functional consequence in the appropriate circumstances, and, percolating up through the functional hierarchy, that must contribute to purpose attainment under appropriate circumstances. Explicating this brings in all of the variables of the function statement—B(i), S, E, P, T, and C. This is the primary criterion. But there are additional heuristic criteria that are extremely useful. They usually do not identify a function, but make it likely that something has a function, or—for (3)—place some constraints on what can be identified as the function. Thus:

1. the complexity of a system is assumed to correlate with the complexity either of the task it is to perform, or the complexity of the way and conditions under which it is to be employed—roughly, with the internal and external complexity of the performance.

And, since it takes some effort to maintain complex systems,

2. the elaboration and persistence of a complex system indicates an importance commensurate with the resource commitment necessary for maintaining it.

And, since we assume that any complex system will be constructed piecemeal,

3. the story told for the function of the system must be consistent with the piecemeal elaboration of that function, and most commonly with its elaboration and successive co-option—as a repeated exaptation through a succession of other initially exaptive functions (see Section 22 below);
4. persistence of traits through diverse variations in related families over extended periods of evolutionary time makes it more likely that they or their causal antecedents have widespread downstream functional consequences dependent on them. This inference is strengthened for traits that are vestigial, not clearly functional, or whether obviously functional or not, that occur earlier in development.

Note that, on this analysis (unlike that of Wright 1973 or Millikan 1984), these are heuristic and pragmatic requirements—not logical ones. And they are cumulative in their effects, indeed in a non-linear fashion—for example, a complex (criterion 1) vestigial trait gets more credit under criterion (4) than a simpler one.

Criteria (3) and (4) are related to processes of generative entrenchment, which are virtually unavoidable and important consequences of the piecemeal and exaptive means of assembly of complex adaptive systems. This has deep consequences for the architecture of complex adaptive systems, and limitations in the manner and degree to which they can be said to be optimal. See Wimsatt (1986, 1999, and especially 2001), Schank and Wimsatt (1988, 2000), Wimsatt and Schank (1988, 2002). For the fullest development of arguments involving generative entrenchment in biology, see Arthur (1997).

These four criteria are evidence that functions are served by the features in question, but give no clues about which functions are served. We frequently attribute specific functions on the basis of analogy in the broadest sense. This includes both homology and analogy as those terms are used by functional morphologists, and also simple physical similarity. These suggest specific functions on the basis of knowledge we may have of other systems. They are particularly important for cross-species, cross-phylum, or even broader comparisons of functional similarity. These analogy-based pragmatic criteria for the attribution of functions are:

5. similarity of the given object, set of objects, behavior, or set of behaviors with another object, set of objects, behavior, or set of behaviors with respect to *physical* features. Where the function of the second entity is known, the same or a similar function is attributed to the first.
6. similarity of the given object, etc., with another object, etc., with respect to *functional* features, where the function of the second entity is known, the same or a similar function is attributed to the first entity.

These last two criteria are partially independent and can conflict. Darwin (1876) describes a species of orchid in which an interior part of the flower resembles the female reproductive apparatus of the species of bee responsible for its pollination. In his abortive attempt to mate with the flower, the male bee picks up pollen, which he carries to the site of his next deception. We disregard the similarity of parts of the flower and female bee on what are ultimately functional grounds. (Bee–flower matings do not produce fertile offspring.) A functional analogy, the fact that this adaptation attracts bees (as other flowers do with nectar or special color markings), is employed to place this adaptation in the orchid's functional hierarchy, rather than the bee's, and to identify it as connected with the function of pollen dispersal. Visual similarity of this part of the flower with anything else to be found in nature gives misleading clues as to its function.

Are physical or functional similarities more conservative in evolution? This is an unanswerable question at this level of abstraction. Traditional focus on deep functional requirements, and more modern focus on generative entrenchment and developmental constraints suggest functional roles as more conservative, but, as these become more superficial, the questions transform to questions about the stability of niche relations and the answers are more conditional. Gould's emphasis on exaption might suggest lower stability of functional roles in evolution. This question is more important to consider than to answer, since it is strongly context dependent and without a general answer. The degree of generative entrenchment of the elements in question is probably more important, and here relative ages are more often the given data we work with than the answers to be found out.

I have nothing more to say about physical similarity, but more should be said about functional similarity—especially of the practical variety employed at stages where the functional hierarchies of the various systems being considered are relatively incompletely elaborated.

13. Varieties of Functional Similarity

We talk freely and almost interchangeably about functional identity, equivalence, isomorphism, correspondence, and analogy. But a number of concepts of functional similarity can be defined in terms of functional hierarchies. Without giving all possible combinations, I outline the kind of considerations that can act as distinguishing dimensions of these modes of comparison. I then discuss the weaker forms of functional similarity that seem to be applied in practical cases.[23]

[23] For simplicity of exposition, I sometimes talk as if the functional hierarchy has a tree structure, even when it does not. Most discussions below are not affected by this. Of those affected by the presence of closed functional loops, most are modifiable in relatively simple ways.

In general, the entities said to exhibit functional similarities are entities whose functional structure is represented by *parts* of the functional hierarchy. A *part* of the functional hierarchy is a subset (not necessarily a *proper* subset) of the nodes in the hierarchy together with some or all of the arrows connecting these nodes such that sufficient arrows remain in the part to connect all of the nodes in that part.

These 'parts' might correspond to behaviors, systems of behavior, physical parts or systems of physical parts, functional consequences and sets of functional consequences, or non-homogeneous combinations of these. They will show at least some integration because of the connectivity condition. I consider environments and sets of environments as well, for these also correspond to 'parts' in this sense, even though they might not generally be considered to be parts of the functionally organized system on spatial criteria.

All varieties of functional similarity as defined here share one thing in common: they are at least partially defined in terms of isomorphisms—between the parts in question, between parts of the parts, between other parts of the hierarchy of which the parts are parts, or between other hierarchies or parts of hierarchies of which the parts could be parts. By *isomorphism* I mean that the nodes and arrows of the two structures said to be isomorphic can be placed in one–one correspondence such that each has the topological properties of the other when the arrows are interpreted as *directed* arrows.

Four kinds of considerations are involved in talking about types of functional similarity.[24] These are:

1. conditions of isomorphism;
2. conditions on the similarity of purpose;
3. conditions on the scope of isomorphisms;
4. conditions on partial functional similarity.

These are four *classes* of conditions. One condition in each class must be met for each type of functional similarity. With the possible exception of some conditions in class 3, all alternatives within a given condition are mutually exclusive. Except as noted below, all are also logically independent. With x, y, z, and w independent conditions in the four classes, we have (x.y.z.w) possible distinguishable types of functional similarity—a large number. But the conditions themselves are of more interest than a lexicon of their combinations.

[24] A fifth class of conditions is naturally suggested by the function statement schema—namely, conditions on the isomorphism of theories used in constructing the functional hierarchies. For cultural evolution, ideas have been compared to conceptual viruses, assuming that both can be treated as kinds of parasites, whose horizontal spread is well modeled by epidemiological equations. (Note added, 1997: Given elaborating theories of cultural evolution, this may become a richer source of functional analogies than many of the others (particularly at more molar levels, since most sets of formal equations are relatively non-specific).)

14. Conditions of Isomorphism

Isomorphisms can be studied under two kinds of conditions. The first and stronger condition is that the two parts in question be *intersubstitutable* in a set of hierarchies, a given hierarchy, or part of a given hierarchy without disturbing the structure of that set, hierarchy, or part. This involves the isomorphism of a set of hierarchies, a hierarchy, or a part of a hierarchy with itself before and after the substitution. It is appropriate when considering spare or replacement parts in engineering contexts, or in cases where there is internal redundancy of organs or genes, or functional equivalence of different alternative courses of behavior.

Alternatively, and more commonly in natural biological contexts, one can ask whether two sets of hierarchies, two hierarchies, or two parts of the same or of different hierarchies are isomorphic to each other. This might be called *comparative* isomorphism, and is the sense of functional equivalence or similarity appropriate in most evolutionary contexts, and essentially all cases of 'comparative functional morphology'.

Two entities can be comparatively isomorphic without being isomorphic under substitution, but not conversely. Thus, consider two computers of identical logical structure. This logical structure in the first is realized via electronic switching elements and in the second by hydraulic 'fluid logic' components. Presumably, the functional hierarchies of these two computers would be comparatively isomorphic. Experimental 'fluid logic' computers were considered and simple partial prototypes built in the late 1950s by the US military because they would be more resistant to radiation effects that could 'fry' electronic computers in missile nose cones. The IBM 709 and 7090 had a similar relationship: they had the same circuit logic diagrams, which the 709 realized with tubes and the 7090 with transistors. The greater volume, heat production, and power and voltage requirements of the first entailed different provisions for space, power, and air conditioning. The greater reliability and lower cost of the transistors produced different maintenance schedules and purchasers, and the faster switching speeds of the transistors meant that the 7090 could accomplish tasks that the 709 could not if online prediction was required. Collectively, these meant that the whole functional architecture of the computer installations were not identical, however similar their logic circuits. In any case, one could not simply replace a worn-out part in one computer by the corresponding part in the other computer and expect the new part to function in the same way as the part it 'replaced'. (One could of course include special 'translating' devices—electrical/hydraulic and hydraulic/electrical 'transformers' and add the relevant power supplies, but (as anyone with an artificial heart or kidney knows), this would require a substantial change in the supporting functional organization of the whole installation, and thus also destroys full isomorphism under substitution for that case.)

If two entities are functionally intersubstitutable under a set of conditions, they are also comparatively isomorphic in corresponding systems. Mass-produced parts are designed to be intersubstitutable and two physical systems of the same type produced on an assembly line with the same corresponding intersubstitutable parts will clearly be comparatively isomorphic.

15. Conditions on the Similarity of Purposes

Comparison of structure within a hierarchy places no explicit constraints on the nature of the purpose associated with the hierarchy.[25] Yet the nature of the purpose clearly has some relevance in comparative judgments of functional similarity. Among possible constraints on two purposes being compared, three come readily to mind.

1. The purposes at the top of the hierarchies being compared are *identical.*
2. The purposes at the top of the hierarchies being compared are *similar in respects* P_i.
3. The purposes at the top of the hierarchies being compared are *dissimilar in all respects.*

These three conditions are arranged in order of decreasing functional similarity, but the apparent precision of this list is misleading. I have not discussed general criteria for the identity or similarity of purposes, and do so here only in rough terms, illustrated with a few examples.

Purposes might be characterized either extensionally (listing the set of states that count as purpose-attainment states) or intensionally (by listing the criteria for purpose-attainment states.) The second way seems generally preferable. I suppose that two identical purposes should have criteria for purpose attainment that are identical in meaning.[26] For scientific applications, one might weaken this requirement somewhat, to say that two purposes are identical if and only if their criteria for purpose-attainment mutually entail each other according to all of the relevant scientific theories for all ranges of conditions within the scope of these theories (an approximation to saying that they have the same 'scientific meaning?').

[25] This set of conditions is not entirely independent of the first set. Isomorphism under substitution would seem to imply that the purpose(s) associated with the hierarchy/ies in which the substitution is performed are identical before and after the substitution. So the full range of these conditions would apply only for isomorphism under comparison.

[26] Mere extensional equivalence of purpose-attainment states will not do. I doubt that extensional equivalence would even guarantee that the purposes are even intuitively similar.

Indexical conditions matter. Consider two persons playing a game that only one of them can win. In some sense, each has the *same* purpose—each wants to win the game. But it is problematic to treat two purposes as identical if they can conflict. So identity of purposes has an indexical component. They both have the same purpose only if they both want the *same* one of them to win the game (though not too obviously, as a father and son might when they are playing chess), and naturally then both of their plays are parts of the same functional hierarchy. Similarly, two evolutionary units, X and Y, may have 'the' purpose of maximizing their probability of survival, but these are two distinct (though similar) purposes.

Situations where purposes appear without an implied indexical reference might be called *general* purposes. When someone says that the purpose of *a* car is to provide transportation, of *a* lathe is to machine metals in axially symmetric ways, of *a* game-player is to win the game, or of *an* evolutionary unit is to maximize its probability of survival, this purpose is held to be the same for different cars, lathes, game-players, and evolutionary units. Such general purposes refer to what is common to a class of indexical purposes defined by the characteristic that if they had the same indices they would be the same indexical purpose. General purposes are types of purposes that are characteristic of certain kinds of objects.[27]

There are various ways in which purposes can be similar. One important way could be that they are both indexical specializations of the same purpose-type. Other modes of similarity (applying both to indexical and to general purposes) might include:

(*a*) for complex purposes (see Section 8 above), sharing some of the criteria for purpose attainment, so 'seeking advancement and power' and 'seeking advancement and pleasure' have 'seeking advancement' in common;

(*b*) for either complex or simple purposes, criteria for attainment of two purposes may share key concepts, thus, for example, 'the survival of the organism', 'the survival of an evolutionary unit', and 'the survival of the state' all make use of the concept of survival. 'Homeostasis', 'adaptation', and 'utility' represent other key 'portable' concepts commonly found in purpose specifications.

This last type of similarity is a sort of higher-level generalization of general purposes, where the types of systems are no longer 'organisms', 'evolutionary units', or 'states' but something like, for example, 'definable units that tend to

[27] Presumably, criteria for identity and similarity of two purpose-types would also be unpacked in terms of logical relations between their sets of criteria, but with no indexical references.

organize themselves in such a way as to increase their probability of survival'. This is a narrow interpretation of (*b*), however, for key concepts need not occur in parallel ways in the different purposes. Thus 'maximizing homeostasis', 'minimizing homeostasis', and 'maintaining the degree of homeostasis at a constant level' are all possible purposes using the concept of homeostasis.

This list of ways in which purposes may be similar is meant to be suggestive, not exhaustive. A closer examination of anthropomorphic and anthropocentric reasoning could turn up other ways in which purposes can be similar.

16. Conditions on the Scope of Isomorphisms

One obvious way of generating different concepts of functional similarity is to individuate different requirements for the scope of the isomorphisms that must be present. In some of these requirements, the definitions of scope will be treated as if the functional hierarchy had a tree structure. Where appropriate, modifications necessary to generalize the definitions to take account of the presence of functional loops are indicated.

The first six requirements of scope (three basic requirements, applied to isomorphism under substitution and to comparative isomorphism) are as follows:

(S.1) *Absolute substitutional isomorphism.* Two parts are intersubstitutable in any hierarchy in which *either could* occur without changing the structure of that hierarchy.

(S.2) *Absolute substitutional homomorphism.* One of the parts is intersubstitutable in any hierarchy in which the *other could* occur without changing the structure of the hierarchy.

(S.3) *Substitutional isomorphism relative to hierarchy H_i.* Two parts are intersubstitutable in hierarchy H_i without changing the structure of that hierarchy.

(S.4) *Absolute comparative isomorphism.* For each hierarchy in which *either* part could occur, there is a corresponding hierarchy that is isomorphic with the first and in which the other part occurs as a corresponding part.

(S.5) *Absolute comparative homomorphism.* For each hierarchy in which *one* of the parts could occur, there is a corresponding hierarchy that is isomorphic with the first and in which the other part occurs as a corresponding part.

(S.6) *Comparative isomorphism relative to hierarchies H_i and H_j.* Hierarchies H_i and H_j are isomorphic, and the two parts are corresponding parts of these hierarchies.

These requirements of scope are so strong that most are of little use in practice,[28] or met only in certain limiting cases.

'Exactly similar' parts of the same materials on a production line presumably approximate to the ideal of absolute substitutional isomorphism, though minor variations from piece to piece belie the terminology. The practice of specifying tolerances for holes, diameters, and crucial dimensions for parts that must fit together increases ease of fit. So does the further practice of having a box full of the relevant parts being assembled—not every nut has to fit every bolt; there just has to be enough overlap in the size distributions that a tolerable fraction of the part pairs fit together on the first try, and there are not too many that do not fit anything. Since a cost–benefit calculation determines the tolerances to which the parts are manufactured, increases in labor costs can lead to the institution of finer tolerances for part manufacturing. Such probabilistic tolerances and cost–benefit trade-offs are endemic to evolutionary biology as well.

Various weakened versions of this constraint (for example, 'exactly similar' machine parts made of different materials, such as steel and brass, or the 'corresponding' parts of the electronic and hydraulic computers) might meet some of the weaker conditions on this list, though this depends in part on what is taken as the functional hierarchy. (A brass part will fail under mechanical and temperature loads that a steel part will sustain, and the two metals have different electrical conductivities. A hydraulic computer is slower than the corresponding electronic computer, but might be used in the absence of an electrical power supply. If the hierarchy is taken for a relatively restricted range of environments, these differences might not show up.)

Below, I include formulations for isomorphism under substitution and comparative isomorphism so that their differences could be seen, but generally I discuss only isomorphism under substitution since the definitions are simpler. Since a corresponding sense of isomorphism under comparison applies and could be given, I 'double number' each definition.

17. Conditions on Partial Functional Similarity

Assume that the functional hierarchy has a tree structure. (If whole cycles are treated as nodes, hierarchies with cycles assume a tree structure, though, if

[28] Several things of formal interest are worth noting. (i) (S.1) implies (S.2), which implies (S.3) and similarly for (S.4), (S.5), and (S.6). None of the converses holds, however. Also, (S.1) implies (S.4), (S.2) implies (S.5), and (S.3) implies (S.6), but not conversely: 'substitutional' isomorphism implies but is not implied by the corresponding 'comparative' isomorphism. (ii) The strongest condition, (S.1), certainly deserves to be christened 'structural functional *identity*'. If the quantitative indices (see Section 21) also remain equal before and after the substitution of parts, it should be named 'functional identity'. This last sense amounts to the satisfaction of Leibniz's law of the identity of indiscernibles for all *functional* properties.

there are many cycles, or any very important ones, most of the functional complexity will be collapsed to a single (very complex) node.) This fact allows defining structural properties of hierarchies with cycles relatively simply in ways suggested by definitions of corresponding structural properties of trees.

The necessary concepts are that of a path between two nodes, the distance between two nodes, the topmost node, above, below, and betweenness for nodes, the branch and branch complement of a node, and a super-branch, super-branch complement, sub-branch, and sub-branch complement of a node. All of these concepts are informally characterized below. (These and the following may be more formally defined using graph theory. See Harary *et al.* 1965, or the definitions in Wimsatt 1971.)

A *path* is an ordered sequence of connected alternating nodes and arrows traversed in the direction indicated by the arrows. A node or arrow is *between* two other nodes or arrows if there exists a path from one of these latter nodes or arrows to the other that includes the given node or arrow. The (or a) *topmost node* of the hierarchy is the (or a) node that has no paths leaving it. A node or arrow is *below* a second node or arrow if a path from the first node or arrow to the (or a) topmost node contains the second node or arrow. The second node or arrow is then *above* the first.

The *distance* between two nodes or arrows is the number of arrows in the shortest path between them. The *length* of a path is the number of arrows in that path. The *level* of a node is the number of arrows in the path from that node to its topmost node.

The *branch* of a node is that node together with all nodes and arrows below it. The branch of a node is a *super-branch* of the branch of any node below it, and the latter branch is a *sub-branch* of the first. The *branch complement* of a node is what remains of the hierarchy when that node's branch is removed. The *order* of a sub-branch or super-branch is the distance between the branch-node and the super- or sub-branch node. A *super-branch (or sub-branch) complement of order n* of a node is what remains of the super-branch of order n (or branch) when that branch (or sub-branch of order n) is removed.

18. Internal, External, and Total Functional Similarity

A natural way of classifying isomorphisms in the same hierarchy (under substitution) or in different hierarchies (under comparison) is to ask where they occur relative to a certain node or set of nodes. A given node in the hierarchy corresponds to a functional system in a class of functionally equivalent environments. The nodes below it correspond to its parts at various levels of functional analysis. So isomorphisms in the branch of that node refer to *internal* functional features of that system, and isomorphisms elsewhere in the hierarchy or hierarchies refer to *external* functional features. Isomorphisms of internal functional features of systems imply similarities in the manner in which they produce their functional consequences. Isomorphisms of external functional features imply similarities in the role their functional consequences play in larger systems of which they are parts.

A physical part will in general correspond to a set of nodes in a functional hierarchy, with one distinct node for each of its functional behaviors. Examples of internal and external functional similarity for physical parts can be drawn from an examination of the design of different internal combustion engines. Two- and four-cycle reciprocating engines, Wankel (rotary) engines, turbojet, ramjet, and pulsejet engines, and rocket and diesel engines can all be designed in such a way as to use spark plugs for ignition. These spark plugs could all be internally functionally similar (they 'operate' in the same way) but not externally functionally isomorphic in these different types of engines. On the other hand, any of these different types of engines could be designed to use surface electrode or gap electrode spark plugs or 'glow' plugs. These three different types of plugs are not internally functionally similar (they operate in different ways) but would play the same external functional role in any given type of internal combustion engine.

If two entities are both internally and externally functionally similar, they will be said to be *totally* functionally similar. This obviously corresponds to (S.3) for substitutional isomorphism and to (S.6) for comparative isomorphism.

More explicitly:

(S.7, 8) *Internal functional similarity.* A hierarchy is *internally* functionally isomorphic under substitution with respect to a node or set of nodes if it is self-isomorphic before and after the substitution in all of the branches of that node or set of nodes.

(S.9, 10) *External functional similarity.* A hierarchy is *externally* functionally isomorphic under substitution with respect to a node or set of nodes if it is self-isomorphic before and after the substitution in the branch-complement of that node or in the intersection of the branch-complements of that set of nodes.

There are at least two essential ways in which it may be important to consider a set of nodes in these judgments of functional similarity. First, we may want to talk about functional similarities of two physical parts in a given environment or set of functionally equivalent environments. Because this part may exhibit several functional behaviors in different functional systems, it is necessary to consider the nodes corresponding to these behaviors. Secondly, it might be desirable to consider the isomorphisms of two given functional behaviors over a range of functionally different environments. This again brings in a set of nodes. These two investigations may also be combined to investigate functional similarities of parts over a range of functionally different environments.

Internal and external functional similarity are still extremely strong requirements, but various weaker forms of internal and external functional similarity can be defined.

19. Relative Internal and External Functional Similarity

There are richer ways to elaborate kinds of functional similarity that bear centrally on notions of functional equivalence and multiple realizability. Using the concepts of a sub- and super-branch, sub- and super-branch complementation, and the order of a sub- or super-branch of sub- or super-branch complement, we can define various forms of relative internal and external functional similarity within a given hierarchy. These are defined in Wimsatt (1971), and I give only an informal discussion of them here.

These types of functional similarity are all ways of ignoring the whole functional hierarchy and looking at functional similarities at places defined by their location relative to the node in question. The senses of relative internal functional similarity can be viewed as the reapplication of external and internal functional similarity respectively, within the branch of a node (instead of within the whole hierarchy) relative to a sub-branch of that node (instead of relative to the branch of that node). The two senses of external functional similarity apply total and external functional similarity, respectively, within a super-branch of that node instead of within the whole hierarchy.

Intuitively, functional similarity refers to the form of the branch complement of a node—external functional similarity, or similarity of functional role. Functionally equivalent things are things that can be plugged in at a node and make no external difference. If the similarity is 'higher level' or 'less detailed' or 'more abstract', the similarities may be in the form of a super-branch complement of the node. Each super-branch has a sub-branch with which it covers the whole hierarchy, so a higher-level functional abstraction is ignoring functional detail in a larger number of branches or functional subsystems of the hierarchy.

Multiple realizability in this context is the recognition that there may be functional equivalents. Functional equivalents may realize their functions differently as long as they interface with the higher-level functional systems in the same way. So their internal functional organizations may differ. But functional equivalence is almost always approximate, or partial—only along limited dimensions of comparison, in the real world. So-called open architecture computer systems may have functional equivalents at various levels—or fail to be quite 'open' (or have components that are not quite functionally equivalent) because of failures that may occur at different levels—exploring the depths of sub-branch complements of a node. Computer magazines test new operating systems by running different applications written for the older operating system, putting each through its paces to determine whether it breaks down, and under what conditions. What results is a variety of co-evolutionary 'patches' or fixes—published alike by hardware and software applications manufacturers, and by the computer manufacturer in later 'bug fixes' for that release of the operating system. And there are some bugs—transients—that never recur reproduceably enough to find, or to analyze. Happily, most of these are also rare enough that they can be ignored.

We act as if there is frequent functional equivalence, and that external functional structure places no constraints on the internal functional structure (so we can mix and match, and are constrained only at the interface). But this is true—even approximately—only in simple cases. A branch is a functional subsystem, and the operation of separating a system into a branch and its complement is naturally decomposing a system along functional lines at one particular functional joint. But, in complex systems, functional organization rarely corresponds to the organization of physical components (Wimsatt 1974), and parts of physical components tend to interact relatively strongly with one another. So, the more complex the functional organization, the more interactions tie internal and external functional structure together. (See Bechtel and Richardson 1992 on complex localization.) Functional equivalence exists for parts in complex systems too, but is usually a product of detailed fine-tuning and co-evolution, and still usually only partial at that. Everyone trying to configure a new computer system wrestles with functional equivalence, near-equivalence, or non-equivalence at various levels. The development of 'compatibility', 'transportability', and 'transparency' in computer software and hardware are some of the finest engineering examples of this available today: a closer look shows just how difficult this ease is to generate. A recent review by Bechtel and Mundale (1999) raises interesting and important questions about the extent of supposed functional equivalences and multiple realizability in the same and different brains, and how these decisions are presupposed by and impinged upon by our very methodologies for studying them.

20. Other and Weaker Forms of Functional Similarity

Biologists doing comparative studies of organisms or anthropologists comparing societies would hardly ever have any chances to apply concepts of functional similarity stronger than the various relative senses of internal and external functional similarity. But even these concepts are usually still too strong in several respects. First, in real cases, judgments of functional similarity take place when the various functional hierarchies and branches are still only partially known. Secondly, such judgments often seem to ignore the structure of whole portions of the tree in ways that fail to match the distinctions made above. In what ways can one afford to ignore portions of the tree in making these judgments: is there any systematic way of describing the 'pruning' of nodes and branches implicit in making these comparisons? At least three patterns emerge. These might be called 'path similarity', 'level similarity', and 'opportunistic pruning'.

1. *Path similarity.* Most cases of claimed functional similarity in biology (where based upon functional and not directly upon physical similarity) involve nothing more than isomorphism of the *paths* (or parts of them) from the node in question to the topmost node in the hierarchy or from a lower node to that node, or both. Such criteria are often strengthened by judgments or presuppositions that several or many of the paths in the corresponding hierarchies are isomorphic. They may be weakened by 'opportunistic pruning' (see below) to make the paths isomorphic.
2. *Level similarity.* One could also consider isomorphism (before or after opportunistic pruning) of the hierarchies only at, above, or below a certain *level* of the hierarchy, where level could be defined either in terms of the hierarchy or relative to the node in question.[29] This presupposes that the tree is sufficiently well elaborated for the levels of various nodes to be defined.

There are also criteria for assigning levels in a hierarchy which do *not* stem from its formal organization, and that sometimes correspond to biological practice. (I discuss levels of organization (and another crucial mode of molar organization, the niche-like 'perspective') in detail, involving relative frequency and time scale of interaction as well as size scale, in Wimsatt 1994.) The first of these criteria associates different levels with the various distance scales characteristically associated with a set of phenomena. Successive levels might be indicated by a sodium ion, inorganic molecules, various small protein molecules (such as insulin), sub-cellular parts, such as ribosomes, golgi apparatus, and mitochondria, and so on up through cells, tissues, organs, system of organs, organisms, populations, and the like.

Different levels are characterized by changes of scale of an order of magnitude or more, though not all parts and effects at a given level involve the same distances, and the 'levels' deal with continuously distributed overlapping scales for objects and effects. Cell sizes may vary over several orders of magnitude. These inconsistencies may arise because biological parts are classified according to type on functional criteria, or because they have similar parts, with no

[29] These can all be characterized more precisely in graph-theoretic terms. See Wimsatt (1971).

intrinsic reference to physical size. Frequently, too, they are classified according to physical inclusion in physical systems. Thus, all cells are regarded as being at the same level, as are all sub-cellular parts, such as ribosomes and mitochondria. The first reflects the influence of functional criteria and the second those of physical inclusion.

This kind of talk about levels is also prejudicial to talking about functional systems as physical objects, since sizes are more readily associated with objects than with behaviors. (However, one can associate sizes with behaviors by considering their 'ranges of effect', as we do when we talk about 'close interaction forces' or 'contact adhesion' or, less obviously, heat- and stretch- sensitivity or phototaxis.) Classification by levels is also useful at early stages in a functional investigation because it involves little by way of theoretical commitments, and one cannot have theoretical commitments before one has theories!

Perhaps more sophisticated assignments of levels to phenomena could be derived by classifying phenomena according to the level of theory most appropriate to their description. Sodium ions are most appropriately studied at classical chemical and the lower biochemical levels, hemoglobin and DNA molecules at levels involving more complex interactions, and tissues at the more classical anatomical, histological, and physiological levels. For reasons given in my 1994, I doubt that this approach will work generally: there is here also interpenetration of levels, and the various theories (where one can even say what they are) are not always in well-defined relations in terms of level. It would be easy to suppose that, if one theory is *reducible* to another, then the first is at a higher level than the second. But as yet we have no formal reductions of one theory to another in this realm. There is still a wide diversity of opinion as to what is to be meant by reduction (see Wimsatt 1976, 2003; Sarkar 1998). In effect, *talk* about levels of organization in terms of this criterion is often accompanied by an *actual* use of criteria of relative size, functional similarity, or physical inclusion.

Phenomena can also be classified by combinations of criteria reflecting the limitations of a technique. Thus, starch or gel electrophoresis classifies proteins according to their 'electrophoretic mobility', where this is, to a first approximation, a function only of their size and net charge. To be 'visible' in an electron microscope (of the transmission type), an object must be within a certain size range, of no greater than a certain thickness (a function of the material involved), and exhibit a certain minimum stability when bombarded by electrons of the energy and intensity used, and when treated with the various heavy metals used in 'shadowing' electron microscope specimens. But this would rarely yield a classification by level. Another means more likely to pick out levels is to classify the level of a phenomenon by what modes of organization you are permitted to 'disrupt' while studying it (Wimsatt 1997*b*).

3. *Opportunistic pruning.* Even with the degrees of freedom inherent in these different forms of functional analogy, I suspect that many judgments of functional analogy cannot be analyzed in terms of any single *a priori* systematic rule for pruning nodes or branches or for considering just various sub-portions of the hierarchy. In many cases, functional hierarchies or parts of them are *not* carefully compared on a point-for-point basis. There is just a 'feeling' or 'gestalt' that two functional hierarchies 'look the same' in some part or parts of their structure—even if closer inspection were to show that none of the senses of functional similarity defined so far seem to apply. There is no direct way to get a handle on such judgments, but their *effect* is just as if two functional hierarchies were selectively pruned until they or some portions of them were isomorphic. But almost any functional hierarchies, if 'cut back' enough, would exhibit isomorphisms on this criterion. So there must be informal and tacit criteria concerning the specific comparisons at hand and what portions of the hierarchies can be ignored or pruned.

Judgments as to level (both theory and size relative) probably enter, as do judgments of the relative functional importance of various branches and nodes. (This latter idea is important and will be treated in the next section.) Ultimately some of these judgments must be laid simply at the doorstep of the intuitions that come with expertise. Helmer and Rescher (1958) have made at least a start in the analysis of the role of 'judgments of expertise' in science.

21. Quantitative Criteria for Functional Similarity

It is profoundly misleading to talk about functional similarity just in terms of discrete categories. Isomorphisms of structure, the conditions of these isomorphisms, and the similarity of purposes associated with these hierarchies obviously have quantitative dimensions. Things are not only functional, nonfunctional, and dysfunctional, but they clearly are more or less functional and dysfunctional.

The structure of the functional hierarchy is determined by interrelations between behaviors regarded as functional, but some are so central to the operation of a system that their absence or malfunctioning would, under almost any conditions, prevent purpose-attainment by that system. (Any loss of these functions is an unconditional lethal.) Deeply generatively entrenched features usually have this status. Other behaviors might have so little importance that they make no difference at all under most conditions and only a small contribution in the remainder. Should the normal operation of the heart be given the same status as the normal growth of a fingernail?

We must be able to make comparative judgments about relative magnitudes of functional contributions. Sometimes selection coefficients may adequately measure 'degree of functional contribution', but other times—more commonly—no such simple standard is to be had.[30] Some problems in defining such a measure or making the relevant functional comparisons are considered in Wimsatt (1971 and 1972). Suppose these problems are solved. The functional hierarchy includes only functional behaviors and systems of behavior, so suppose that each node in the hierarchy is assigned a value of an index of functionality between 0 and 1. Consider two corresponding nodes in three distinct, but otherwise similar hierarchies. In the first, they have values of 0.9999 and 0.0001. In the second, they both have values of 0.5000, and, in the third, values of 0.0001. (Fisher treated selection coefficients down to this value as significant.) Could corresponding nodes with such divergent values represent very similar components? Not a chance! These three systems must be quite dissimilar in functional architecture, in spite of their formal isomorphisms. So we cannot ignore quantitative considerations.

Are there further constraints secondary to structural isomorphism? No. Quantitative relations should clearly in some cases *overwhelm* judgments of structural isomorphism in deciding to which of two parts a third is more similar. The magnitude of effects matters: large differences in selection intensities acting on a trait can have a variety of effects on other linked and unlinked traits, and may entirely change the course of evolution for that system (Wade 1996).

And we should not merely be concerned to delete nodes. If we can delete a node of 0.001 in one hierarchy to make a comparison, we could just as well add a node of 0.001 in the other. Without clear historical or functional markers

[30] A measure could plausibly be defined in terms of probability of purpose-attainment, and degree of functionality as the probability of purpose-attainment if the behavior or object is present minus that probability in its absence. This would yield values between +1 and −1 for this index, and makes plausible definitions for functional, non-functional, and dysfunctional behaviors.

There are problems, however.

First, probability of purpose-attainment without a functional trait cannot be evaluated simply by removing it. Simply extirpating the functional item will usually preturb the system in diverse ways, which will prejudice the result. This is a corollary to criticisms made by Gregory and Lorenz of ablation and deprivation experiments (see note 17 above). Might the degree of functionality be *defined* in counterfactual terms as the difference between the actual probability of purpose-attainment (with the entity present) and the imagined probability of purpose-attainment in a thought experiment where *just* the entity in question is affected? But counterfactuals are consistent with an arbitrarily large number of states of affairs. If the thought experiment is not filled out in great detail, ambiguities result. Nor is the 'possible worlds' gambit much help here. See Mikkelson (1997) for an elaboration of the use of counterfactuals in ecology and situations of comparable complexity. This measure should have other desirable metrical properties. Thus its maximum value should be found at the topmost node of the hierarchy under certain variations of the values of this index at lower nodes, and should equal the sum for the nodes immediately below it and one level down, corresponding to its 'next smallest' subsystems. No such measure has been defined.

indicating a preferred direction, we should consider transformations symmetrically. Pruning or adding nodes may seem *ad hoc*, but may sometimes be more fruitful. Defense of categorical structural isomorphisms would often be far more *ad hoc* for real measurements on real systems.

1. Data of different quality or degree of detail or of accuracy for the two systems being compared may produce artifactual differences. Small functional contributions or small differences in them may go unnoticed or be impossible to detect in one system, while greater experimental tractibility or background data on another may yield apparently greater functional elaboration.
2. We should try different resolutions or resolving powers on the functional systems being compared, to see whether focusing or defocusing yields greater similarities, and, if so, in what functional systems or subsystems.
3. Selection looks at functional hierarchies also. Nodes with small effect are more easily lost owing to drift or not protected in evolutionary changes from interactions with other systems that destroy their effects.

So for a variety of reasons, defocusing—not looking with too much detail and precision—may be advantageous in studying functional hierarchies. This is not always so, however: focusing on details can sometimes lead to significant reassessments of function. If, from a 'God's-eye' view, we knew that the functional effects of an adaptation were very small, we might well be able to ignore it, but we do not have that kind of knowledge. And today's nearly irrelevant trait may be tomorrow's pivotal exaptation.

Quantitative similarity could be characterized more precisely, but I see no compelling reason to try. Precise distinctions would suffer from the same arbitrariness as does calling any Chi-square level greater than 95 per cent 'significant' and all others not. Applying names and making distinctions like this may be a useful guide in establishing a common procedure (much as agreeing upon units in which measurements are to be expressed), but does not cut any relevant conceptual ice here. Similar remarks would apply to quantitative criteria for when to prune nodes in considering functional organization or isomorphisms. 'Should a node be pruned when its index is one order of magnitude less than that of the node in the same system with the next smallest index—or two orders of magnitude, or one-half?' That depends upon the purposes of the investigator and other features of the situation. It should not be decided by fiat.

22. Concluding Thoughts: Exaptations, Kluged Adaptations, and Functional Architectures

We have given the God's-eye view of functional organization. Now for the real thing. Our pictures of functional hierarchies and functional organization, as well as the things themselves, emerge out of 'measurements' on and improvements to the performance of real systems in the real world under conscious or natural selection forces. This is true for natural selection no less than for our engineering and naturalistic tests. Functional architectures are crisp in engineering drawings and design proposals. Any structure we come up with through observation—of systems in the real world—is a curve-fitting approximation to real world performance. For our engineered artifacts, the designed architecture we started with is blurred and muddied by the kluged fixes made to get it to work at least approximately as it should in the real world. (Ask any software programmer whether the program as executed actually matches the stated specifications at any level we would write down and explain with commented code.) But at least the plan that we started with, and major modifications made to it, give a reference source for the functional organization.

For organic selected systems, the situation is worse: in a relevant sense, the architecture our studies reveal, if accurate, is of the real world. So it is based there. But given that our estimates of systems that we did not design are derived from studying how they break down, we must ask generally what the relation is between a functional organization based on lesions, deletions, and stimulations and the 'real' functional architecture. This is a big problem, and we can say generally that we know how to map from one to the other only when we already know (to some degree) how the system works and what it is for. This sounds like a vicious circularity, but it is not. It bespeaks a dialectical feedback process of investigation.

But there is a deeper problem. Our reifications of natural systems into parts, processes, mechanisms, and systems is somewhat misleading: nature cares only for results after all, and no one said that it must be assembled in the way that we say it is. And, of course, the only way that any design architecture was really assembled by evolution was through a tortuously long layered history of successive kluges—many or most of which are not accessible to us. It is not clear what relation this history has to currently perceivable functional architecture. To evaluate current performance, we must infer function on the basis of currently perceivable architecture. Generative entrenchment means that past history will be a rich source of constraints and clues on its present organization. With increased entrenchment, those aspects of functional architecture are stabler through history, and should become more crisply defined. But we must remember that the architecture is itself constantly in process.

This points to a bigger problem concerning relations between current architecture and evolutionary history. And generative entrenchment provides, right now, the only tool we have for relating them. As the main evolutionary source of stability, it is the appropriate theoretical tool for delineating and filling in with successively decreasing stability and precision, the outlines of a natural functional architecture. But this is a matter for another occasion.

References

Arbib, M. A. (1965), *Brains, Machines, and Meta-Mathematics* (New York: Macmillan).

Arthur, W. (1997), *The Origins of Animal Body Plans* (Cambridge: Cambridge University Press).

Bechtel, W., and Mundale, J. (1999), 'Multiple Realizability Revisited: Linking Cognitive and Neural States', *Philosophy of Science*, 66/2 (June), 175–207.

——and Richardson, R. C. (1992), *Discovering Complexity: Decomposition and Localization as Scientific Research Strategies* (Princeton: Princeton University Press).

Beckner, M. (1959), *The Biological Way of Thought* (New York: Columbia University Press).

——(1968), 'Teleology', in Paul Edwards (ed.), *The Encyclopedia of Philosophy* (New York: Macmillan), viii. 88–91.

——(1969), 'Function and Teleology', *Journal for the History of Biology*, 2/1: 151–64.

Brandon, R. N., and Burian, R. (eds.) (1985), *Genes, Organisms, Populations* (Cambridge, Mass.: MIT. Press).

Buss, L. (1987), *The Evolution of Individuality* (Princeton: Princeton University Press).

Campbell, D. T. (1974), 'Evolutionary Epistemology', in P. A. Schilpp (ed.), *The Philosophy of Karl Popper* (La Salle, Ill.: Open Court), ii. 413–63.

Cannon, W. (1939), *The Wisdom of the Body* (New York: W. W. Norton).

Cummins, R. (1975), 'Functional Analysis', *Journal of Philosophy*, 72/20: 741–65.

Darwin, C. (1859), *On the Origin of Species* (London: John Murray).

——(1876), *On the Fertilization of Orchids* (London: John Murray).

Dawkins, R. (1976), *The Selfish Gene* (Oxford: Oxford University Press).

Dobzhansky, T. (1968), 'On Some Fundamental Concepts of Darwinian Biology', in T. Dobzhansky, M. K. Hecht, and W. C. Steere (eds.), *Evolutionary Biology* (New York: Appleton-Century-Crofts), ii. 1–34.

Durham, W. (1991), *Coevolution: Genes, Culture and Human Diversity* (Palo Alto, Calif.: Stanford University Press).

Fodor, J. (1965), 'Functional Explanations in Psychology', in Max Black (ed.), *Philosophy in America* (London: Routledge & Kegan Paul).

——(1968), *Psychological Explanation* (New York: Random House).

Gould, S. J., and Vrba, E. (1982), 'Exaptation—a Missing Term in the Science of Form', *Paleobiology*, 8/1: 4–15.

Gregory, R. L. (1962), 'The Logic of the Localization of Function in the Central Nervous System', in E. E. Barnard and M. R. Kare (eds.), *Biological Prototypes and Synthetic Systems* (New York: Plenum Press), i. 51–3.

Harary, F., Norman, R. Z., and Cartwright, D. (1965), *Structural Models: An Introduction to the Theory of Directed Graphs* (New York: Wiley).

Helmer, O., and Rescher, N. (1958), 'On the Epistemology of the Inexact Sciences', Rand Report P-1513, The Rand Corporation.

Hempel, C. G. (1959), 'The Logic of Functional Analysis', repr. in C. G. Hempel, *Aspects of Scientific Explanation, and Other Essays in the Philosophy of Science* (New York: Macmillan, 1965), 297–330.

Hubby, J. and Lewontin, R. C. (1966), 'A Molecular Approach to the Study of Genic Heterozygosity in Natural Populations I: The Number of Alleles at Different Loci in *Drosophila pseudoobscura*', *Genetics*, 54: 577–94.

Jardine, N. (1967), 'The Concept of Homology in Biology', *British Journal for the Philosophy of Science*, 18: 125–39.

Jukes, T. H. (1966), *Molecules and Evolution* (New York: Columbia University Press).

Kauffman, S. A. (1971), 'Articulation of Parts Explanation in Biology and the Rational Search for Them', in R. C. Buck and R. S. Cohen (eds.), *PSA 1970* (East Lansing, Mich.: Philosophy of Science Association), 257–72.

Levins, R. (1968), *Evolution in Changing Environments* (Princeton: Princeton University Press).

Lewontin, R. C. (1961), 'Evolution and the Theory of Games', *Journal for Theoretical Biology*, 1: 382–403.

——(1978), 'Adaptation', *Scientific American*, 239: 212–30.

——(1974), *The Genetic Basis of Evolutionary Change* (New York: Columbia University Press).

——and Hubby, J. (1966), 'A Molecular Approach to the Study of Genic Heterozygosity in Natural Populations II: Amount of Variation and Degree of Heterozygosity in Natural Populations of *Drosophila pseudoobscura*', *Genetics*, 54: 595–609.

Lloyd, E. (1988), *The Structure and Confirmation of Evolutionary Theory*, repr. with a new introduction (Princeton University Press 1994).

Lorenz, Konrad Z. (1965), *Evolution and Modification of Behavior* (Chicago: University of Chicago Press).

Mikkelson, G. (1997), 'Other Things Being Equal: Counterfactuals, Natural Laws, and Scientific Models; with Case Studies from Ecology', Ph.D. dissertation, Committee on the Conceptual Foundations of Science, the University of Chicago.

Millikan, R. G. (1984), *Language, Thought, and Other Biological Categories: New Foundations for Realism* (Cambridge, Mass.: MIT. Press).

Nagel, E. (1961), *The Structure of Science: Problems in the Logic of Scientific Explanation* (New York: Harcourt, Brace, & World).

Putnam, H. (1967), 'The Mental Life of Some Machines', in H. N. Castaneda (ed.), *Intentionality, Minds, and Perception* (Detroit: Wayne State University Press), 177–200.

Raff, R. (1996), *The Shape of Life: Genes, Development and the Evolution of Animal Form* (Chicago: University of Chicago Press).

Riedl, R. (1978), *Order in Living Organisms: A Systems Analysis of Evolution* (New York: Wiley).

Sarkar, S. (1998), *Genetics and Reductionism* (Cambridge: Cambridge University Press).

Schank, J. C., and Wimsatt, W. C. (1988), 'Generative Entrenchment and Evolution', in A. Fine and P. K. Machamer (eds.), *PSA 1986*, vol. 2 (East Lansing, Mich.: Philosophy of Science Association), 33–60.

———(2000), 'Evolvability: Adaptation, and Modularity', in R. Singh, K. Krimbas, D. Paul, and J. Beatty (eds.), *Thinking about Evolution: Historical, Philosophical and Political Perspectives: Festschrift for Richard Lewontin* (Cambridge, Cambridge University Press), 322–35.

Simon, H. A. (1969), *The Sciences of the Artificial* (Cambridge, Mass.: MIT Press), 3rd edn., 1996.

Thoday, J. M. (1953), 'Components of Fitness', *Symposia of the Society for Experimental Biology*, 7: 96–113.

Waddington, C. H. (1969), 'Paradigm for an Evolutionary Process', in C. H. Waddington (ed.), *Towards a Theoretical Biology* (Edinburgh: Edinburgh University Press), ii. 106–24.

Wade, M. J. (1996), 'Adaptation in Subdivided Populations: Kin Selection and Interdemic Selection', in Michael Rose and George Lauder (eds.), *Adaptation* (New York: Academic Press), 381–405.

Williams, G. C. (1966), *Adaptation and Natural Selection: A Critique of some Current Evolutionary Thought* (Princeton: Princeton University Press).

Wimsatt, W. C. (1971), 'The Concepual Foundations of Functional Analysis', Ph.D. dissertation, Department of Philosophy, University of Pittsburgh.

——(1972), 'Teleology and the Logical Structure of Function Statements', *Studies in History and Philosophy of Science*, 3: 1–80.

——(1974), 'Complexity and Organization', in K. F. Schaffner and R. S. Cohen (eds.), *PSA 1972* (Dordrecht: Reidel), 67–86.

——(1976), 'Reductive Explanation: A Functional Account', in A. C. Michalos, C. A. Hooker, G. Pearce, and R. S. Cohen (eds.), *PSA 1974* (Boston Studies in the Philosophy of Science, 30; Dordrecht: Reidel), 671–710.

——(1980*a*), 'Randomness and Perceived-Randomness in Evolutionary Biology', *Synthese*, 43/3: 287–329.

——(1980*b*), 'Reductionistic Research Strategies and their Biases in the Units of Selection Controversy', in T. Nickles (ed.), *Scientific Discovery*, ii. *Case Studies* (Dordrecht: Reidel), 213–59.

——(1981*a*), 'Robustness, Reliability, and Overdetermination', in M. Brewer and B. Collins (eds.), *Scientific Inquiry and the Social Sciences* (San Francisco: Jossey-Bass), 124–63.

——(1981*b*), 'Units of Selection and the Structure of the Multi-Level Genome', in P. D. Asquith and R. N. Giere (eds.), *PSA 1980*, vol. 2. (East Lansing, Mich.: Philosophy of Science Association), 122–83.

——(1986), 'Developmental Constraints, Generative Entrenchment, and the Innate-Acquired Distinction', in W. Bechtel (ed.), *Integrating Scientific Disciplines* (Dordrecht: Martinus Nijhoff), 185–208.

——(1994), 'The Ontology of Complex Systems: Levels, Perspectives, and Causal Thickets', in M. Matthen and R. Ware (eds.), *Biology and Society: Reflections on Methodology. Canadian Journal of Philosophy*, supp. vol. 20: 207–74.

——(1997*a*), 'Functional Organization, Functional Analogy, and Functional Inference', *Evolution and Cognition*, 3/2: 2–32.
——(1997*b*), 'Aggregativity: Reductive Heuristics for Finding Emergence', in L. Darden (ed.), *PSA 1996*, vol. 2 (East Lansing, Mich.: Philosophy of Science Association), supp. vol. 2, S372–S384.
——(1999), 'Generativity, Entrenchment, Evolution, and Innateness', in V. Hardcastle (ed.), *Biology Meets Psychology: Philosphical Essays* (Cambridge, Mass.: MIT Press), 139–79.
——(2001), 'Generative Entrenchment and the Developmental Systems Approach to Evolutionary Processes', in S. Oyama, R. Gray, and P. Griffiths (eds.), *Cycles of Contingency: Developmental Systems and Evolution* (Cambridge, Mass.: MIT Press), 219–37.
——and Schank, J. C. (2002), Generative Entrenchment, Modularity, and Evolvability: When Genic Selection Meets the Whole Organism, in G. Schlosser and G. Wagner (eds.), *Modularity in Development and Evolution* (Chicago: University of Chicago Press).
——(2002), *Re-Engineering Philosophy for Limited Beings: Piecewise Approximations to Reality* (Cambridge, Mass.: Harvard University Press).
——and Schank, J. C. (1988), 'Two Constraints on the Evolution of Complex Adaptations and the Means for their Avoidance', in M. Nitecki (ed.), *Evolutionary Progress* (Chicago: University of Chicago Press), 231–73.
Wright, L. (1973), 'Functions', *Philosophical Review*, 82: 139–68.
——(1976), *Teleological Explanation* (Berkeley and Los Angeles: University of California Press).

2002 Epilogue to a 1970s Architecture

If I were to change any single thing in this paper now it would be to the conceptual scheme represented in figure 7.3. There the functional hierarchy is represented as rooted fundamentally in the behaviours of genes. Pleiotropy is represented here, but not polgeny. I had assumed that would be adequately captured in higher level epistatic interactions hidden in the functional hierarchy. That is just the start of the trouble. Developmental genetics, evo-devo, developmental systems theory, and eco-evo-devo have changed the conceptual geography of genotype-phenotype relations. It isn't all bottom up. Maternal effect and epigenetics have forced on us (see Jablonka 2001, and other papers there) the recognition that genes are not all that is inherited and that talking about the behaviour of genes in isolation is about as useless as trying to figure out what the behavior of a screw in the armrest of my car has to tell us about the dynamics of its own motion. (It does after all sometimes contribute (redundantly, for one screw out of four) to bracing my arm, which holds the steering wheel, which . . . .) Analogous problems may arise for more 'bottom up' functional hierarchies in other functional domains with lots of modularity. In the US we now replace starter motors, rather than clean their armatures or change their brushes.

Joblonka, E., (2001), 'The Systems of Inheritance', in S. Oyama, R. Gray, and P. Griffiths (eds.), *Cycles of Contingency: Developmental systems and evolution*, (Cambridge, Mass.: MIT. Press), 99–116.

8. Function and Design Revisited

David J. Buller

ABSTRACT

Several analyses of biological function (for example, those of Williams, Millikan, and Kitcher) identify an item's function with what natural selection designed it to do. Allen and Bekoff claim, in contrast, that design by selection is a special case of biological function. I argue that Allen and Bekoff's account of natural design is unduly restrictive and that it fails to mark a principled distinction between function and design. I distinguish two approaches to natural design—the 'trait-centered' approach of Allen and Bekoff and the 'organism-centered' approach—and defend the latter, according to which function and design are coinstantiated phenomena.

1. Introduction

The concept of *design* has played a central role in several prominent theories of biological function. Williams, for example, argues that the demonstration of design is both necessary and sufficient for the demonstration of function (1966: 209). Millikan says: 'Having a proper function is a matter of having been "designed to" . . . perform a certain function' (1984: 17). And Kitcher echoes: 'the function of *S* is what *S* is designed to do' (1993: 380). None of these authors, however, provides an analysis of design. Williams assumes, instead, that 'design is something that can be intuitively comprehended' by analogy with human engineering (1966: 260; see also 1992: 40). And, given the fact that Kitcher analyzes function in terms of design, while illustrating design with engineering examples, he also appears to take design to be 'intuitively comprehended'. In contrast, Millikan treats 'function' and 'design' as synonyms, which are mutually defined by her version of the etiological theory of functions; for she says that the goal of her 'theory of proper functions is to define this sense of "designed to" . . . in naturalist, nonnormative, and nonmysterious terms' (1984: 17).

I am grateful to Colin Allen, Harold Brown, David Hull, the editors of this volume, and (especially) Karen Neander for helpful comments and input on beta versions of this chapter.

Allen and Bekoff have recently provided an account of design that removes any need for 'intuitive comprehension' and makes design a property of traits that can be objectively ascertained through phylogenetic analysis (Allen and Bekoff 1995*a*, *b*; Bekoff and Allen 1995). In providing this account, Allen and Bekoff argue that there are non-biological cases of function without design that have biological analogues and that, consequently, function and design are distinct phenomena in the biological domain as well. Biological design, or *natural design*, they contend, is a matter of having a function *plus* having undergone a history of modification under selection for better performance of that function. This entails not only that the concepts of function and design are not synonymous (contra Millikan), but that they are not even coextensive, which in turn entails that the concept of design cannot be employed to analyze the notion of biological function (contra Williams and Kitcher).

In the next section, I argue that Allen and Bekoff's account of natural design is unduly restrictive and that it fails to mark a principled distinction between biological function and natural design. I diagnose these problems as a by-product of their non-biological examples, which are crucially disanalogous to biological cases of function, and I articulate and defend an alternative conception of natural design that avoids these problems. Before developing that alternative conception of design, however, it is necessary, in Section 3, to explore in more detail the conception of biological function to which both Allen and Bekoff and I are committed. Then, in Section 4, I sketch two distinct ways of approaching the phenomenon of natural design—the 'trait-centered' approach, characteristic of Allen and Bekoff, and the 'organism-centered' approach—and defend the latter. With that alternative conception of design in place, I show that the concepts of biological function and natural design, while not synonymous, are nonetheless coextensive (as per the theories of Williams and Kitcher). I conclude by discussing some implications for the concept of adaptation.

2. Biological Function versus Natural Design?

Allen and Bekoff begin with examples to show that, in the artifactual case, although design is sufficient for function, it is not necessary:

> In *The Dixie Chicken* in College Station and many other down-home drinking establishments in Texas, stags' heads function as wall decorations. They are clearly not designed for that purpose. (Although the stags' heads were presumably put on the wall intentionally, hence by intent-design.) Likewise, the function of a rock on a desk may be to hold down loose papers, but unless the rock has been modified by, e.g., having a flat base chiselled into it, it is not appropriate to say that this object was designed for the purpose of holding down papers. Thus, having a function does not entail being designed for that function. (Allen and Bekoff 1995*b*: 33)

In these artifactual cases, design involves not only having a function, but being *modified* in some way in order *better* to perform that function. In short, artifact design = artifact function + modification for better performance of that function.

Since natural selection frequently modifies traits in the direction of greater 'perfection' for their functional roles, Allen and Bekoff contend that an analysis of natural design should parallel their analysis of artifact design. Thus, they propose, a trait *T* is *naturally designed* for *X* if and only if:

> i. X is a biological function of T and
> ii. T is the result of a process of change of (anatomical or behavioral) structure due to natural selection that has resulted in T being more optimal (or better adapted) for X than ancestral versions of T. (1995*b*: 34)

So, 'natural design entails both possession of biological function and a history of progressive structural modification under natural selection for improved performance of that function' (1995*b*: 3).

Allen and Bekoff intend the conception of biological function in clause (i) as a version of the etiological theory of functions: 'the functions of a given trait are those effects the trait had in the past that contributed to the selection of organisms with that trait' (1995*b*: 26). Clause (ii) is thus intended to build on this conception of function in the following way (see 1995*b*: 38; Bekoff and Allen 1995: 254): a (current) version *T* of some trait is *naturally designed* to do *X* if and only if it is the (etiological) function of *T* to do *X* and

(ii.*a*) there was some earlier form of *T*, such that *T* arose as a modification of that earlier form,
(ii.*b*) the earlier form of the trait also had the function of doing *X*, and
(ii.*c*) a comparison of *T* with the earlier form of the trait shows *T* to be more effective than that earlier form in doing *X*.

To illustrate, consider the example to which Allen and Bekoff themselves apply their distinction: the progressive modifications to the forelimb structures of land-bound saurians that resulted in wings (1995*a*: 615–16). Their idea is that, if we consider the series of these modifications to forelimbs, we would find a form of the forelimbs that did not allow flight and a subsequent modification to them that did enable flight. The modified forelimbs thus functioned to produce flight, but were not designed for flight, since the earlier form of the forelimbs did not function to produce flight—that is, condition (ii.*b*) is not satisfied. However, when the modified forelimbs were further modified by selection to produce more effective flight, the secondarily modified forelimbs were also designed for flight; for they arose as a modification of an earlier form of forelimbs (so condition (ii.*a*) is satisfied), that earlier form of forelimbs (that is, the primarily modified form) had the function of producing flight (so condition (ii.*b*) is satisfied), and a comparison of the later and earlier forms shows the later form to be more effective in the performance of their shared function (so condition (ii.*c*) is satisfied). Thus, the secondarily modified forelimbs represented natural selection's modifying the forelimbs in the direction of greater optimality for flight, and is for that reason *designed* for flight.

There are, I think, two problems with this conception of natural design. First, it is unduly restrictive, for it counts a trait as an instance of natural design only if it has undergone directional selection, in which selection modified the trait towards greater optimality. But some traits are simply maintained by selection in a population, with active selection against any emerging variants. (Stabilizing selection is one form of such maintenance (see Endler 1986).) Since traits that have been maintained in a population by selection will not have undergone modification in the direction of enhanced performance of function, Allen and Bekoff will not classify them as instances of natural design. But this is problematic. If selection has maintained a trait in its initial form, selecting against any variants that arose, that is because the trait was already as optimal as possible and no modification could have made it more optimal (relative, of course, to the constraints on the trait). An account of natural design should allow us to say that such traits are instances of *good design*, regardless of whether they have undergone a history of modification. Not to recognize such traits as instances of natural design is to employ an account of natural design that is too tied to only one way in which selection operates (cf. Reeve and Sherman 1993: 7).

Secondly, Allen and Bekoff's distinction between natural design and mere function (that is, function without design) is not principled. To see this, begin by considering that Allen and Bekoff take it to be an advantage of their conception of natural design that it accords with the traditional, though by no means universal, biological usage of the concept of adaptation. According to this traditional conception, 'the external world sets certain "problems" that organisms need to "solve"' and adaptation is 'the process of evolutionary change by which the organism provides a better and better "solution" to the "problem"' (Lewontin 1978: 213). A trait is an adaptation if it is a product of this process. Thus, like Allen and Bekoff's conception of natural design, the traditional conception of adaptation incorporates the criterion of modification by selection for more effective performance of some function (better solution of some problem). This results, Allen and Bekoff believe, in a simple relation between natural design and adaptation: adaptations, and only adaptations, are products of natural design (1995*b*: 38; see also 1995*a*: 612–13).

But not all adaptations are the result of progressively better solutions to relatively stable environmental problems. Many adaptive problems an organism faces are set not by the organism's physical environment, but by other organisms in evolving lineages—particularly parasites, predators, and conspecifics that are competing for the same resources. In such cases, the very problems an organism must solve are themselves undergoing evolution. Consideration of these facts has led to a growing treatment of some adaptations as moves within *evolutionary arms races*, rather than as solutions to relatively stable environmental problems (Dawkins 1982: ch. 4; 1991: ch. 7; Vermeij 1987: chs. 1, 15; Weis *et al.* 1989; Davies and Brooke 1991; Endler 1991; Krebs and Davies 1993: ch. 4; Lotem and Rothstein 1995; Ridley 1996: ch. 22). In an evolutionary arms race, there is progressive modification not only in the solutions to an adaptive problem, but in the adaptive problems themselves. 'Improving' solutions are continually matched by ever more 'difficult' problems, with evolution in either precipitating the evolution of the other. The result is the 'Red Queen effect' (van Valen 1973): lineages evolve as fast as they can merely to be equally effective as they once were in solving adaptive problems; so there is no cumulative progress in their solutions to adaptive problems. While Allen and Bekoff's distinction between natural design and mere function appears plausible given a more traditional conception of adaptation, when viewed within the context of arms races, which are a significant aspect of evolution, it is not a principled distinction.

To see why, consider the following very simplified sketch of an evolutionary arms race, in which progressively modified forms of a trait co-evolve with progressively modified forms of an adaptive problem. A trait (*T*) evolves under selection in a population as a response to an adaptive problem (*P*); thus, it is the (etiological) function of the trait to solve that problem. Over time, mutation introduces a modified form of the trait (call it *T**), which provides a more effective solution to the adaptive problem. The modified trait thus increases in frequency in the population, eventually replacing the unmodified form. This precipitates evolution of the adaptive problem, with a modified form of the problem (call it *P**) coming to replace the earlier form (*P*). Although the modified trait did not evolve as a response to the modified problem, it nonetheless adequately solves it (though not as effectively as it had solved the unmodified form of the problem); so selection maintains the modified trait in the population and it thus becomes the (etiological) function of that trait (*T**) to solve the new adaptive problem (*P**). Over time, mutation further modifies the trait, introducing yet another form of the trait (*T***), which is more effective than the earlier form (*T**) in solving the new adaptive problem (*P**). This precipitates the evolution of the adaptive problem (into *P***), and so on.

Now, given Allen and Bekoff's conception of natural design, is *T**, the first modified form of the trait, an instance of natural design? There are several conflicting ways of answering this question, and there is no genuinely non-arbitrary way of choosing among them. Consider just a few main options.

(1) *Yes.* T^* is an instance of natural design, since it arose as a modification of T with which it shared the function of solving P, and it performed that function more effectively than T.

But things are not so simple. For (1) ignores the fact that the adaptive problem itself evolved (from P to P^*) during the tenure of T^* and T^* also functioned to solve the new form of adaptive problem, which it was not a function of T, the earlier form of the trait, to solve. And there is no reason why the two traits should be compared with respect to their functional effectiveness in solving the earlier adaptive problem (P) rather than the later (P^*). We could, then, opt for a more nuanced answer:

(2) *Yes and no.* T^* was designed for P for the reasons given in (1), but it was not designed for P^*, since condition (ii.*b*) above is not satisfied (that is, the earlier form of the trait from which T^* arose as modification did not share the function of solving P^*). This would make it the mere function of T^* to solve P^*.

(2), however, depends on treating successively modified forms of adaptive problems in an arms race as genuinely distinct problems (such that solving P is a different function from solving P^*). But there is nothing in the nature of things to force this interpretation. We could, instead, view such successive adaptive problems as different versions of the *same* problem, which differ only with respect to degree of difficulty. Suppose the successively modified forms of the trait (the Ts) are progressively increasing running speeds of cheetahs, and the successively modified forms of the adaptive problem (the Ps) are progressively increasing running speeds of gazelles. Then, instead of considering catching gazelles running at n m.p.h. a distinct problem from catching gazelles running at $k > n$ m.p.h., we could consider both the same problem for cheetahs—namely, catching gazelles—and the various speeds of gazelles as merely different degrees of difficulty of that problem. This would allow treating successive forms of a trait as solutions to the same problem, but it would leave open how to compare their effectiveness in solving that problem at different degrees of difficulty. Thus, comparing the effectiveness of T and T^* in solving the P degree of difficulty will generate a modified version of (1). Comparing their effectiveness in solving P, while taking into account the fact that T^* also solved the P^* degree of difficulty, will generate a modified version of (2). Finally, comparing T's solution of P with T^*'s solution of P^* will generate

(3) *No.* T^* is not an instance of natural design, since condition (ii.*c*) above is not satisfied because of the Red Queen effect (that is, T^* was not more

effective at solving P^* than T was at solving P). Again, this would make it the mere function of T^* to solve P^*.

The problem is that there are no principled grounds for choosing among these various ways of comparing the functional effectiveness of successive forms of a trait in an evolutionary arms race. And, since these different modes of comparison generate different classifications of the same trait given Allen and Bekoff's definition of natural design (some classifying it as an instance of natural design and others as an instance of mere function), the absence of principled grounds for choosing among these modes of comparison entails that Allen and Bekoff's distinction between function and natural design is not principled.

Allen and Bekoff could respond by stipulating that the functional effectiveness of successive traits should be compared with respect to that point in time at which there was selection for the later form over the earlier. This stipulation would entail option (1) above and would thus resolve the ambiguity inherent in Allen and Bekoff's definition of natural design. But, again, things are not so simple. Within the clean lines of the above extremely simplified model there is *steady* replacement of an early form of a trait with its modified form and, consequently, a fairly straightforward sense in which there was a 'time at which there was selection for' the latter over the former. 'In the wild,' however, there is rarely such steady replacement of one trait by another, and this undermines the proposed solution.

To see why, consider a population of a bird species in which some birds have broad beaks and others have slender ones. In drought conditions, the broad-beaked birds enjoy a selective advantage, since they are better able to feed on the available dry seeds; in rainy conditions, the slender-beaked birds enjoy a selective advantage, since they are better able to feed on the insects in tree bark. In a reasonably extended drought, the broad-beaked birds will increase in frequency relative to the slender-beaked birds. But this extended drought may be followed by an extended wet period in which the slender-beaked birds stage a comeback and increase in frequency relative to the broad-beaked birds. Climate being what it is, of course, there will be fluctuations in the frequencies of the two beak types over an extended period of time—that is, over a long time span there will be periodic reversals in the direction of selection. But, when measured in a long enough time span, there may be a clear net direction to the changing frequencies. Perhaps rainy conditions are more frequent than droughts in the long term, so that slender-beaked birds show an overall increased frequency over broad-beaked birds in the very long haul.

Now let us modify the above arms-race scenario to reflect a process of selection that proceeds by such fits and starts, rather than steadily, and see what becomes of the proposed revision of Allen and Bekoff's account of natural design. Suppose that it takes 250 years for a modified trait completely to replace its ancestral form in the population. Over that 250-year period, we can say that there was selection for the modified form over the ancestral form; consequently, we can compare the two forms of the trait with respect to how effectively they performed their shared function of solving an adaptive problem during that 250-year period (as per option (1) above). But there will not have been *steady* selection for the modified form over the ancestral form during this entire 250-year period. This long period will show some fluctuations, including short periods in which the ancestral form actually increased in frequency relative to the modified form owing to fluctuations in the conditions under which the two traits solve the adaptive problem.

At this point, we confront the same difficulty as before. We could take the entire 250-year period as the 'time at which there was selection for' the modified trait over its ancestral version, and then we will classify the modified trait (T^*) as an instance of natural design, since it solved the adaptive problem (P) better than the ancestral form of the trait (T) did in the long haul. But there is nothing in the nature of things forcing us to measure the relative frequencies of the two traits with respect to the entire period in which one replaced the other. We could just as justifiably focus on a shorter time span. Since the frequencies of the two traits fluctuated during the 250-year period, there will be at least one shorter time span s in which T increased in frequency under selection relative to T^*. What are we to say about T^* during s? We seem to be faced with the following dilemma in applying Allen and Bekoff's revised criteria. On the one hand, we could insist that s be ignored and that the two versions of the trait should be compared only with respect to those periods in which T^* is increasing in frequency relative to T. But this would be wholly arbitrary. On the other hand, we could take s into account. But then T^* will not count as an instance of natural design during s, since T will be performing their shared function better than T^* (so T^* will not satisfy condition (ii.*c*) during s). Thus, within the longer 250-year period in which T^* replaces T, there will be short time spans in which T^* is an instance of natural design by Allen and Bekoff's criteria and other short time spans in which it will not be an instance of natural design, but instead will have the mere function of solving the adaptive problem P. And, again, there are no principled grounds for comparing the functional effectiveness of the two versions of the trait in one of these time spans rather than another, or in the long time span rather than the shorter ones. But, again, this entails that there is no principled distinction between a trait's being designed for solving an adaptive problem and its having only the mere function of solving that problem.

3. A Closer Look at Functions

The distinction between design and mere function was very clear, however, in Allen and Bekoff's artifactual examples. For instance, although a rock of sufficient mass can function as a paperweight, that rock is not designed to be a paperweight unless it has been modified to 'perfect' it for the function of being a paperweight. So how can a distinction that is so clear in the artifactual cases that Allen and Bekoff mention be so unprincipled in the biological case? Because Allen and Bekoff's examples of functional artifacts are strongly disanalogous to functional traits of organisms in lineages undergoing evolution. One might think this disanalogy derives from a difference between artifact function and biological function. There is, indeed, a difference between these, but the source of the disanalogy lies elsewhere. To see this, however, it is necessary first to be very clear about the conception of biological function to which Allen and Bekoff are committed; for I share their conception and will employ it in Section 4.

In the biological case, Allen and Bekoff claim to hold a fairly standard version of the etiological theory of functions. Up to this point, I have treated the etiological theory as ascribing a function to a trait when it has been selected for producing some effect. But elsewhere I have shown that there has been a systematic, yet undetected, ambiguity in the various formulations of the etiological theory (Buller 1998). It has generally been assumed that the various formulations of the theory share the same fundamental commitment and that they are, consequently, merely stylistic variants of that same essential commitment. But the formulations are, in fact, stylistic variants of two non-equivalent versions of the etiological theory.

One frequent formulation of the etiological theory is the following, which I will call the 'strong etiological theory':

> A current token of a trait *T* in an organism *O* has the function of producing an effect of type *E* just in case past tokens of *T* contributed to the fitness of *O*'s ancestors by producing *E and* were selected for (over alternative items) *because of* this contribution to the fitness of *O*'s ancestors.

This formulation is to be found in Millikan (1993: 35–6), in which the parenthetical clause is explicit, and in Neander (1991*a*: 173; 1991*b*: 459) and Godfrey-Smith (1994: 359), in which it is not. Regardless of whether the parenthetical clause is explicit, however, any formulation that invokes selection for a trait in specifying its function is committed not only to the parenthetical clause, but to much stronger conditions as well, as I will show in a moment.

In contrast to the above, the etiological theory is sometimes articulated without appeal to selection for the functional trait, but with an emphasis only on the requirement that the functional trait be a reproduction of items that had the same effect(s) (see e.g. Millikan 1989*b*: 289). This idea is encapsulated in the following, which I will call the 'weak etiological theory':

> A current token of a trait *T* in an organism *O* has the function of producing an effect of type *E* just in case past tokens of *T* contributed to the fitness of *O*'s ancestors by producing *E*, and thereby causally contributed to the reproduction of *T*s in *O*'s lineage.

Like the strong etiological theory, the weak etiological theory defines the function of a current token of a trait in terms of the role played by ancestral tokens. Nonetheless, the two theories are not equivalent; for, while the strong etiological theory entails the weak, the weak does not entail the strong.

To see why, consider first the strong etiological theory. By defining the function of a current token of a trait as the production of an effect because of which there was selection for past tokens of that trait, the strong etiological theory attributes a function to a trait *T* only when the following three conditions are satisfied, each of which is a necessary condition for there being selection for a trait: (1) *T* is hereditary, (2) there has been variation in *T* within a common selective environment (see Brandon 1990), and (3) the bearers of *T* had greater fitness within that common environment than bearers of *T*'s variants (at least partly) because of possessing *T*. The weak etiological theory, in contrast, does not define the function of a trait in terms of selection for it, and thus does not make it a necessary condition of a trait's having a function that there was variation in that trait. And, since variation is not required, it is also not necessary that the bearers of that trait had greater fitness than any other organisms. The weak etiological theory requires only that a trait contributed to the fitness of the ancestors of its current bearers (and thereby contributed to the reproduction of tokens of that trait). These conditions are satisfied provided that (1) *T* is hereditary and (2) *T* causally contributed to the fitness of ancestral bearers of the trait, for any trait that satisfies both of these conditions will have causally contributed to its own reproduction via genetic mechanisms of inheritance and development.

The weak etiological theory thus attributes functions to traits more liberally than the strong etiological theory. Indeed, the strong theory's function attributions constitute a proper subset of the weak theory's function attributions, since the weak theory counts as functional any hereditary trait of an organism that causally contributed to the fitness of that organism's ancestors, while the strong theory requires in addition that there was selection for that trait. (It is sometimes assumed (for example, by Millikan 1989*a*) that the notion of a trait's 'contribution to fitness' can be made sense of only if the trait has been selected for making that specific contribution. But it is just the opposite: the notion of a trait's being selected for can only be made sense of in terms of its contribution to fitness (see Buller 1998: 510–11; see also Sober 1984: 100)).

To illustrate the difference between the strong and weak theories, suppose that *T* is a hereditary physiological trait that plays a causal role in the process of gamete production, but that either (*a*) through genetic happenstance the necessary mutation(s) never occurred in the population to produce an alternative to *T* (cf. Kitcher 1993: 388), or (*b*) mutation produced alternatives to *T* in the population, but no two of the alternatives occurred within a common selective environment. In either case, there cannot have been selection for *T*. But, since *ex hypothesi T* does play a causal role in gamete production, it does causally contribute to the fitness of its bearers. Under these conditions, the strong theory would withhold a function attribution from *T*, since there has never been selection for it. But, since *T* does make a hereditary contribution to the fitness of its bearers, it does have some effect in virtue of which it gets reproduced across generations; so the weak theory would attribute a function to it. Thus, while the strong theory looks for a history of selection for a trait, the weak theory looks only for a history of hereditary contribution to fitness.

While not acknowledging the distinction between the strong and weak etiological theories, Allen and Bekoff clearly seem committed only to the weak etiological theory, as evidenced by a number of passages. For example, they formulate the etiological theory as holding that 'the functions of a given trait are those effects the trait had in the past that contributed to the selection of organisms with that trait' (1995*b*: 26). Here the function of a trait is not defined in terms of *selection for the trait*, but rather in terms of the trait's contribution to the *selection of organisms* with that trait, where the selection of organisms is a matter of overall fitness differences among organisms, which may ultimately be due to fitness differences in other traits. Consequently, this formulation does not require that a functional trait has a history of variation within a common selective environment, but only that it has a history of contributing to the fitness of its bearers. 'Thus,' Allen and Bekoff continue, 'the claim that "a function of a bow in a population of canids is to communicate that what follows is play" means that the past tendency of bows to communicate that what follows is play contributed to the reproductive success of ancestors of the present population' (1995*b*: 26). In other places as well they formulate the etiological theory in a way that does not require a history of selection for a functional trait, but only a history of contribution to ancestral fitness leading to the reproduction of tokens of the functional trait: 'a function of a trait is an effect of the trait that has contributed (in ancestral populations) to the preservation of the trait (in descendant populations) via the differential survival and reproduction of organisms with that trait' (1995*b*: 28). Note here that it is not the differential success *of traits* that matters, but the differential success *of organisms* bearing the traits, which again may be due to fitness differences in other traits. And, finally, in characterizing the 'historical notion of function', Allen and Bekoff say, 'a trait's function is the specific contribution that ancestral versions of the trait made to individual fitness' (Bekoff and Allen 1995: 253).

I have belabored these points because it is commonly assumed that the etiological theory requires selection for a trait in order for that trait to have a function. But there is a clearly historical—hence etiological—conception of function according to which a trait itself need not have been selected for, but need only have made a contribution to the success of organisms undergoing selection. This is what I have called the 'weak etiological theory' and, if we are to take Allen and Bekoff's explicit statements literally (which is what I propose to do), Allen and Bekoff are committed only to the weak etiological theory, not the strong. Since I am also committed to the weak etiological theory (Buller 1998), I am in agreement with Allen and Bekoff about what it is for a trait to have a function. So none of my disagreement with Allen and Bekoff concerning the relation between function and design turns on a difference in our conceptions of biological function.

To return, then, to the distinction between artifact functions and biological functions, there is a clear difference between the two. For example, the effect of weighting papers is not an effect of rocks that contributes to the reproduction of rocks (in fact, rocks are simply not reproduced from other rocks at all). Similarly, decorating walls is not an effect of stags' heads that contributes to the reproduction of stags' heads. So there is a sharp difference between biological and artifact functions. Nonetheless, I will now argue that it is not this difference that accounts for the disanalogy between Allen and Bekoff's artifactual examples and the functional traits of organisms. Rather, the source of the disanalogy lies in a difference between the types of *item* that possess the functions: each example of a functional artifact that Allen and Bekoff cite is radically unlike a functional trait of an organism. And this difference involves a different relation between the phenomena of function and design in the case of organismic traits (but not only organismic traits) than in the artifactual cases. For, I will now argue, every trait that has a function in an organism as per the weak etiological theory (which, I have argued, is how we should interpret Allen and Bekoff's clause (i)) is also designed for its role (contra Allen and Bekoff's clause (ii)).

4. System Design, Constraints, and Component Design

Begin by considering two different paths by which one can approach the phenomenon of natural design, where these are a function of different perspectives on the operation of natural selection. First the two perspectives on selection.

On the one hand, natural selection can be seen as working to create *adaptations* via accumulated small modifications to traits over evolutionary time.

Indeed, historically, one of the central explanatory functions of the theory of evolution by natural selection has been explaining how 'organs of extreme perfection' come to be. This trait-centered perspective on natural selection is the perspective of optimization theory, which typically models a trait of an organism in relative isolation from the rest of the organism. Insofar as other traits of the organism enter such models, they appear as historical, developmental, or trade-off constraints on the optimality of the trait under study (see e.g. Seger and Stubblefield 1996). Given this abstraction of a trait from its organismic context, optimization theory studies how natural selection may have optimized the trait for a particular function by retaining each mutation that moved the trait closer to optimal functioning.

On the other hand, natural selection can be seen as working to create *complex organisms*, or other living things, which exhibit functional interdependence of parts and overall adaptedness to the environment. This organism-centered perspective on natural selection corresponds to another primary explanatory function of the theory of evolution by natural selection—that of answering the Big Question of the *origins* of complex living things. Answering the Big Question was one of the major triumphs of Darwin's principle of natural selection over Paley's argument from design, and forcefully articulating and defending this explanation is the principal point of works such as Dawkins's *The Blind Watchmaker*. As Dawkins (1991: 43) says, 'living things are too improbable and too beautifully "designed" to have come into existence by chance. How, then, did they come into existence? The answer, Darwin's answer, is by gradual, step-by-step transformations from simple beginnings, from primordial entities sufficiently simple to have come into existence by chance.' From this perspective, the theory of evolution by natural selection can be seen as having 'ushered in a new way of thinking about the *design of organisms*' (Lauder 1996: 58; emphasis added).

Of course, these two perspectives on natural selection are not in the least incompatible. Indeed, it is one of the strengths of the theory of evolution by natural selection that it explains the origins of both complex living things and their 'organs of extreme perfection'. But these two ways of looking at the products of natural selection are avenues to very different conceptions of natural design. If the phenomenon of natural design is approached from the trait-centered perspective on natural selection, design will appear in the first instance as a property of *traits*, and attempts to analyze the concept of design will take the form of specifying the conditions under which it is appropriate to say that some trait is 'designed for' some particular role. Given this approach to design, these conditions will be specified in terms of a particular type of history of natural selection acting directly on a trait, independently of how that trait fits into the overall structure of the organism possessing it. And this, in fact, is precisely the approach to design exemplified by Allen and Bekoff's theory. From the organism-centered perspective, in contrast, design will appear in the first instance as a property of *organisms as wholes* (as per the above quoted passages from Dawkins and Lauder). From this vantage point, selection is seen as 'choosing' among competing designs in each generation, preserving those designs that exhibit the highest overall adaptedness to the environment (Lauder 1996: 60). Traits then appear as 'components of design' (Lauder 1996: 56) and are considered 'designed for' whatever contributions they make to the success of organismal designs under selection. In this case, it will not be necessary for selection to have acted directly on a trait in order for it to be designed for its role, and *ipso facto* not necessary that the trait underwent modification by selection for its role; it is sufficient simply that it contribute to the overall design of the organism possessing it.

It is important to note that both conceptions of design agree in taking design to be a product of, or created by, the operation of natural selection. So both conceptions of design see natural design only where natural selection has done some work to create it. They differ only with respect to what they emphasize natural selection as creating—traits (adaptations) or organisms. This, in turn, makes the views differ with respect to how direct a link there must be between the operation of selection and what a trait is designed for. In the trait-centered approach, selection must have operated directly on the trait. In the organism-centered approach, the link between the design of a trait and the operation of selection can be very indirect: selection must only create overall organismic design and a trait must contribute to that design, independently of whether there was direct selection for the trait's role in that design. Precisely how a trait can be designed by natural selection in this sense is what I will now explain and defend.

Consider again Allen and Bekoff's rock–paperweight example. First, the rock–paperweight is *unarticulated*—that is, it does not possess an internal articulation into parts that themselves produce effects that are involved in, or required for, the rock's performing its function as a paperweight. Secondly, the rock–paperweight is *unintegrated*—that is, it is not itself a component part of a larger paperweight system and, hence, does not produce an effect that plays a causal role within a more inclusive system that is a paperweight (rather, the rock itself is the entire paperweight). Thus, there are no constraints on the rock's function deriving from either processes that it contains or processes in which it is contained. The rock performs its function as a paperweight just by directly 'engaging' the world with a single effect, holding down papers, which it produces whenever it is placed on top of papers.

In contrast to the rock–paperweight, functional traits of organisms (that is, traits that satisfy the weak etiological theory of functions) are *articulated*, since they are internally complex, decomposable into substructures that themselves make a causal contribution to the production of the functional trait's proper effect. A favorite functional example of philosophers, the heart, clearly illustrates such internal articulation. Such traits are also *integrated*, since they have functions in virtue of making (or having made in the past) a contribution to fitness via a causal contribution to a component of fitness—to viability, fertility, fecundity, or mating ability. But no trait makes a causal contribution to fitness in isolation from the other traits of the organism possessing it. Rather, and properly speaking, traits are *components* within complex functionally integrated organismic *systems*. In all such complex systems (natural or artifactual), a component makes a contribution to features of the entire system in virtue of causal interactions with the other system components to which it is connected, where the sum of the actions of this network of connected subsystems produces the features of the system as a whole. The heart, again, illustrates such integration; for it makes a contribution to fitness only in virtue of being properly connected with the rest of the circulatory system, which in turn makes a contribution to fitness only in virtue of being connected with the rest of an organism's vital systems (for example, the respiratory system).

When a functional item is, like an organismic trait, both internally articulated and externally integrated, it is nested within a functional *hierarchy* of subsystems. There are thus subsystems 'below' the functional item in the hierarchy whose effects are causally necessary for its producing its proper effect—in particular, those subsystems that constitute its internal articulation. The heart, for example, is internally articulated into the atria and ventricles (among other things), the contractions of which are causally necessary for the heart's pumping. In addition, there are (sub)systems 'above' the functional item in the hierarchy to whose proper effects its own proper effect is causally necessary—in particular, those (sub)systems with which it is externally integrated. The heart's pumping, for example, is causally necessary for the circulatory system's producing its proper effect of transporting nutrients to the cells of the body.

Being so nested in a system hierarchy, there are *constraints* on an item's function that stem both from the subsystems below it and the (sub)systems above it in the hierarchy. That is, only an item with a highly specific set of properties could occupy a locus in a system hierarchy in such a way as to effectively interact causally with other system components and thereby contribute to the overall success of the system. These constraints on a functional item in effect constitute a set of system 'needs' for the locus occupied by that item; and only an item that can satisfy those 'needs' will come to occupy that locus. This contrasts sharply with artifacts such as the rock–paperweight; for there are no constraints on the rock–paperweight deriving from subsystems to which it is connected, since it is neither articulated nor integrated. There is thus a strong disanalogy between organismic traits and artifacts such as the rock–paperweight.

But what does this disanalogy have to do with function and natural design? Well, since there are no systemic constraints on the function of the rock–

paperweight, it can perform its function of paperweighting just by being placed on top of papers. Design for paperweighting is in no way necessary for the rock to perform this function. Even more than this, however, since the rock can perform its paperweighting function in the absence of design for that function, there is no room for design to enter into the rock's function *except* by way of modification to better perform the function that it already performs in the absence of design. In contrast, the systemic 'needs' that a functional item must satisfy in order to occupy a locus within a complex hierarchical system constitute a set of *design constraints* on that functional item. That is, the design of the system as a whole—its design as a functionally integrated system—requires an item with a highly specific set of properties to occupy a particular locus within that system in such a way as to interact effectively with the other system components. An item that possesses that set of properties, and effectively interacts with other system components so as to satisfy the design constraints on the locus it occupies, is in a strong sense *designed for* its role within the system. Thus, design enters into the functions of system components in virtue of the causal requirements placed on those components by the overall design of the system of which they are parts. And a component can be designed for the locus it occupies without ever having been modified to better perform the function required of it.

Let me sharpen these points with an artifactual example that is analogous to an organismic trait and, thus, more appropriate than the artifactual examples offered by Allen and Bekoff. Consider the prototype of the gasoline-powered internal combustion engine. The engine as a whole had to be designed to convert chemical energy into mechanical energy with a certain minimum degree of efficiency. (Analogously, natural selection designs organisms *as wholes* to solve the adaptive problems posed by their environments.) The overall design requirements on the internal combustion engine in turn required a component that would perform the function of vaporizing gasoline and delivering the resulting gas-air mixture to the combustion chamber. This design requirement was satisfied by the carburetor, which had to be constructed in a highly specific way so as to satisfy the design constraints that the system as a whole placed on the locus the carburetor was to occupy within it. (Analogously, the adaptive problems that natural selection designs organisms to solve place demands on those parts of organisms that can be involved in solving those problems, and these demands create 'a hierarchy of ever more specific selection pressures' on the subsystems that compose those parts (Kitcher 1993: 390).) Of course, internal combustion engines underwent tremendous evolution after the prototype, and part of that evolution involved modifications to the carburetor to make it more efficient in performing its function. But the crucial question is: was the carburetor of the prototype—the first version of the carburetor, prior to any subsequent modification—designed to vaporize gasoline or not?

Clearly it was. So it is not true that only modified versions of the first carburetor were designed to vaporize gasoline, while it was merely the function of the first carburetor to do so. The modified versions of the carburetor were merely *better designed*.

The difference between the rock–paperweight and the carburetor is that the rock–paperweight can perform the function of paperweighting in the absence of design, whereas nothing can perform the function of vaporizing gasoline in an internal combustion engine without satisfying a set of design constraints derived from the overall design of the engine within which that function is performed. Indeed, in general, *no* component performing a function in a complex system would even *be* a component of that system if it did not satisfy the design constraints placed on it by the system; components that do not satisfy such design constraints are eliminated (by natural selection in the biological case and human designers in the artifactual case). In other words, causally integrating within the overall design of the complex system that contains a component is a necessary condition for that component's performing a function in that system, since selection processes (either natural or artificial) weed out components that fail to integrate within overall system design. Thus, being designed for its role within a complex (artifactual or natural) system is a necessary condition of a component's performing a function within that system, since, if the component were not designed for its role, it would not be there in the system performing that function.

When the focus is thus shifted from Allen and Bekoff's artifactual examples to artifactual examples that actually are analogous to organismic traits, it can be seen that proper design is a necessary condition for component function in a complex system, even if that component has never undergone modification for its function. And, since traits are properly speaking components of complex organismic systems, proper natural design of a trait is a necessary condition for biological function, and modification for that function is not necessary for design. The design of components of complex systems (including traits of organisms) is a function of the role those components play within overall system design. Design, as it were, 'trickles in' from the system as a whole to the components that contribute to the system's satisfaction of the design requirements that are placed on it as a whole. Thus, in designing organisms as wholes to respond to certain adaptive problems, natural selection thereby designs the conditions that parts of organisms must satisfy in order to help the organism respond to environmental demands; and, when a part satisfies those conditions, it is designed for its role within the organism.

It may seem, however, that this goes too far; for the above discussion may appear to make natural design a wholly non-historical phenomenon, thereby divorcing it radically from the phenomenon of function, which is essentially a historical notion according to the weak etiological theory. In particular, it may appear that an item must first be designed for its role within an organismic system and only subsequently be reproduced because of its playing that role. To put this graphically, a trait *T* that contributes to the fitness of the 'first' organism *O* in a lineage to possess *T* does not have a function in that organism, since a trait has a function only if it is a reproduction of earlier items that contributed to fitness in the same way. And it may appear that the conception of design I have been urging would count *T* as designed for its contribution to the fitness of *O*, since it may appear that *T* causally interacts with other components of *O* in such a way as to contribute to *O*'s response to its selective environment.

But this appearance is deceptive; for traits that have no history under a selection process are not actually instances of natural design. The reason is the same as the reason they do not have functions under the etiological conception of function (see Neander 1991*b*: 460–1). Each token trait that has a function in an organism is embedded in a causal nexus where it produces numerous effects, many of which are not produced by other tokens of its (homologous) type. Which of these numerous effects it is the *function* of that token to produce is a matter of the contribution of the trait *qua type* to the fitness of organisms possessing tokens of it—that is, a matter of which effect serves to contribute to fitness when averaged over the class of tokens of that type. Such average contributions to fitness can be assessed only transgenerationally, however, since they are determined by the operation of selection on a lineage. So, under the etiological conception, a trait has the function of producing some effect only when a transgenerationally stable relationship between that *type of trait* and a *type of effect* emerges through the process of natural selection. In exactly the same way, and for exactly the same reasons, design ascriptions are also grounded in a transgenerationally stable relationship between a type of trait and the types of design constraint the trait satisfies in the relevant organismic systems; and this transgenerationally stable relationship emerges only through the process of natural selection. Thus, a history under selection (of design constraint satisfaction) is just as necessary for a trait's being an instance of natural design as it is for its having an etiological function. (Note, in both cases, the relevant selective history consists in the operation of selection on a lineage, not on a trait. Traits have (weak etiological) functions and design because of their history of contributing to the success under selection of organisms in a lineage.)

Drawing all these considerations together, it is now possible to formulate succinctly an alternative to Allen and Bekoff's conception of natural design. I will call the following formulation the *systemic conception*: a token of a trait *T* in an organism *O* is *naturally designed* for *X* if and only if

(i) *O*'s ancestor/ancestors faced some adaptive problem *P*, the solution of which was necessary for its/their reproductive success;

(ii) there was some organismic (sub)system *S* of *O*'s ancestor/ancestors that solved *P*;

(iii) the demands that *P* placed on *S* required a causally integrated response from components of *S* and this, in turn, required a component occupying locus *L* in *S* causally to interact with the other components of *S* by performing task *X*; and

(iv) *T* occupied *L* in *S* and successfully did *X*.

To briefly illustrate the systemic conception of natural design: my kidneys are designed for filtering metabolic wastes from my blood, since (i) my ancestors faced the problem of combating toxicity, which (ii) their urinary systems solved by removing metabolic wastes from the body, which (iii) required a causally integrated response from components of the urinary system (for example, renal arteries, ureters, bladder), and this, in turn, required a component occupying a position with both access to the blood and an outlet to the bladder to perform the removal of metabolic wastes from the blood, and (iv) the kidneys occupied this position and successfully removed metabolic wastes from the blood.

The systemic conception of natural design is consequently more inclusive than the conception of design defended by Allen and Bekoff, since it does not make modification a necessary condition of design. But it is thereby coextensive, although not synonymous, with the weak etiological concept of function accepted by both Allen and Bekoff and me. Any trait that satisfies the systemic conception will have contributed to an ancestral solution to an adaptive problem and thereby will have contributed to ancestral reproductive success; such a trait will thus also satisfy the etiological conception of function and be ascribed the function of making that particular contribution to reproductive success.

5. Conclusion: Adaptation Again

The systemic conception of design thus restores the connection between function and natural design, but it loosens the connection between design and adaptation, *if* adaptation is conceived as requiring modification by selection for greater effectiveness in functional performance (the traditional conception of adaptation, which Allen and Bekoff accept). As the arms-race argument of Section 2 showed, however, there is a systematic ambiguity in the notion of 'modification for greater functional effectiveness'. And this ambiguity infects the traditional conception of adaptation just as much as it infects Allen and Bekoff's conception of natural design. So an account of adaptation lacking this systematic ambiguity may restore a tighter connection between adaptation and natural design.

There are, at this point, two ways in which a tighter connection could be restored. First, one could adopt something like Sober's definition of 'adaptation': '*A* is an adaptation for task *T* in population *P* if and only if *A* became prevalent in *P* because there was selection for *A*, where the selective advantage of *A* was due to the fact that *A* helped perform task *T*' (1984: 208). This makes 'adaptation' synonymous with 'function' as defined by the strong etiological theory, and consequently makes both instances of 'natural design' under the systemic conception. The disadvantage of this account of adaptation is that it departs rather widely from the common idea that adaptations involve a history of modification.

Secondly, one could retain the requirement of modification, but divorce it from the notion of greater effectiveness in the performance of some function. This could be accomplished as follows: an adaptation is a trait that has undergone *design improvement* under selection. Under the systemic conception of natural design, the notion of design improvement would require only that a trait be modified to more efficiently satisfy the design constraints placed on it by the organismic system of which it is a part. That is, a modification would be judged to be a design improvement wholly in terms of *intrasystemic* engineering standards, and not in terms of extrasystemic effectiveness in solving a specific adaptive problem: a modification to a system component that results in more efficient overall system design is a design improvement, regardless of whether it results in greater effectiveness in solving an adaptive problem.

Suppose, for example, that mutation produced a modification to the leg muscles of cheetahs, which made the muscles more efficient in converting chemical energy into contractile force, resulting not in an increase in contractile force, but only in a decrease in the amount of chemical energy used by the muscles in producing the same contractile force as before the modification. Presumably, the leg muscles of cheetahs have the function of enabling cheetahs to catch prey. But this modification would not result in greater running speed, hence would not result in greater effectiveness with respect to solving the problem of catching prey. Nonetheless, the modification would reduce the energy burned by cheetahs in chasing prey and thus reduce the frequency at which feeding is necessary. In short, the modification would produce a more efficient overall organismic design without resulting in 'greater functional effectiveness' with respect to running speed and the catching of prey. If adaptations are understood in this way, as traits that have undergone intrasystemic design improvements, the cheetah's modified muscles would count as an adaptation.

This account of adaptation thus avoids the systematic ambiguity involved in the notion of 'modification for greater functional effectiveness'. At the same time it accords very well with the conception of adaptation that is at work in discussions of evolutionary arms races. As Dawkins describes evolutionary arms races, for example, there is 'progress in design, but no progress in accomplishment', because an arms race is a process of '*equipment* improving while its net *effectiveness* stands still' (1991: 186). No account of adaptation that defines improvement in terms of increased effectiveness in solving a specific adaptive problem can apply to these sorts of adaptive change. But the adaptive changes that typify arms races fit very naturally within an account of adaptations as traits that have undergone intrasystemic design improvement, when this is understood in accordance with the systemic conception of natural design. In sum, then, the current proposal can accept Allen and Bekoff's distinction between merely functional traits and adaptations; it simply shows both to be instances of natural design, while showing adaptations to be special cases of natural design.

When the focus is thus shifted away from Allen and Bekoff's artifactual examples to artifactual examples that are articulated and integrated functional components of complex systems, proper design emerges as closely related to function. And functional traits of organisms are precisely such components of complex functionally integrated systems. Of course, Allen and Bekoff's artifactual examples do show that, if we are concerned to provide an analysis of the concept of function in *all* its uses (if we are concerned to provide something like a set of necessary and sufficient conditions for the fully general employment of the concept), design is not a necessary condition for function. But the considerations that motivate this conclusion are not relevant to cases of *biological* function. In the biological case, function and design are inseparable.

References

Allen, Colin, and Bekoff, Marc (1995*a*), 'Biological Function, Adaptation, and Natural Design', *Philosophy of Science*, 62: 609–22.

————(1995*b*), 'Function, Natural Design, and Animal Behavior: Philosophical and Ethological Considerations', in N. S. Thompson (ed.), *Perspectives in Ethology 11: Behavioral Design* (New York: Plenum), 1–46.

Bekoff, Marc, and Allen, Colin (1995), 'Teleology, Function, Design, and the Evolution of Animal Behavior', *Trends in Ecology and Evolution*, 10: 253–63.

Brandon, Robert N. (1990), *Adaptation and Environment* (Princeton: Princeton University Press).

Buller, David J. (1998), 'Etiological Theories of Function: A Geographical Survey', *Biology and Philosophy*, 13: 505–27.

Davies, Nicholas B., and Brooke, Michael (1991), 'Coevolution of the Cuckoo and its Hosts', *Scientific American*, 264/1: 92–8.

Dawkins, Richard (1982), *The Extended Phenotype* (Oxford: Oxford University Press).

——(1991), *The Blind Watchmaker* (London: Penguin).

Endler, John A. (1986), *Natural Selection in the Wild* (Princeton: Princeton University Press).

——(1991), 'Interactions between Predators and Prey', in J. R. Krebs and N. B. Davies (eds.), *Behavioural Ecology: An Evolutionary Approach* (Oxford: Blackwell Scientific), 169–96.

Godfrey-Smith, Peter (1994), 'A Modern History Theory of Functions', *Noûs*, 28: 344–62.

Kitcher, Philip (1993), 'Function and Design', *Midwest Studies in Philosophy*, 18: 379–97.

Krebs, J. R., and Davies, N. B. (1993), *An Introduction to Behavioural Ecology*, 3rd edn. (Oxford: Blackwell Scientific).

Lauder, George V. (1996), 'The Argument from Design', in M. R. Rose and G. V. Lauder (eds.), *Adaptation* (San Diego, Calif.: Academic Press), 55–91.

Lewontin, Richard (1978), 'Adaptation', *Scientific American*, 239/3: 212–30.

Lotem, Arnon, and Rothstein, Stephen I. (1995), 'Cuckoo–Host Coevolution: From Snapshots of an Arms Race to the Documentation of Microevolution', *Trends in Ecology and Evolution*, 10: 436–7.

Millikan, Ruth Garrett (1984), *Language, Thought, and Other Biological Categories: New Foundations for Realism* (Cambridge, Mass.: MIT Press).

——(1989*a*), 'An Ambiguity in the Notion "Function"', *Biology and Philosophy*, 4/2: 172–6.

——(1989*b*), 'In Defense of Proper Functions', *Philosophy of Science*, 56/2: 288–302.

——(1993), 'Propensities, Exaptations, and the Brain', in *White Queen Psychology and Other Essays for Alice* (Cambridge, Mass.: MIT Press), 31–50.

Neander, Karen (1991*a*), 'Functions as Selected Effects: The Conceptual Analyst's Defense', *Philosophy of Science*, 58: 168–84.

——(1991*b*), 'The Teleological Notion of "Function"', *Australasian Journal of Philosophy*, 69: 454–68.

Reeve, Hudson Kern, and Sherman, Paul W. (1993), 'Adaptation and the Goals of Evolutionary Research', *Quarterly Review of Biology*, 68/1: 1–32.

Ridley, Mark (1996), *Evolution*, 2nd edn. (Cambridge, Mass.: Blackwell Science).

Seger, Jon, and Stubblefield, J. William (1996), 'Optimization and Adaptation', in M. R. Rose and G. V. Lauder (eds.), *Adaptation* (San Diego, Calif.: Academic Press), 93–123.

Sober, Elliott (1984), *The Nature of Selection: Evolutionary Theory in Philosophical Focus* (Cambridge, Mass.: MIT Press).

van Valen, Leigh (1973), 'A New Evolutionary Law', *Evolutionary Theory*, 1: 1–30.

Vermeij, Geerat J. (1987), *Evolution and Escalation: An Ecological History of Life* (Princeton: Princeton University Press).

Weis, Arthur E., McCrea, Kenneth D., and Abrahamson, Warren G. (1989), 'Can there be an Escalating Arms Race without Coevolution? Implications from a Host-Parasitoid Simulation', *Evolutionary Ecology*, 3: 361–70.

Williams, George C. (1966), *Adaptation and Natural Selection: A Critique of Some Current Evolutionary Thought* (Princeton: Princeton University Press).

——(1992), *Natural Selection: Domains, Levels, and Challenges* (New York: Oxford University Press).

9. The Continuing Usefulness Account of Proper Function

PETER H. SCHWARTZ

ABSTRACT

'Modern History' views claim that in order for a trait X to have the proper function F, X must have been *recently* favored by natural selection for doing F (Griffiths 1992, 1993; Godfrey-Smith 1994). For many traits with prototypical proper functions, however, such recent selection may not have occurred, since traits may have been maintained owing to lack of variation or selection for other effects. I explore this flaw in Modern History accounts and offer an alternative etiological theory, which I call the 'Continuing Usefulness' account. According to my view, a trait has the proper function F if and only if, first, the trait was favored by selection for doing F *at some point* (perhaps far in the past), and, secondly, the trait has *recently* contributed to survival and reproduction by doing F. This separates the requirement involving natural selection from the one involving the recent past, two issues that the Modern History account conflates. The clear separation allows a detailed analysis of the causal judgments and form of adaptationism that ground the etiological approach to function.

In the last quarter century, etiological accounts have dominated much of the debate over the concept of function. Following Larry Wright (1973), these accounts say that a trait has a function F only if it has been favored by natural selection for doing F. Even critics now agree that this approach is needed to explicate at least one of the concepts at work in biology, which some have termed 'proper function' (cf. Millikan 1984, 1989*b*). Attempting to work out the details, Griffiths (1992, 1993) and Godfrey-Smith (1994) have presented convincing arguments that an etiological account should require that the relevant natural selection occurred *recently*, often in the form of maintenance selection. I describe the resulting 'Modern History' view and the reasons for its adoption in Section 1.

My deepest thanks go to Gary Hatfield and Gary Ebbs for their guidance and encouragement. I also thank Zoltan Domotor, David Magnus, Ruth Millikan, Karen Neander, Shari Rudavsky, and Neil Shubin for their help. I presented an earlier version of this work at the Sixteenth Biennial Meeting of the Philosophy of Science Association, which appeared in the proceedings of that meeting (Schwartz 1999*b*). I am grateful to the Philosophy of Science Association for permission to reprint sections of that paper here.

However, there is an important problem for this theory that has not been adequately recognized. For many traits with prototypical proper functions, the essential sort of recent selection may not have occurred. Traits may have been maintained in a species owing to lack of variation or selection for other effects. In Section 2, I explain how this can happen and give examples from the biological literature. In fact, if biologists were abiding by the Modern History theory's requirements, they could not assign proper functions to many prototypically functional traits, since they often do not know whether the necessary form of maintenance selection occurred. In Section 3, I argue that this and other related problems pose a serious challenge to the Modern History theory. Because this theory has stood as the best supported etiological approach, and because the arguments of Griffiths (1992, 1993) and Godfrey-Smith (1994) so successfully undermine all other etiological accounts, the problem threatens the entire etiological approach.

But there is a way out. In Section 4, I describe an etiological account that avoids the problems Griffiths and Godfrey-Smith point out for other views, as well as the problem I emphasize for theirs. According to my 'Continuing Usefulness' account, a trait has the proper function F if and only if, first, the trait was favored by selection for doing F *at some point* (perhaps far in the past), and, secondly, the trait has *recently* contributed to survival and reproduction by doing F. This separates the requirement involving natural selection from the one involving the recent past, two issues that the Modern History account conflates. The clear separation allows a detailed analysis of the causal judgments and form of adaptationism behind the etiological approach to function. My discussion of these topics in the chapter's final sections branches out of the function debate into philosophy of biology more generally.

1. The Modern History Account of Proper Functions

Etiological theorists generally accept that two concepts of function are at work in biology. The concept of 'proper function', which they hope to explicate, has certain normative and explanatory implications. Saying that a trait-type[1] has the proper function F implies that it has the purpose of doing F (Millikan 1989*a*, *b*, 1993), that its doing F is in some way normative (Neander 1991*a*, *b*), or that its doing F helps explain the natural selection of the trait (Griffiths 1992, 1993; Godfrey-Smith 1994). The other concept of function at work in biology carries none of these implications. An item has its 'causal role function' (to use Neander's term (1991*a*)) based on the causal contribution it makes to the production of some effect (cf. Cummins 1975); there is no implication that this is

[1] Throughout this chapter, 'trait' will mean trait-type, unless otherwise specified.

what the item should do or has done in the past. Proper function, by contrast, requires a certain sort of causal history, in particular one involving natural selection. Neander (1991*a*: 174) writes: 'It is the/a proper function of an item (X) of an organism (O) to do that which items of X's type did to contribute to the inclusive fitness of O's ancestors, and which caused the genotype, of which X is the phenotypic expression, to be selected by natural selection.' Millikan (1984, 1989*b*) and Neander (1991*a*, *b*) have played key roles in delineating the concept of proper function and arguing for the etiological approach to explicating it.

Neither of them places a restriction on when the crucial action of natural selection must have occurred and thus their accounts provide no way for a trait to lose its proper function. Griffiths (1992, 1993) and Godfrey-Smith (1994) argue that this aspect undermines the theories since in fact traits do lose their proper functions, and thus the accounts fail to fit one important aspect of the concept's use in biology. As Godfrey-Smith (1994) emphasizes, a trait may arise under selection pressure for carrying out one effect, but then may be modified in various ways and come to have a new proper function.[2] For example, although feathers may have arisen under selection for thermoregulation (Ostrom 1979), the wing feathers of modern flying birds may now only have the proper function of aiding flight. Or, as Griffiths (1993) emphasizes, a trait may become 'vestigial': it may lose a proper function it once had without acquiring a new one. The sightless eyes of cave-dwelling species, for example, no longer allow sight, although they did so in the ancestral species that lived in the light.

Both Griffiths (1992, 1993) and Godfrey-Smith (1994) modify the etiological approach to mesh with this aspect of biology by adding the requirement that the relevant selection occurred *recently*. Thus, if a trait stops being selected for doing F, then after a certain amount of time it will no longer count as having the proper function F. In this way, the wing feathers of modern birds may lose the proper function of thermoregulation, and the sightless eyes of cave-dwelling species may lose the proper function of aiding sight.

The two Modern History theories differ somewhat. Godfrey-Smith (1994), following Millikan (1984, 1989*b*), uses the idea of a 'reproductively established family' which is (roughly) a group of entities united because at least one of their properties is 'copied' from one individual to the next. Thus, all human hearts are members of a 'family' since they have key properties that are copies of properties of past hearts.[3] Using this terminology, Godfrey-Smith (1994: 359) states his definition of proper function as follows:

[2] Gould and Vrba's discussion (1982) of 'exaptations' has been crucial to drawing this sort of example to the attention of philosophers of biology.

[3] As mentioned, this terminology stems from Millikan (1984) and there is simply not enough space here to provide a full recounting of its use. Please refer to Millikan (1984, 1989*b*) and Godfrey-Smith (1994) for more details.

(F3) The function of m is to F iff:
(i) m is a member of family T,
(ii) members of family T are components of biologically real systems of type S,
(iii) among the properties copied between members of T is property or property cluster C,
(iv) one reason members of T such as m exist now is the fact that past members of T were successful under selection *in the recent past*, through positively contributing to the fitness of systems of type S, and
(v) members of T were selected because they did F, through having C. (emphasis added)

Godfrey-Smith's two main novelties are his requirement, in (ii), that the members of the family T 'are components of biologically real systems' and, in (iv), that they have been selected 'in the recent past'.

Griffiths (1992, 1993) states his version slightly differently; he emphasizes the importance of 'regressive evolution'—that is, changes that result when a trait no longer plays a key adaptive role, such as when a cave-dwelling species becomes sightless. But he wants to allow traits to become vestigial even before such modification occurs, and thus he constructs a more complex account. He (1993: 417) defines the notion of 'an evolutionarily significant time period' for a trait as 'a period such that, given the mutation rate at the loci controlling T and the population size, we would expect sufficient variants for T to have occurred to allow significant regressive evolution if the trait was making no contribution to fitness'. A 'proximal selective explanation' is 'one that involves the action of selective forces during the last evolutionarily significant period, or would have involved such action during that period had the mutation rate not fallen below expectation' (1993: 417–18). Using these terms, he states his formal account of proper function: 'Where i is a trait of systems of type S, a proper function of i in S's is F iff a proximal selective explanation of the current non-zero proportion of S's with i must cite F as a component in the fitness conferred by i' (1993: 418). Griffiths's definition is slightly more liberal than Godfrey-Smith's since it does not require *actual* recent activity by natural selection: a trait can have a proper function if there were no selection because the mutation rate dropped 'below expectation', a distinction to which I return later.

2. How Maintenance Selection may Fail to Act as Expected

2.1. The Two Possibilities

Despite the Modern History theory's advantages, a basic fact about biology interferes: many traits with prototypical proper functions may not have been recently favored by natural selection for carrying them out. We must remember that, for selection to occur, there must be 'heritable variation in fitness' (Lewontin 1970; Endler 1986: 4; Sober 1993: 9). This means that, in order for trait X to have been favored by selection for doing F during some period P, there must have been more than one form of X that was *heritable*, where these various forms performed F to different degrees, and where this difference in doing F resulted in a difference in the bearers' *propensity to survive and reproduce.*[4] Even if tokens of X contributed to the survival and reproduction of their bearers during this period by doing F, there are at least two ways that X (or its tokens) could have failed to have been selected for doing F: either (*a*) there may not have been suitable variation in X for natural selection to occur at all, or (*b*) selection that maintained X in the population may have favored some effect other than F.

And the simple fact is that, for most traits in most populations, biologists do not have the data to rule out these two alternatives. There are many barriers to demonstrating the necessary heritable variation and fitness differences even in current populations, and studies doing so are few and far between (cf. Endler 1986). The problems are even greater when the analysis concerns what happened in the past. Therefore, if discerning proper function depended on a judgment about the recent action of natural selection, biologists could not confidently assign proper functions to traits. This indicates that there is something seriously wrong with the Modern History approach to explicating the concept.

Let me describe the two possibilities at greater length. Although the universal occurrence of mutations guarantees some genetic variation in traits, there is good evidence that various mechanisms reduce the range of resulting phenotypic variation. For example, in a process called 'canalization', selection may favor genetic arrangements that reduce the risk of a trait's failure to be expressed (Waddington 1959). Given a trait X that makes a crucial contribution to survival and reproduction by doing F, natural selection will favor individuals with the lowest percentage of offspring and grandoffspring lacking the trait (all else being equal). Computer models predict the development of mechanisms that reduce the phenotypic variation created by genetic mutations (Wagner 1996), and experimental data in *Drosophila* suggest that traits that are more important to fitness are correspondingly less likely to be altered by random genetic changes in the genome (Stearns and Kawecki 1994).

[4] This way of explicating 'heritable variation in fitness' depends on the widely accepted 'propensity account of fitness' (Mills and Beatty 1979; Sober 1993). In addition, in what follows I adopt Sober's simplifying assumption (1993: 10) that a trait is heritable if 'offspring "tend" to resemble their parents' in their expression of this trait. As Sober (1993) acknowledges, this sense of 'heritable' applies to predictable similarities between parents and offspring that are non-genetic, including ones that result from parents and offspring eating similar things, for example. A more formal definition says that a trait is heritable only if offspring have a higher probability of resembling their parents than of resembling other adults in the population in regard to this trait. By this definition, a fixed and invariable trait in a species is not heritable. I will use the less formal definition—which allows fixed traits to be heritable—although the difference is not crucial: the cases that matter to my argument are ones where variation exists.

In addition, even if some heritable variation exists in a trait's ability to carry out its useful effect, it may be favored by selection for other reasons. For example, consider the feathers of flying birds. Imagine that wing feathers of type p1 are ubiquitous in some species. Because of a mutation, feathers of type p2 arise, where from an engineering standpoint these feathers are not as efficient producing flight as feathers of type p1 (in this species). Assume in addition that small decreases in flight efficiency translate into decreases in survival and reproduction. It might seem unavoidable that p2 feathers will be selected against because of their inferior flight-allowing properties. But p2 feathers may also make their bearers more susceptible to hypothermia, or less efficient at eliciting feeding from parents, and these effects could kill off such individuals before they even have a chance to fly. Under these scenarios, p2 feathers will be selected against because of these other effects, rather than because of their failure to allow efficient flight.

Similarly, 'pleiotropic' effects of genetic changes may eliminate some heritable variation. Imagine now that p2 feathers carry no unfortunate side effects such as hypothermia or starvation; the feathers are just relatively inefficient for flight. But it is possible that the genetic mutation (let us call it g2) that would lead to feathers of type p2 has effects on other traits—that is, pleiotropic effects—that are maladaptive. For example, perhaps g2 causes a malformation of some other ectodermal tissue. Once again, if individuals with mutation g2 die before they have a chance to attempt inefficient flight, phenotype p1 would be maintained because of these other effects of g2, not because of the inefficiency of p2 feathers for flight.[5]

2.2. *Specific Cases from Biology*

In cases where biologists investigate a gene that they think plays a certain key role, they often find that it plays other roles that are even more crucial. Cheng *et al.* (1995) and Turner *et al.* (1995) set out to confirm that the Syk gene in mice plays a role in the development of the B cells of the immune system. They attempted to establish this with a standard 'knockout' experiment, producing mice lacking the gene and looking for defects in B-cell development. The only problem is that most knockout mice died before birth or shortly after, apparently because of the absence of Syk's usual contribution to the development of blood vessels. Thus, Syk may be maintained by selection because of its effect on the circulation system rather than because of its effect on the immune system.

[5] Buller (1998) has described another way that a trait can make contributions to survival and reproduction without being favored by natural selection for doing so. In his cases there is genetic variation in the trait, but the individuals expressing that variation do not live in a common environment (1998: 508).

Biologists also discuss how an apparently adaptive trait may be maintained because of effects other than its most obvious effect. One example involves the 'helpers-at-the-nest' phenomenon of some bird species, where some adults help raise offspring that are not their own progeny. Many biologists have proposed that this behavior is favored by selection because it increases the successful breeding of close relatives (kin selection), improves access to resources controlled by the breeding pair (reciprocal altruism), or just provides experience that aids the helper when it attempts to raise its own offspring (training).

Jamieson (1989, 1991) points out that, even if helping at the nest has any of these positive effects, selection may not have maintained the behavior for this reason. First of all, he notes that no studies have established the existence of heritable variation in the behavior or the tendency to perform it (1989: 397). Secondly, he (1991) says that, even if there is variation in the behavior, selection might maintain the trait by favoring some other effect. In particular, the same changes in neural organization that result in a decreased proclivity to help others raise their offspring could also result in a decreased proclivity to feed one's own offspring. The latter effect may be the one that natural selection discriminates against. This possibility shows that we simply cannot be sure, given the information we now have, that selection has recently maintained the helping behavior (Jamieson 1991: 278).

Similar debates have concerned the evolution and maintenance of structures, as well as behaviors. One of the racier ones centers on female clitoris and orgasm. Gould (1987) argues that the clitoris and the good feelings it engenders may not themselves have been favored by natural selection. Like male nipples, the clitoris could be the non-adaptive result of developmental processes selected for their effect on the other gender's analogous structure. To be precise, and to answer critics such as Sherman (1988, 1989), explanations of this sort must postulate either (i) an absence of relevant heritable variation, or (ii) selection on the heritable variation that does not involve benefits stemming from sexual sensitivity. Gould or his followers might hypothesize that females without a clitoris were selected against not because they failed to pursue matings as fervently, for example, but because their sons (and brothers, and uncles, and nephews, etc.) had some increased prevalence of sexual malfunction.

3. Consequences for the Modern History Theories

These examples establish, I believe, that a trait with an effect that aids survival and reproduction may persist in a population for reasons other than natural selection's favoring its accomplishment of this effect. I'll call this the possibility of 'Non-Selective Maintenance', even though it includes situations where selection plays a role—that is, favoring some other effect. The possibility of Non-Selective Maintenance and the fact that biologists usually cannot rule it out greatly diminishes the attractiveness of the Modern History interpretation of proper function.

Griffiths (1992) and Godfrey-Smith (1994) acknowledge this problem in passing, but provide little defense against it. Godfrey-Smith (1994: 356–7) writes,

> The modern history view does, we must recognize, involve substantial biological commitments. *Perhaps traits are, as a matter of biological fact, retained largely through various kinds of inertia. Perhaps there is not constant phenotypic variation in many characters, or new variants are eliminated primarily for non-selective reasons.* That is, perhaps many traits around now are not around because of things they have been doing. Then many modern-historical function statements will be false. If functions are to be understood as explanatory, in Wright's sense, there is no avoiding risks of this sort. (emphasis added)

In some sense, he is correct to say that any etiological theory takes 'risks of this sort', but he ignores how drastically magnified such risks are after the addition of his requirement. We may have good reasons to believe that in most cases of proper function the necessary sort of selection occurred at some point, even though we cannot be sure that it continued into the recent past. As I will discuss below (Sect. 4.3), we can be confident, for example, that the wing feathers of modern flying birds were favored *at some point* for their contribution to flight, even though within many given species we cannot be confident that such selection has occurred in the recent past.

The existence of this uncertainty means that his theory cannot fulfill the goal he sets for it. Godfrey-Smith (1994: 345) says that he wants to provide a conceptual analysis of proper function that is 'guided more by the demands imposed by the role the concept of function plays in science, the real weight it bears, than by informal intuitions about the term's application'. One of proper function's key roles in biology is to place objects into functional categories such as 'heart': an item is a heart only if it has the proper function of pumping blood (Millikan 1984, 1989*a*, *b*; Neander 1991*a*, *b*; Godfrey-Smith 1994: 347). But, if the concept of proper function was defined according to the Modern History view, biologists would only rarely be confident that an organ counts as a heart, since they could only rarely rule out the possibility of Non-Selective Maintenance.

Griffiths (1992) also recognizes the possibility of Non-Selective Maintenance and then discounts it for unconvincing reasons. He repeatedly acknowledges that a trait that stops contributing to survival and reproduction may be maintained for some other reason, such as its playing a key role in embryology or because 'there is no genetic variation' (1992: 127; see also pp. 122, 123, 125, 129). But, although he recognizes the possibility, he offers little response to the problem it suggests. He concludes his initial acknowledgment of the difficulty by writing, 'How common this phenomena is must be determined by empirical research, rather than philosophical speculation' (1992: 123). While this is certainly true—only biological research can establish the prevalence of Non-Selective Maintenance—it is unclear why this fact helps his account. A philosophical explication of a concept *currently* at work in biology should not depict biologists as relying on assumptions that they cannot *currently* justify.[6]

Many philosophers have questioned the general project of conceptual analysis, and Millikan (1989*b*) has argued that the search for an explication of proper function should not be judged through this lens. One prominent alternative would be to consider explications as proposals for how to define this concept in the future, not as analyses of anything about current biological concepts. But, taken in this way, the Modern History account fails again, for basically the same reason. Since in most cases biologists are not able to judge whether a trait has satisfied the Modern History requirement, in most cases they will be unable to determine the proper function of given traits. Thus a concept whose definition includes the requirement would often have little use.

One final possible defense of the Modern History accounts holds that, although Non-Selective Maintenance *is* a real possibility, biologists ignore it when assigning proper functions. In other words, someone could argue, biologists employ a concept of function defined just as the Modern History theory says, even though in doing so they rely on the flawed assumption that selection must have occurred recently.[7] This defense would be strengthened if there were any direct evidence that biologists discount the possibility of Non-Selective Maintenance in assigning proper functions. Godfrey-Smith and Griffiths do not present such evidence. In addition, I have described specific cases where biologists do discuss these possibilities, often in conjunction with functional descriptions of traits.

Admittedly, if the Modern History view was the only workable etiological account available, then we might infer that biologists are making mistaken assumptions about recent selection. But there is an alternative etiological account, one that does not ascribe any false beliefs to biologists. I present, explain, and defend this view in the next section.

[6] Griffiths's theory does allow one way that a trait can have the proper function F even if it has not been favored by selection for doing F. This is the case where there are certain variants that could have been expected to arise, and that would have been selected against because of their failure to do F, but which did not actually arise since the mutation rate fell 'below expectation'. But notice that Non-Selective Maintenance includes many other possible ways for a trait to be maintained without being favored recently for carrying out its proper function, so Griffiths's theory does not avoid the general problem.

[7] David Magnus and Ruth Millikan have suggested this possible response for the Modern History theorists to me.

4. The Continuing Usefulness Account

4.1. Statement of the Theory

The Continuing Usefulness account has two necessary conditions, one dealing with the action of natural selection and one with the recent effects of a trait on survival and reproduction:

A trait-type X has the proper function F (at time t) if and only if
- C1 X has arisen, been modified, or been maintained by natural selection at some point (prior to t) because its doing F contributed to the fitness of individuals with X, and
- C2 X's doing F has recently and importantly (before t) causally contributed to the survival and reproduction of organisms in this species with this trait.

The theory stands firmly within the etiological school because of the requirement in (C1) that X was favored by selection for doing F. Condition (C2), in contrast, is not meant to involve selection at all: X's doing F has contributed to survival and reproduction if trait-tokens of X have been a relevant causal factor in the survival and reproduction of their bearers by doing F.[8] I will discuss these two requirements in the next two sections.

4.2. Condition (C2) and the Idea of Causal Contribution

The theory allows facts about recent contributions to survival and reproduction to do what facts about recent selection did for the Modern History theory—that is, explain how traits can lose their proper functions. For example, the eyes of cave-dwelling species will lose their proper function of allowing sight if they do not do this any more or if their doing it no longer contributes to survival and reproduction. The notion of 'recently' in (C2) plays a key role. If the Continuing Usefulness account required that the trait *currently* contributes to survival and reproduction by doing F, the theory would be vulnerable to attacks Neander (1991*a*) introduced against the theories that inspired (C2). She points out that such theories have the following unacceptable consequence: if a disease that stops X from doing F spreads through all members of a species, then F would suddenly no longer count as X's proper

[8] Condition (C2) most closely resembles a condition Neander (1991*a*: 182–3, n. 12) mentions as having been suggested by Christopher Boorse and William Lycan in personal communication. Neither she nor they consider combining this condition with an etiological requirement, as I do here. Wright (1973, 1976) combined a causal condition with the etiological one he emphasized, although in a very different way.

function. For example, if a virus causing blindness spread to every member of a species, eyes would no longer actually allow sight and thus could not have this proper function. In this case, eyes would no longer be correctly described as 'dysfunctional' or 'diseased', which Neander rightly rejects as 'nonsense' (1991*a*: 182). According to the Continuing Usefulness theory, in contrast, only after a trait has stopped carrying out its proper function for a long time will it become vestigial for this function.[9]

From this perspective, part of the Modern History requirement's attraction was its power to keep a trait from losing its proper function just because it failed to carry out the function for some short period of time. The Continuing Usefulness account accomplishes the same thing by using a notion of recent contribution, thus avoiding the requirement of recent selection. I will not try to define the notion of 'recently' further, beyond noting that it is probably best calculated in terms of a number of generations, rather than years, in order to explain differences between bacteria or fruit flies and primates or elephants. My use of 'recently' here introduces no more imprecision than in the Modern History accounts: Godfrey-Smith (1994) did not explain how to delimit 'the recent past', and Griffiths's definition (1992, 1993) of an 'evolutionary significant period' relies on the notion of 'significant' regressive evolution. Although there always will be confusing borderline cases, there will be many more clear cases.[10]

The idea of 'importantly' in (C2) is necessary to prevent something like a single event from restoring a vestigial trait's functionality. Imagine that because of some mutation one naked mole rat is born with rudimentary vision, which contributes just slightly to its survival and reproduction. If (C2) did not include a notion of 'importantly', then suddenly naked mole rat eyes (the entire trait-type) would regain the proper function of aiding sight. I resist defining 'importantly' as implying any specific requirements, such as that the trait accomplishes its effect a majority of the time. In his classic non-etiological account of function, Boorse (1976) required this, an aspect of his theory that has been convincingly criticized. Millikan (1989*b*) points out that many traits fulfill their proper functions rarely (think of sperm fertilizing ova, or mating

[9] Neander (1991*a*: 183, n. 12) reports that Boorse and Lycan proposed that the assignment of proper function be made based on contributions to survival and reproduction 'relative to a larger time-slice of a population (such as the past millennium or two)'. As mentioned above, she does not describe herself or them as considering combining such a condition with an etiological criterion.

[10] A more complete account of proper function would discuss how various factors can affect the determination of the length of 'recently'. For example, if the trait has stopped making its usual contribution to survival and reproduction because of a widespread infection, perhaps it will take longer until we consider the trait vestigial than if it stopped making this contribution because of a change in lifestyle or the environment. This raises questions about the concept of 'normal environment' that are beyond the scope of this chapter, and that I discuss in my dissertation (Schwartz 1999*a*).

dances attracting mates), and Neander (1991*a*) drives the point home with her example of universal disease (discussed above).

The relevant aspects to determine whether X's contribution to survival and reproduction has been 'important' include the percentage of the time an average token accomplishes the effect, the percentage of tokens that ever do, and the magnitude of the effect. There seems to be no problem with knowing 'importantly' when we see it, even though (again) there will always be borderline cases. Note that the mutation restoring sight to naked mole rats could spread in the population if it conferred an advantage to individuals with the mutation. In this case naked mole rat eyes would eventually reacquire the proper function of allowing sight. The Continuing Usefulness and the Modern History accounts both have this consequence.

It is important to clarify that the notion of cause in (C2) is meant to be quite liberal, allowing a wide range of ways that a trait token can be taken to 'causally contribute' to survival and reproduction.[11] Take the eyes of zebras: we know that tokens of this trait-type have contributed to survival and reproduction by allowing sight—for example, by allowing individuals to see approaching predators. But this organ has also contributed to survival and reproduction in other ways—for example, by conducting blood from the optic artery to the optic vein: if we removed a zebra's eyes, the animal would run a high risk of bleeding to death. Eyes may even contribute to survival and reproduction in other ways—for example, by keeping parasites from settling in the eye sockets. Another way of putting this is to say that we can assign the eye all these roles in a 'functional analysis' (in Cummins's sense (1975)) of the animal's ability to survive and reproduce. And the notion of 'cause' at work in (C2) is not meant to keep any of these effects from possibly counting as a contribution to survival and reproduction. In fact, it is unclear that there is any way to restrict the notion from applying to these cases without invoking facts about natural selection.[12] Any such amendment would make the Continuing Usefulness account basically identical with the Modern History account.

Instead, (C2) is meant to range widely—many, many effects can count as putative contributions to survival and reproduction—while (C1) blocks most of these from counting as proper functions. So, even if we decide that it is true that the zebra's eye contributes to survival and reproduction by transporting blood from the optic artery to the optic vein or by blocking parasites, neither of these will count as a proper function of this organ unless we believe that natural selection favored these effects. And we do not: we have no reason to believe that there were ever individuals with heritably different eyes who were selected against because of increased parasitic infection. In contrast, there is no question

[11] I have been greatly aided by discussions with Gary Ebbs about topics covered in this section.

[12] Ruth Millikan (1989*b*: 174; 1993: 39–40) and Godfrey-Smith (1994: 352) have made this argument against the non-etiological approaches to function that inspired condition (C2).

that selection has favored eyes that provide eyesight (a weak-adaptationist claim I will defend below). Non-etiological approaches, such as in Boorse (1976) and Bigelow and Pargetter (1987), cannot rely on an etiological condition to play this pruning role, and thus these theories run into the dilemma of assigning proper functions too liberally or appealing to the very facts of natural selection they hope to avoid.

The only way that liberality in (C2) could cause problems for the Continuing Usefulness theory would be in cases where the condition must act as a restraint on assigning proper functions—for example, when discussing vestigial traits. But there does not seem to be any serious problem knowing when a trait no longer makes the causal contribution that was once favored by selection: this just means that no reasonable analysis of the animal's recent ability to survive and reproduce would appeal to X's doing F.[13] This is particularly obvious in cases where X does not even do F any more, such as the case of the sightless eyes of a cave-dwelling species. In cases where X still does F but F no longer contributes to survival and reproduction, the explanation of why we think that X no longer has the proper function F involves explaining why we think that X's doing F no longer contributes to survival and reproduction.

4.3. *Weak Adaptationism and Criterion (C1)*

Condition (C1) preserves the explanatory implications of proper function ascription, widely prized by etiological theorists since Wright (1973, 1976), because saying that X has the proper function F implies that a complete explanation of X's prevalence or form must mention X's being selected at some point for doing F. But, as Godfrey-Smith (1994: 356–7) points out (in the paragraph quoted in section 3 above), this requires 'substantial biological commitments' for any etiological theory—that is, that every proper function F of every trait X was favored by natural selection *at some point.*

This claim has special significance given my attack on the Modern History view. I claim that, while in many cases we cannot be confident that a prototypical trait with proper function F has been *recently* favored by selection for doing F, we can be confident that it was favored by selection for doing F *at some point.* I call this a commitment to 'weak adaptationism': we can be confident that *some* aspects of *some* traits were favored by selection for carrying out their useful effects. Although 'stronger' forms of adaptationism, ones that infer the action of natural selection behind every beneficial effect of every trait, face serious problems (cf. Gould and Lewontin 1979), this weaker form is tenable, I will argue, and in fact indispensable in biology.

[13] Ruth Millikan (personal communication) suggested this last way of putting it.

Let us return to the example of feathers discussed above. Even though we do not have enough information to know that the feathers of most species of birds have *recently* been favored by natural selection for their contribution to flight (as I argued in Section 2), I believe that we *do* have enough information to know that feathers were favored by natural selection *at least at some point* for their contribution to flight. My confidence stems from the qualities of feathers that are so well suited to flight. Feduccia (1996: 13) presents a particularly gripping list of these aspects:

> Almost every structural detail of feathers has aerodynamic significance. Feathers are a near-perfect aerodynamic design. They are lightweight . . . possess an unusually high strength-to-weight ratio, have graded flexibility, possess a ventral reinforcing furrow on the ventral aspect of the shaft, and have great resilience—when broken, they come back together in original aerodynamic form, and their modular construction limits damage. Feathers create individual airfoil cross-sections, slotted wings, and wings with overall airfoil cross-sections. Their smooth, aerodynamic contours produce laminar flow on the body, thus reducing drag.

The interaction of microscopic 'hooklets' and 'barbules' allow the wing to come back together 'like Velcro' immediately after being separated, quickly recreating the flying surface.

As Dawkins (1986) emphasizes, some facts about living things demand a certain sort of explanation because of the 'appearance of design'. The wing feathers of modern flying birds are a case in point. In cases like this, an appeal to natural selection helps explain the impressive fit between trait and role. A complete story of the evolution of these structures certainly involves many non-selectional factors, such as genetic drift, random mutation, and genetic, physiologic, and developmental constraints. But, in the absence of selection for the trait's carrying out the crucial effect—in this case flying—any explanation for the existence of so many useful aspects evaporates. Modern biologists should not too quickly assume the action of natural selection, and there must be ways to test selection hypotheses. But they also cannot easily give up their belief that natural selection acted at some point in the selection of wing feathers of modern birds, at a cost of giving up any hope of explaining their many unusual aspects.[14]

We do *not* know that *every* facet of feather structure was favored by natural selection for making a contribution to flight. It is always possible that some feature that is apparently well suited to flying was favored for some other reason or is simply a by-product of some other selected feature. For example,

[14] This argument is inspired by Dawkins (1986) and Dennett (1995), although I reject some aspects of their accounts of natural selection. In particular, I believe that natural selection is not literally a designer (contra Dennett 1995) and not literally a problem-solver (contra Dawkins 1986). I discuss this issue at greater length in Schwartz (1999*a*: ch. 2).

perhaps the 'ventral reinforcing furrow' on the shaft is simply a by-product of some developmental process: perhaps there have never been individuals whose feathers lacked this furrow and were selected against because their feathers were too flexible. But here is my weak-adaptationist claim: it is completely beyond reasonable biological thinking to claim that *no aspect of feathers was ever favored by natural selection for contributing to flight*. If this was true, we would have no explanation how such an apparently well-designed object could have come into existence. Biologists can be confident that at least some of the features of this incredibly efficient item were selected for their contribution to survival and reproduction.

And this sort of reasoning applies to most of the 'apparently designed' traits of organisms. If we compile all the aspects of a bird's physiology that make flight possible—including the musculature associated with the wings, the light bones, the feathers, and so on—we can imagine that *some* of them were not favored by selection for their contribution to flight, but not that *none* of them was. Similarly, within a complex organ like the human eye, for example, we can imagine that some features may not have been favored by selection, but not that none of them was. And a similar comment applies to subparts of complex traits: we must believe that at least some aspects of the lens of the human eye were favored for their ability to focus images on the retina, even though we may not be sure exactly which ones were and which were not.[15]

Thus we see that biologists can be more confident about certain aspects of long-ago natural selection, which created and modified a trait, than they are about the trait's recent maintenance selection. If there has been maintenance selection of X for doing F, then condition (C1) is satisfied, and, if the selection occurred recently, condition (C2) is satisfied as well. Therefore, according to the Continuing Usefulness account (and other etiological accounts), recent maintenance selection of X for doing F will be *sufficient* for X's having the proper function F. The Modern History view went wrong in converting this useful sufficient condition for X's having the proper function F into a necessary condition. A new necessary condition decreases epistemological confidence, and the Modern History account pays the price of this dearly. The Continuing Usefulness account avoids this difficulty, even as it answers the problems that faced earlier etiological theories.

5. Conclusion

In this chapter, I have first pointed out a crucial problem for some of the most important recent accounts of proper function in biology. Secondly, I

[15] I discuss the consequences of this epistemological fact on the explication of various concepts of function at work in biology in Schwartz (1999*a*: ch. 4).

have proposed a new etiological account that solves all the problems identified, without relying on questionable assumptions about the recent natural selection of traits. Thirdly, this account has allowed me to clarify and defend ideas about causation and natural selection that ground the etiological approach. In the future, etiological accounts of function must grapple with the many ways in which the natural selection of traits may vary, in order to provide a complete account of notions of function and their use in modern biology. Nothing less than our understanding of the basis and place of teleological thinking in biology is at stake.

References

Amundson, Ron, and Lauder, George V. (1994), 'Function without Purpose: The Uses of Causal Role Function in Evolutionary Biology', *Biology and Philosophy*, 9: 443–69.

Bigelow, J., and Pargetter, R. (1987), 'Functions', *Journal of Philosophy*, 84: 181–96.

Boorse, Christopher (1976), 'Wright on Functions', *Philosophical Review*, 85: 70–86.

Buller, David J. (1998), 'Etiological Theories of Function: A Geographical Survey', *Biology and Philosophy*, 13: 505–27.

Cheng, Alec M., Rowley, Bruce, Pao, William, Hayday, Adrian, Bolen, Joseph B., and Tony Pawson (1995), 'Syk Tyrosine Kinase Required for Mouse Viability and B-Cell Development', *Nature*, 378: 303–6.

Cummins, Robert (1975), 'Functional Analysis', *Journal of Philosophy*, 72/20: 741–65.

Dawkins, Richard (1986), *The Blind Watchmaker* (New York: W. W. Norton & Co.).

Dennett, Daniel C. (1995), *Darwin's Dangerous Idea: Evolution and the Meanings of Life* (New York: Simon & Schuster).

Endler, John (1986), *Natural Selection in the Wild* (Princeton: Princeton University Press).

Feduccia, Alan (1996), *The Origin and Evolution of Birds* (New Haven: Yale University Press).

Godfrey-Smith, Peter (1994), 'A Modern History Theory of Functions', *Noûs*, 28: 344–62.

Gould, Stephen Jay (1987), 'Freudian Slip', *Natural History*, 96: 14–21; repr. as 'Male Nipples and Clitoral Ripples', in *Bully for Brontosaurus: Reflections in Natural History* (New York: W. W. Norton & Co., 1991), 124–38.

——and Lewontin, Richard (1979), 'The Spandrels of San Marco and the Panglossian Paradigm: A Critique of the Adaptationist Programme', *Proceedings of the Royal Society of London Series B: Biological Sciences*, 205: 581–98.

——and Vrba, Elisabeth S. (1982), 'Exaptation—a Missing Term in the Science of Form', *Paleobiology*, 8/1: 4–15.

Griffiths, Paul E. (1992), 'Adaptive Explanation and the Concept of a Vestige', in Paul R. Griffiths (ed.), *Trees of Life* (Dordrecht: Kluwer), 111–31.

——(1993), 'Functional Analysis and Proper Functions', *British Journal for the Philosophy of Science*, 44: 409–22.

Griffiths, Paul E. (1996), 'The Historical Turn in the Study of Adaptation', *British Journal for the Philosophy of Science*, 47: 511–32.

Jamieson, Ian G. (1989), 'Behavioral Heterochrony and the Evolution of Birds' Helping at the Nest: An Unselected Consequence of Communal Breeding', *American Naturalist*, 133: 394–406.

——(1991), 'The Unselected Hypothesis for the Evolution of Helping Behavior: Too Much or Too Little Emphasis on Natural Selection', *American Naturalist*, 138: 271–82.

Lewontin, Richard (1970), 'The Units of Selection', *Annual Review of Ecology and Systematics*, 1: 1–14.

——(1983), 'Gene, Organism and Environment', in D. S. Bendall (ed.), *Evolution from Molecules to Men* (Cambridge: Cambridge University Press), 273–85.

Millikan, Ruth Garrett (1984), *Language, Thought, and Other Biological Categories: New Foundations for Realism* (Cambridge, Mass.: MIT Press).

——(1989*a*), 'An Ambiguity in the Notion "Function"', *Biology and Philosophy*, 4/2: 172–6.

——(1989*b*), 'In Defense of Proper Functions', *Philosophy of Science*, 56/2: 288–302.

——(1993), 'Propensities, Exaptations, and the Brain', in *White Queen Psychology and Other Essays for Alice* (Cambridge, Mass.: MIT Press), 31–50.

Mills, S., and Beatty, J. (1979), 'The Propensity Interpretation of Fitness', *Philosophy of Science*, 46: 263–86.

Neander, Karen (1991*a*), 'Functions as Selected Effects: The Conceptual Analyst's Defense', *Philosophy of Science*, 58: 168–84.

——(1991*b*), 'The Teleological Notion of "Function"', *Australasian Journal of Philosophy*, 69: 454–68.

Ostrom, J. H. (1979), 'Bird Flight: How Did It Begin?', *American Scientist*, 67: 46–56.

Schwartz, Peter H. (1999*a*), 'Function, Dysfunction, and Disease in Biology and Medicine', Ph.D. dissertation, Philosophy Dept., University of Pennsylvania.

——(1999*b*), 'Proper Function and Recent Selection', *Philosophy of Science*, 66 (Proceedings), S210–S222.

Sherman, Paul (1988), 'The Levels of Analysis', *Animal Behaviour*, 36: 616–19.

——(1989), 'The Clitoris Debate and Levels of Analysis', *Animal Behaviour*, 37: 697–8.

Sober, Elliot (1993), *Philosophy of Biology* (Boulder, Colo.: Westview Press).

Stearns, Stephen C., and Kawecki, Tadeusz J. (1994), 'Fitness Sensitivity and the Canalization of Life-History Traits', *Evolution*, 48: 1438–50.

Turner, Martin, Mee, P. Joseph, Costello, Patrick S., Williams, Owen, Price, Abigail A., Duddy, Linda P., Furlong, Michael T., Geahlen, Robert L., and Tybulewicz, Victor L. J. (1995), 'Perinatal Lethality and Blocked B-Cell Development in Mice Lacking the Tyrosine Kinase Syk', *Nature*, 378: 298–302.

Waddington, C. H. (1959), 'Canalisation of Development and the Inheritance of Acquired Characters', *Nature*, 183: 1654–5.

Wagner, Andreas (1996), 'Does Evolutionary Plasticity Evolve?', *Evolution*, 50: 1008–23.

Wright, Larry (1973), 'Functions', *Philosophical Review*, 82: 139–68.

——(1976), *Teleological Explanations* (Berkeley and Los Angeles: University of California Press).

Part Three

Teleosemantics

10. Pagan Teleology: Adaptational Role and the Philosophy of Mind

Mark Perlman

ABSTRACT

Though teleology seems a promising source of help in explaining mental representation and mental content, it will not solve the main problem of mental content: misrepresentation. Thus biological functions cannot adequately provide the basis of mental content. I examine three attempts to ground the content of mental representations on biological functions, Millikan, Dretske, and Papineau, and show how they fail to meet the conditions that must be satisfied for functions to solve the problem of mental content. Other related theories fail for related reasons. I will also discuss Neander's distinction between two forms of teleology, and explain why I will insist on a third alternative. This is not a diatribe against teleology *per se*, and does in fact admit that we can assign clear functions to mental representations, as well as mental processes and mechanisms. But these functions will not have the content needed to underwrite misrepresentation.

Since the 1980s, philosophers of mind have increasingly looked to biology, and biological functions, to supply a stable basis for mental content. Philosophy of mind has long been in need of naturalizing, to free it from what was the metaphysical approach to the mind that has dominated philosophy since at least Descartes. The central issues were the mind–body problem, freedom of the will, and the nature of ideas. Even among metaphysical problems, these are notoriously vague and seemingly insoluble. Moreover, they cut off the mind from the brain and the physical world in a way that increasingly clashed with the ever-increasing reach of science. The first big fix came in the 1920s and 1930s from psychology, specifically from behaviorism. This 'solved' the

Many of these ideas are also defended in my book *Conceptual Flux: Mental Representation, Misrepresentation, and Concept Change* (2000), but are expanded here. I want to thank my colleagues and co-editors Robert Cummins and André Ariew for very helpful comments. In fact, much of the material on use theories and on Millikan was highly influenced by years of discussions with Cummins—so much so that I am no longer certain where his views end and mine begin.

problems mostly by eliminating them, as a by-product of eliminating the mind altogether and reducing the mental to conditioned responses. But behaviorism turned out to be hopeless. In the 1950s and 1960s it became easy to spot phenomena that could not be reduced to stimulus and response (language—for example Chomsky 1957). So the search was on for a new way to keep the mind (and mental phenomena) out of the dark grasp of metaphysics. Modern sources come from psychology (Goldman 1976, 1986), artificial intelligence and connectionism (Churchland 1979, 1981, 1990), and biology (beginning with Dennett (1969: esp sect 9), Cummins (1975), and flourishing with Millikan (1984, 1986, 1993), Dretske (1986), and Papineau (1987)). Of interest here is how considerations of evolutionary role or adaptational role have been put to use to shed light on the problem of mental content. The older question, 'What makes ideas mean what they do?', has been changed into the question: 'What gives mental representations their content?' The biological answer is: their evolutionary function (or the function of the mechanisms that produce them or use them in cognition).

In much of the twentieth century, *teleology*, that is, having a function, had been deemed hopelessly metaphysical and unsuitable for a scientific world view. It also seemed hopelessly religious, immediately inviting one to ask who gave those functions to those objects, organisms, processes, and so on. Yet modern biology has thrived while never having stopped talking about the functions of organisms, organs, genes, processes, behaviors, and so on. So, as Logical Positivism faded away, it became more and more tempting to take functions seriously. Ruth Garrett Millikan's 1984 book *Language, Thought, and Other Biological Concepts* was an elaborate and detailed account of how teleological functions could be seen as the basis of human cognition, and, in particular, mental representation.

The feature that makes functions an inviting source of mental content is that an object can have a function but not actually perform that function on all occasions. The function of a book may well be to carry information, but it can also be used as a paperweight. It still has the function to carry information, but in being used as a paperweight that function is of no importance. Used as a paperweight, a book may indeed still carry the information, and still, in a way, perform its function. But in many other cases, an object may perform a task and no longer perform its function at all—as when a dried-up ink pen is used as a pointer. This split between what an object actually does and what its function is makes functions seem ideal for providing content to mental representations, which must perform an analogous task: representations may be *about* one thing yet sometimes be used to represent some other thing. When a representation points to something not included in its representational content, this is known as misrepresentation. This phenomenon is familiar enough, but notoriously difficult to explain adequately. This mismatch between content and use is what we might explain by appealing to teleological functions.

We ought not to deny the great philosophical advances that have come from examining the mind from the perspective of cognitive psychology, AI, and evolutionary biology. Great strides have been made in making teleology, especially the notion of biological functions, scientifically and methodologically respectable. But, though teleology seems a promising source of help in explaining mental representation and mental content, it will not solve the main problem of mental content: misrepresentation. Thus biological functions cannot adequately provide the basis of mental content.

To show this, I need first to describe both the problem in the philosophy of mind that biological functions have been drafted into service to solve and the conditions which must be satisfied for functions to solve that problem. I will then examine three attempts to ground the content of mental representations on biological functions: Millikan, Dretske, and Papineau. These three views are models for many others, and these related theories will fail for related reasons. I will also discuss Karen Neander's distinction between two forms of teleology, and I will insist on a third alternative. This chapter is not a diatribe against teleology *per se*, and does in fact admit that we can assign clear functions to mental representations, as well as mental processes and mechanisms. What it will argue is that these functions will not have the content needed to underwrite misrepresentation.

1. Mental Representation and the Problem of Misrepresentation

The goal is to find a naturalistic basis for meaning, and, in particular, for the content of mental representations. If we are naturalizing, it will be natural either to think that the mind is identical to the brain or at least that mental states are reducible to or supervenient on brain states. The biological approach typically begins with the most primitive states of the mind, states that we share with animals. If we can give an evolutionary account of a cognitive function we share with many animals, that should serve as a basis for a naturalistic account of more complex and uniquely human mental activity. But the goal is a general theory of content based on natural properties of representations.

To have a theory of content is to have a theory that identifies the objects, states of affairs, or events that a representation type correctly applies to. In a sense, we are looking for a property that divides the universe in half: the portion *meant* by the representation, and the portion not included in its meaning. This division is present in the most 'animal' representations—What is 'FOOD' and what is not?—and in the most complex and intellectual—What is 'ART' and what is not? Notice that in both cases this division brings with it a range of possible misapplications of the representation. If the representation 'FOOD' carves out a certain set of nutritious objects, applying it to a non-nutritious object counts as a misrepresentation. If an idea can correctly represent a horse, then surely it misrepresents when applied to a can-opener. So, in providing a theory of representational content, dividing the universe into what is covered or included in the meaning of the representation and what is not covered or included, one main desideratum is that the theory specify the misrepresentations as well as the representations. As it turns out, this is not so easy.

One way of stating the difficulty in accounting for misrepresentation has been called by Fodor (1987) the 'disjunction problem'. The problem is that proposed cases of misrepresentation can be redescribed as cases of correct but disjunctive representation. Suppose that some non-cat, Fido, causes the concept 'CAT' to be tokened in my head.[1] 'CAT' is supposed to represent the property of being a cat, and Fido is not a cat. But on a crude causal theory of representation, a representation means whatever causes it, so 'CAT' actually means *dog-or-cat.* This disjunctive content makes the representation correct: Fido is indeed a dog-or-cat. If we include any cause (or worse any condition under which the representation is used), in its content, then it will correctly apply to that object, state, or event, whatever it is. Misrepresentation would be impossible, and, since beliefs are often said to be representations processed in particular ways, there could never be a false belief. More sophisticated causal theories try idealization to some set of optimal, typical, or normal conditions, and Fodor (1987, 1990*a*) proposes the ingenious solution of 'asymmetric dependence', according to which 'CAT' means *cat* and not *dog-or-cat* because dogs would not cause 'CAT' unless cats did, but cats would still cause 'CAT' even if dogs did not. But, whatever the theory, the driving force behind contemporary theories of mental content has been this problem: that citing actual uses of a representation yields disjunctive contents, and disjunctive contents make misrepresentation impossible. I agree with Cummins's assessment (1989) of all these attempted solutions—even if they get the conditions for content right, it is not clear that they *explain* anything. They do not indicate *why* it is that those conditions are the ones that correctly identify the content of a representation, and how such conditions could help us understand the role that representations play in our cognitive functioning.

But, more importantly, the disjunction problem really is just the problem of drawing the distinction between meaning-fixing and non-meaning-fixing uses

[1] To help distinguish between mention of words and of concepts (representation types), I use single quotes around lower-case letters to indicate words ('cat'), and single quotes around small capitals to indicate concepts ('CAT'). To clarify the difference between expressions and the meanings attributed to them, I italicize descriptions of meanings: the word 'cat' and the concept 'CAT' mean *cat.*

of a representation. The uses that lead to the disjunctive meaning must somehow be removed from the uses that fix meaning in order to solve the disjunction problem. The game in contemporary theorizing about mental content is to find a property (or properties) of uses of a representation that is possessed only by the uses that make the meaning nice and non-disjunctive. Uses that lack the property will be capable of being misuses, and a theory that finds the right property will have grounded the existence of misrepresentation and error. Exactly which property or properties will do the job is the subject of intense debate. This task is the target for modern evolutionary or adaptational role theories. Perhaps a feature of biological function will correctly divide meaning-fixing and non-meaning-fixing uses of representations, and make room for possible misrepresentations. Adaptational Role theorists (like Millikan, Papineau, and Neander) would not like their views being classified as sub-varieties of use (or conceptual role) theories, but, as I will explain, like it or not, that is what they are.

A 'use' theory of content is a naturalistic theory that equates the content of a mental representation with some function of its use. A formal way of describing the pattern of a representation's use is to refer to the 'conceptual role'. Conceptual role tells us what in the cognitive system causes the representation (stimuli and mental processing), and how that representation causes other representations to be tokened, as well as how it leads to behavior.[2] Conceptual role theories identify the content of the representation with this pattern of use, as the particular representation's position in the system of other representations that mediate its occurrence. Use theories of representational content can thus be equated with conceptual role theories. Conceptual role is just a complex way of describing the use of the representation, identifying a pattern of use with the causes and effects of tokening of the representation, as well as its connections to other representations.

A simple use theory has no room for misrepresentation. If every actual use of a mental representation or concept is constitutive of its meaning, then there can be no misuse. If one is tempted to say that misuses of mental representations can be part of the meaning as well, the problem is that we cannot call the application of a concept a misuse until we know the meaning of the representation or concept, and that is just what we are trying to determine. The most common way to introduce the possibility of misrepresentation is to limit the uses of a representation that count in determining its meaning. Rather than say that meaning is a function of all use (leaving no room for misuse), one can say,

[2] Functionalist theories of content are often described as basing meaning on *inferential role* instead of *conceptual role* (see Block 1986, 1993, Fodor 1987, Fodor and Lepore 1992, Boghossian 1993). *Inferential role* includes the purely linguistic aspect of an expression's overall causal role, while conceptual role includes relations to non-linguistic stimuli or objects in addition to linguistic aspects of causal role.

somewhat imprecisely, that meaning is a function of 'special' use, or 'good' use. Accordingly, no special use is misuse, some non-special uses can be correct uses, and other non-special uses can be incorrect uses, or misrepresentations. Every misrepresentation must be a non-special use. The specification of which uses are the special uses differs from theory to theory. Fodor (1987, 1990*a*) and Dretske (1981, 1986) identify special uses by causal features (uses by detectors under optimal conditions, or that exhibit asymmetric dependence). Harman (1982) identifies special uses as those in 'normal contexts'. Papineau (1987) identifies special uses as the ones adapted to satisfy desires, and Millikan (1984, 1986) identifies special uses as the ones that maximize biological fitness. To have misrepresentation, there must be some uses that do not determine meaning (or, if you like, representational content), and only they can possibly be misuses.

In the search for a non-arbitrary way to distinguish causes and uses as meaning-fixing or non-meaning-fixing, it seemed natural at a certain point to appeal to biology. After all, the distinction would not be arbitrary if it were built-in, selected by evolution. The attraction is heightened by the depth of actual empirical data about the design of organisms and the well-confirmed scientific theory of evolution as an explanation of the source of this structure and design. If humans are highly evolved animals, and our brains are only different in degree from those of many different animal species, not only can we view our representational mechanisms as a product of natural selection, but we can utilize our knowledge of more primitive kinds of representation and indication to shed light on human intentionality. So, teleology and biological functions are now at the center of a new and promising attempt to establish this required split between meaning-fixing and non-meaning-fixing uses of a representation, and, since this split is what is needed to allow misrepresentation, we see function invoked to ground misrepresentation. The question is: does it succeed?

2. Millikan's Adaptational/Teleological Theory

Millikan (1984) proposes a biological/evolutionary/historical account, of how cognitive mechanisms evolve, as a source of a theory of mental content intended to allow for misrepresentation and escape the disjunction problem. Her key insight is this: a thing can have a function and not actually fulfill it. The function of a typewriter is to type, but I can also use it as a paperweight or a doorstop. These alternative uses do not change the typewriter's original function—to type. This distinction seems just what is needed to establish the content of a representation and allow misrepresentation. A representation (or cognitive mechanism) can have the function to represent one thing, but in particular instances might represent some other thing, hence failing to function correctly. If we define a representation's content by its function, failure to perform that function will be a misrepresentation. The connection to use theories is that adding the function provides a source of content other than the entire set of actual uses. Appeal to the function of the representation is really a way of attempting to specify the ideal use of the representation, and, when actual use deviates from ideal use, we have misrepresentation. This is much the same as Dretske's and Fodor's appeals to ideal use, but Millikan's method of specifying which uses are ideal is different.

Millikan's version of ideal use is teleological, based on a biological view of function that is adaptationally advantageous. She develops a technical notion of *Proper Function*, where a *Proper Function* is a function needed by the organism for survival in *Normal* conditions—that is, in conditions in which the majority of organisms of that kind have lived and evolved. *X* performs its Proper Function when it does the sort of thing that has been historically responsible for replication of *X*s. Proper Functions can fail to be fulfilled fairly often and still be to an adaptational advantage: they have to be fulfilled enough times for survival. So a Normal *X* need not fulfill its Proper Function most of the time—just often enough to keep the organism out of danger. A 'PREDATOR' representation need not be reliable in the sense of being tokened only when a predator is around—it can give a lot of false positives and still be a good mechanism, but it cannot give any false negatives, for just one will be almost a sure chance of being eaten. The mouse who represents any bird, even harmless ones, as attacking hawks or owls will scurry more often than necessary, but will run from all the hawks. But if the mouse misrepresents the attacking hawk as a harmless sparrow just once, and as a result fails to hide from the bird, that is once too many. So reliable indicators as such are not needed for a thing to be an evolutionarily advantageous system. The function of the representation is to represent a predator, and it still has this function even when it gives a false positive—that is, when there is no predator.[3]

This fact is the kernel of Millikan's theory of misrepresentation: if a representation has as its Proper Function to represent *X*, then its content is *X* even

[3] This is somewhat of an oversimplification of Millikan's view, since she in fact has two different versions of the teleological account of representational content, and neither of them assigns functions to the individual representations themselves. In *Language, Thought, and Other Biological Categories* (1984) her view is that it is not the representations themselves but the consumers of the representations that have the proper functions, to process representations that represent the environment correctly. On this version, desire content is dependent on belief content, where the function of beliefs is to be true. In 'Thoughts without Laws: Cognitive Science without Content' (1986) she assigns the function to the producers of the representations, the perceptual mechanisms, and makes belief content dependent on desire content, since desire satisfaction is really what the purpose of having the beliefs is. (This is similar to Papineau's view of functions.) We need not go further into this difference because, for our purposes here, the resulting failure of allowing misrepresentation occurs in either view.

if there is no actual *X* around. And, if it represents *X*, a tokening caused by a non-*X* is a misrepresentation. The reason is that the representation and its mechanism are there because, in the history of the organism or species, the representation and mechanism have served to help proliferate the species. Lots of things seldom perform their Proper Function, but a few performances may be enough. Millikan's example of this is human reproduction: the function of sperm is to reach and fertilize an egg, and that is the function of all sperm even if few actually make it. Bee dances are another example that she cites: the orientation of the bee's dance tells the other bees where food is. If the dance is misoriented, its content is still *direction of food*; it just fails to fulfill its Proper Function, and is a misrepresentation of the direction of food.

There are many places where we have only scratched the surface, but even from this short description we can see that Millikan's account of misrepresentation simply will not work. Indeed, teleological accounts in general cannot solve the disjunction problem. Consider Millikan's case (1986: 71) of the toad that snaps at little black moving things. Most of the little black moving things are flies or other bugs, and the toad is sustained by eating bugs, but the toad can be fooled by bee-bees (small metal pellets) tossed in the air nearby, and will snap at them too. Millikan argues that the content of the representation is *bug* and not *bug-or-bee-bee* because bugs are what it is after, and the purpose of the mechanism is to get bugs, not bee-bees.[4] As she puts it: 'We say that the toad thinks the pellets are bugs merely because we take it that the toad's behavior would fulfill its proper function (its 'purposes') Normally only if these were bugs *and* that this behavior occurs Normally . . . only upon encounter with bugs' (Millikan 1986: 71–2). The important point for Millikan is that it is bugs that toads historically have sought, and eating bugs is what has led to reproduction of their representational mechanism. The mechanism is adapted to being an indicator of the presence of bugs because indicating that has helped toads survive and reproduce. The possibility of being fooled by a bee-bee being tossed near the toad does not change the historical tie to bugs, and hence the adaptational theory assigns the content *bug* to the representations. In Normal circumstances, the toad snaps at bugs, so in abNormal circumstances, the content of its representation is what it would have been in Normal circumstances: *bugs* (see Fig. 10.1). This is a straightforward appeal to ideal use, where actual use that deviates from ideal use is misuse, and hence misrepresentation.

[4] In David Israel's hands (1987), the 'bugs' and 'toads' of Millikan's example (1986) turn into 'flies' and 'frogs', and these are the terms that Fodor (1990*a*) uses to discuss the question of whether the content is 'flies' or 'flies-or-bee-bees' that are eaten by the 'frogs'. But, while the point of the examples is the same, Millikan's 'bugs' are a better example. I take it that neither frogs nor toads need flies specifically; most any insect/'bug' will do for good reptile or amphibian nutrition, but bee-bees will not.

Fig. 10.1. Does the toad representation 'r' mean, *bug*, or the disjunctive *bug-or-bee-bee*?

'The trouble is', as Fodor (1990*a*: 71) notes, 'that this doesn't *solve* the disjunction problem; it just begs it'. We can tell the story so that in Normal circumstances the toad snaps at bugs, and so this makes the function of the representations to resonate to *bugs*, and this makes the intentional content *bug*. But there is nothing to prevent us from telling this story so that the content is disjunctive: *bugs-or-bee-bees*, or, even better, *little-black-moving-things* (which covers bugs and bee-bees). The toad's/frog's snapping at those little black moving things is evolutionarily favored because enough of the little black moving things happen to be bugs. As Fodor (1990*a*: 72) says, 'Darwin doesn't care which of these ways you tell the teleological story.' How we describe the snapping is irrelevant, as long as the toad gets enough bugs. Natural Selection treats reliably coextensive representations as synonymous, but psychology and philosophy do not, at least if they are to provide an account of misrepresentation.

Fodor argues that, because we can tell the story of the function of the toad's representational mechanism as picking out either the content *bugs* or the content *bugs-or-bee-bees*, teleology does not give a solution to the disjunction problem, but rather presupposes one in telling us to prefer the more specific content. We could pick either, and we have no principled grounds for choosing the function that makes misrepresentation possible over the function that makes it impossible. But we can give a stronger argument against Millikan than Fodor's. Far from leaving us with no reason to prefer either reading of the function over the other, on adaptational grounds we should prefer the reading of the function to be more conservative, as long as it is accurate enough. So we are not merely in need of some non-question-begging reason to prefer Millikan's reading over the other one. We have reason to think that teleology gives us just the opposite answer from the one that Millikan recommends. In fact, accuracy can actually be an adaptive disadvantage because it costs speed and resources.

The deep problem with Millikan's view is that, even though it is based on the notion of having a function and failing to perform it, as illustrated by the example of false positives like the toad snapping at a bee-bee, she does not appreciate one crucial aspect of these cases. Her adaptational account is based on the picture of success being measured by the number of times the function is precisely performed. This can be a minority of cases, as long as there are enough cases for survival. So her view of adaptive uses of representations is that they are correct uses, and a cognitive mechanism is adaptive if it yields enough correct uses. This is in keeping with the general strategy of explaining success naturalistically, in terms of survival. Millikan ends up equating epistemological error with adaptive error. The fully adaptive cases are those in which the system gets it exactly right, and this need occur only a small portion of the time for the mechanism to be adaptive and for the organism to survive.

What this view ignores is that adaptive success can tolerate unbounded epistemological error.[5] Accuracy is not necessarily adaptive, and is often harmful. All of these cases of adaptation of primitive perceptual mechanisms are cases in which what is indicated is the presence either of food or of predators. Depending on the nature of the food or the predators, these cognitive mechanisms must indicate such presence fast. Bugs fly fast, and, if the toad is to eat one, its cognitive mechanism must immediately produce the representation that leads to snapping. If the cognition is prey representing presence of predators, similar speed is needed, since successful predators pounce quickly. The important point is that there is always a trade-off between accuracy on the one hand and speed and resources on the other. We can reduce false positive readings by increasing the time allowed for the perceptual process, or increasing the cognitive resources in the toad's brain. But, in the real toad-eat-bug world, this is not an option, since the bugs are not about to decrease their flying speed for finicky toads, and toads do not have more computing space to use in bug identification without taking space away from other important processes. The significance of this is that a mechanism could be adaptive even if it never represented things exactly right, as long as it was close enough for the organism to get sufficient food and escape enough predators to survive. The fairly accurate system that always gets things only approximately right while never getting them exactly right is more adaptive than the system that gets it exactly right some of the time. This is because the extra accuracy is not needed; if the size of the error is not fatal, there is no special pay-off for precision. And there is a cost for precision: time and resources. The archer who always comes close to the bull's-eye but never actually hits it is a better shot than the archer who hits the bull's-eye a few

[5] Cummins (1989, 1996) also pushes this line of argument (so well, in fact, that he convinced me of it).

times but who misses badly much of the time. So, as far as adaptive value is concerned, Millikan has just got it wrong: it could be adaptive for no representations to get things exactly right, and hence it could be adaptive for them all to be, in Millikan's sense, misrepresentations. In the example of the toad representing the bugs and bee-bees, it could very well be adaptational to have the intentional content be simply *bee-bee*, in which case all bug-directed representations would be misrepresentations, as long as the toad snapped at them anyway. Thus, adaptation will not provide the distinction between correct and incorrect representation, since it tolerates total misrepresentation. Appeal to adaptational value may pick out pragmatic virtues of having one cognitive system rather than another, but it will not divide up the correct and incorrect uses of representations, and hence it will not serve as a basis for the distinction between meaning-fixing and non-meaning-fixing uses.

3. Dretske's Adaptational Role Theory

Dretske (1986) also invokes functions to solve the problem of misrepresentation. His teleological account of misrepresentation rests on a relatively simple claim: a representation *M* means that *p* iff it is *M's* function to indicate the condition of *p*. Functions are supposed to distinguish the events that are supposed to occur, and that determine content, from those that merely do occur. When *M* is tokened by the conditions it is its function to represent, it correctly represents them. When it is tokened by conditions other than the ones it is its function to indicate, it fails to perform its function and is a misrepresentation.

Dretske (1986: 26) illustrates his point with an oft-cited biological example:

> Some marine bacteria have internal magnets (called magnetosomes) that function like compass needles, aligning themselves (and, as a result, the bacteria) parallel to the earth's magnetic field. Since these magnetic lines incline downwards (toward geomagnetic north) in the northern hemisphere (upwards in the southern hemisphere), bacteria in the northern hemisphere, oriented by their magnetosomes, propel themselves towards geomagnetic north. The survival value of magnetotaxis (as the sensory mechanism is called) is not obvious, but it is reasonable to suppose that it functions so as to enable the bacteria to avoid surface water. Since these organisms are capable of living only in the absence of oxygen, movement towards geomagnetic north will take the bacteria away from oxygen-rich surface water and towards the comparatively oxygen-free sediment at the bottom. Southern hemispheric bacteria have their magnetosomes reversed, allowing them to swim towards geomagnetic south with the same beneficial results. Transplant a southern bacterium in the North Atlantic and it will destroy itself—swimming upwards (towards magnetic south) into the toxic, oxygen-rich surface water.

Dretske would have liked to be able to say that in such a case the transplanted magnetosome is misrepresenting the world, in that its function is to indicate the direction of oxygen-free water, which it does by pointing to magnetic south, and hence the content of the representation is oxygen-free water. Once transplanted, it actually indicates oxygen-rich surface water when it means oxygen-free deep water: it misrepresents the environment.

The problem with saying that the function is to point to oxygen-free water (in the northern hemisphere) is that it also simply points towards magnetic north, and the decision about whether it is representing or misrepresenting depends on which we take its function to be. There is both a liberal reading of the representation's function, indicating oxygen-free water, and a conservative reading of the representation's function, indicating the direction of surrounding magnetic field. On the conservative reading, it is the function of the magnetosome to represent magnetic north, and even the transplanted magnetosome is correctly representing. Unfortunately, transplantation to the wrong hemisphere puts it in an environment where following its correct representations will kill it. Dretske realizes that for the bacteria we cannot find a solid basis for preferring one reading of function to the other, and he concludes that the function is indeterminate. In recognizing the problem about assigning functions, he writes:

> No matter how versatile a detection system we might design, no matter how many routes of informational access we might give an organism, the possibility will always exist of describing its function (and therefore the meaning of its various states) as the detection of some highly disjunctive property of the proximal input. At least, this will always be possible *if* we have a determinate set of disjuncts to which we can retreat. (Dretske 1986: 35)

To deal with this possibility, Dretske appeals to complexity. While function will not give us the desired answer for magenetosomes and similar simple organisms, it will for higher organisms.

Dretske argues that: with more complex organisms, organisms capable of learning, we are entitled to the more liberal function of oxygen-indication, whereas in less complex organisms without the resources for expanding their information gathering resources, the more conservative magnetic-field-indication is the function.[6] Dretske claims that for organisms capable of associative learning there is 'virtually no limit' to the kind of stimuli the organism could be conditioned to substitute for that proximal input, which will then trigger a representation *R*. If this is so, there will be no determinate set of

[6] Curiously, many people misunderstand Dretske to be the father of the indeterminacy argument against functions, when in fact he says functions *do* ground misrepresentation in all but extremely simple organisms. This idea is also present in Millikan: she recognizes that at a very low level there is no difference between an imperative and a description. Is a symbol like FOOD→ a description of what is there, or an order to go get food? For low levels we cannot say which, and it does not really matter. What matters is whether or not the organism has a mechanism that reliably gets it food.

disjuncts to define the disjunctive property that is supposedly infallibly detected. So our only choice in specifying a determinate property would be to make the property non-disjunctive, and hence make the content of *R* non-disjunctive. This is just what is needed to allow for misrepresentation. Since an almost unlimited number of different proximal stimuli could be conditioned to trigger the representation, Dretske claims that none of them can be its content, not even their disjunction. He writes:

> If we are to think of these cognitive mechanisms as having a time-invariant function at all (something that is implied by their continuing—indeed, as a result of learning, more efficient—servicing of the associated need), then we *must* think of their function, not as indicating the nature of proximal (even distal) conditions that trigger positive responses . . . but as indicating the condition (*F*) for which these diverse stimuli are signs. (Dretske 1986: 35–6).

We can see what is really behind this suggestion if we ask this question: if these other stimuli could be substituted for the stimuli that presently control triggering of the representation (in this case magnetosome orientation), why do they not actually control the triggering of the representation now? The answer will be that conditions were not right for the response to become conditioned to those stimuli, and that conditions were right for whatever connection to stimuli—that does exist to develop (or remain, in the case of an innate connection). This claim—that other possible stimuli could be substituted for operating stimuli—boils down to is this: under optimal conditions those other stimuli would be substituted. But this just amounts to creating optimal conditions during the learning period. What is really behind this appeal to functions of the representation and possible stimulus substitution is that, in the optimal and ideal conditions of controlled learning, *R* would precisely covary with the condition *F*. As Cummins (1989: 74–5) remarks:

> It is clear that Dretske accepts the following constraint on the relevant function assignments:
>
> A function of *R* is to indicate *x* only if *r* would covary with *x* under optimal conditions.
>
> This is what does all the work in the arguments; deflationary conservative attributions of content are ruled out solely on the ground that the relevant covariance wouldn't hold 'even under optimal conditions.' The appeal to functions is completely idle here.

If this is what Dretske's solution turns out to be, then it is just the appeal to an infallible learning period that Dretske appealed to earlier in *Knowledge and the Flow of Information* (1981). This earlier view is one that Dretske had already given up in promoting functions as the meaning-fixing property, but we should perhaps review why the earlier causal/informational theory failed.

According to Dretske's causal/informational theory of content (1981), a representation gets its informational content by covariation with properties that would not be present to be perceived unless a particular proposition were true. This by itself leaves us with no misrepresentation: a representation will mean whatever it covaries with, and so its content includes whatever causes it. Dretske's solution in *Knowledge and the Flow of Information* is to appeal to a special kind of ideal conditions: the representational content of *M* is the informational content it *would have* under ideal conditions, which are those during the period in which the representation is learned.[7] During the learning period *M* is a perfect indicator, since the input signals are controlled precisely so that the subject acquires the desired representations and all the inputs count as meaning-fixing causes. After the learning period ends, *wild* tokenings of '*X*' can occur: those not caused by an *X*. But, in Dretske's view, since '*X*'s were caused by *X*'s only during the learning period, '*X*' means *X*, and, after the learning period is over, all non-*X*-caused tokenings of '*X*' are misrepresentations.

Besides the well-known objection that there is no clear cut-off of the learning period[8] (Fodor 1990*a*: Does a whistle blow when training is over?), the important thing to notice is that appeal to the learning period does not do what it is designed to do—specify the content such that misrepresentation is possible. This is seen in Fodor's most serious criticism (1990*a*) of Dretske, that his theory is supposed to utilize counterfactual supporting correlations, and his appeal to the learning period 'ignores relevant counterfactuals'. Suppose that, during the learning period, representation *s* was caused exclusively by property *T*. What *would have happened* if a wild input *F* had occurred during the learning period of representation *s*? If the theory includes counterfactuals, then such a possible wild input contributes to fixing the meaning of *s* even if *F* did not actually occur. And, even if *s* was only actually ever tokened by *T* during the learning period, if *F* would have also caused it, then *s* in fact carries the information that *T-or-F*. Hence *s* means that *T-or-F* and any *F*-caused tokening of *s* after the learning period is hence correct representation, not a misrepresentation.

To rule out such unwanted counterfactuals and allow only the convenient and cooperative counterfactuals to contribute to the meaning would be both *ad hoc* and would seem to beg the question by already assuming that *s* means that *T* and nothing more. If we do not presuppose that *s* is supposed to mean that *T*, how are we to specify which counterfactuals to exclude? We might try

[7] This is similar to Stampe's informational account (1977, 1986), but Stampe appeals simply to ideal conditions: *R* represents *p* only if *R* is such as to enable one to come to know the situation—i.e. that the situation is *p*, should *R* be a faithful representation. So *R* carries the information that *p*. If *R* were faithful (i.e. if conditions were optimal or ideal), one would infallibly know that the situation is *p*.

[8] As we have seen, Dretske later (1986) replaced his device of a learning period with an appeal to functions, a move in part motivated by these kinds of criticisms from Fodor and others, and in part to solve the problem of misrepresentation.

to exclude all of them and have meaning constituted solely by actual tokens of *T* (the actual causes of *s*) during the learning period, and at times it seems that Dretske leans in this direction. But this would make *s* mean too little, since, obviously, not all exemplars of type *T* were causes of *s* during that learning period. Without those counterfactual instances of *T*, the representation *s* would have the content T_1 *&* T_2 *&* T_3 *through* T_n, only for the *n* tokens of objects or events of type *T* that actually did occur during the learning period. In this case *s* would not carry the information that *T*, but merely the information *that* T_1*-or-*T_2*-or-*T_3*-or . . . or-*T_n. It is not clear what the point would be of having a representation that is merely a record of all the actual causes during the learning period. There would be a point if it reliably indicated objects or events of type *T* in the future as well. But, if the representation is to have any such generality at all, any applicability to all tokens of type *T*, this generality can be generated only by having counterfactual instances of *T* count towards the meaning of *s*. After all, Dretske's view is that it is a law that *T*s cause *s*s, for this support of counterfactuals is what makes the representation useful as a reliable indicator of presence of objects or events of type *T* in the future. The account we would end up with is that representation *s* means that *T* iff:

1. *s* reliably indicates *T* during the learning period,
2. possible occurrences of *T* would have caused tokening of *s* during the learning period, and
3. we ignore possible occurrences of anything other than *T* that would have caused tokening of *s*.

Of course such conditions would yield *T* as the meaning of *s*, but the conditions can be satisfied only by specifying ahead-of-time which counterfactuals do not fix meaning, and so it will not serve as a criterion for determining the content of any representation the meaning of which is not already determined at the outset. But if it had already been specified which counterfactuals mattered and which did not, that is, if it had already been specified which counterfactuals were meaning-fixing and which were not, we would not need to look for a source or ground for the representation's content—it would already have to have one. Dretske's causal/informational theory (1981) does not supply a theory of content, it presupposes one. How do we distinguish meaning-fixing from non-meaning-fixing uses and establish misrepresentation? We are told to distinguish meaning-fixing *causes* from non-meaning-fixing ones. And how do we do that? We are then told to do it by distinguishing meaning-fixing and non-meaning-fixing *counterfactuals.* And how do we do that? Dretske simply moves the problem, but does not solve it. This overall strategy is repeated over and over in theories of content, as we will see.

So the old optimal conditions constraint ends up being what distinguishes, supposedly non-arbitrarily, the functional descriptions that fix meaning from those that do not. And this appeal to optimal conditions had already been shown to be both arbitrary and question-begging in virtue of taking some counterfactual learning conditions as content-fixing and others as irrelevant to the determination of meaning. So Dretske's use of biological functions fails as a theory of mental content.

4. Papineau's Teleological Account

Like Millikan, Papineau (1987: 64) recommends that we understand representation as a matter of the biological functions of our beliefs and desires: 'The biological function of any given belief type is to be present when a certain condition obtains: that then is the belief's truth condition. And, correspondingly, the biological function of any given desire type is to give rise to a certain result: that result is then the desire's satisfaction condition.' Papineau also invokes the distinction between *normal* and *abnormal* causes of beliefs and desires to ground the teleology and allow for a gap between the actual causes and the privileged causes that determine the function of the beliefs and desires. As we have seen, such a gap is what is required to allow for misrepresentation. Simply put, the normal causes of a belief are those that the belief is 'supposed to respond to', and the explanation of the presence of such beliefs is in terms of natural selection: 'we have the present disposition to form beliefs of that type because, in the past, its tokens have generally had advantageous behavioural effects, and this has led to its preservation' (Papineau 1987: 65). The objects of desires, their satisfaction conditions, are said to be the effect they are biologically supposed to produce. But there is a problem here that Papineau discusses, in that, biologically speaking, the object of all desires would seem to be the same: passing on genes.

Papineau's specific example, the desire for sweet things, highlights the difficulty: 'Prima facie, this desire is biologically supposed to produce, not only the taste sensation, but also the ingestion of sugar, enhanced metabolic activity, survival, and, eventually, the bequest of the genes behind the desire to the next generation' (Papineau 1987: 67). But this complete string of effects does not always occur in its entirety. Sometimes the ingestion of sugar is detrimental to health and survival. And it seems too much to attribute bequest of genes as the object of the desires of someone with a sweet tooth. They desire sugary foods, nothing more. Papineau's suggestion is that we can take the object of the desire to be the more immediate effect if we think of the brain as being designed or adapted to accomplish the ultimate ends of survival and bequest of genes by focusing on such 'proxy' ends as getting sweet things (or, in other cases, sex or security) and assuming implicitly that these are biologically advantageous. 'Our cognitive mechanisms, although in general designed to ensure gene bequests, are not sensitive to evidence about such things as the connection between eating sweet things and gene bequests. . . . The satisfaction of the desire is the *first* effect (eating sweet things) which is taken to be relevant by natural selection and not by the agent' (Papineau 1987: 69). Thus Papineau tries to focus the object of the desire on the most immediate effect, not the ultimate effect in which the biological advantage really lies.

This account appears to treat beliefs and desires separately, but, ultimately, the content of beliefs is dependent on the logically prior determination of the satisfaction conditions of desires.

> Beliefs will in general have biologically advantageous effects only in so far as they have effects which satisfy desires. So we ought to count the truth conditions of beliefs not simply as circumstances in which they have biologically advantageous effects, but more specifically as the circumstances in which they will have effects that will satisfy the desires they are working in concert with. (Papineau 1987: 70)

The point is that having various true beliefs is, in itself, not biologically advantageous. It is so only if those true beliefs assist in satisfaction of desires, which in the end is biologically advantageous. So Papineau's account has all determination of mental content ultimately based on serving the function of satisfying desires.

In his discussion of Dretske's magnetosome case, Papineau agrees that complexity is the crucial issue, and argues that the bacteria with the magnetosomes are too simple to cut the content down to either *oxygen-free water* or *magnetic north.* Since the bacteria are too simple to have desires, there is nothing to help specify our content assignment. 'The only end that can sensibly be attributed to the bacteria is survival: and from this perspective their magnetosomes simply represent that the indicated direction is conducive to survival' (Papineau 1987: 71). Even though this specification of the content does not choose either of our two candidate contents, *oxygen-free water* or *magnetic north*, the content *direction-conducive-to-survival* is one that, it seems, still can support a notion of misrepresentation. If a southern-hemispheric bacterium is transplanted to the northern hemisphere, its magnetosomes point to the surface and its oxygen-rich water, and lead the organism to its death. If the content of their representations is *direction-conducive-to-survival*, then these transplanted magnetosomes, which now are pointing the wrong way, are misrepresenting the direction conducive to survival.

Several problems arise from Papineau's theory. One obvious one for our purposes involves specifying desire content. On the one hand, simple organisms really do have desires simply to survive, and nothing more specific. But complex intelligent organisms have just the opposite: they have desires for the first effect (such as the sweet taste of sugar) that is relevant to natural selection, and they do not (consciously) desire the more remote effect, such as survival or passing on genes. This difference seems peculiar for a theory that seeks to take advantage of the biological basis of representation generally, across the biological spectrum from humans to bacteria. But, though this asymmetry is perhaps unexpected, it is not in itself a conclusive objection to his account. A more serious problem is saddling individual beliefs, and individual concepts, with typical behavioral consequences. As Sterelny (1990: 130–1) points out,

> That's necessary if a belief, or more exactly a belief-circumstance pair, is to be selected for. For selection on cognitive states will operate via the behaviour those states give rise to. The problem is that it's far from obvious that there are typical behaviours beliefs, or even belief-circumstance pairs, give rise to. The most telling criticism of behaviourism is that it wrongly supposed there to be a signature behavioural type for each mental state type.

Even the perceptual beliefs that Papineau is most concerned with need not have any direct effect on behavior. The behavioral effect of any single mental state comes only, as the functionalist accounts which replaced behaviorism tell us, in relation to its role in the entire cognitive system. Millikan avoids this mistake by invoking the distinction between teleological accounts of the mechanism and teleological accounts of the products of the mechanism, though, as we have seen above, she still attributes functions to individual beliefs and desires. Fodor (1990*a*) argues that the functions should instead be attributed to the belief- and desire-forming mechanisms. Papineau's assigning behavioral consequences to individual beliefs is a serious flaw, yet one he cannot avoid if he wants the individual beliefs themselves to be selected for.

But the most serious shortcoming of Papineau's theory, the one that undermines his attempt to account for misrepresentation, is his basing all representational content on desire-satisfaction. If we already assume that one must have desires to have any representation more complex than *thing-that-will-help-me-survive*, then of course the magnetosomes in the bacteria will have the representational content *direction-of-survival* and not *direction-of-magnetic-north.* But the same alternatives apply here as to Dretske and Millikan. On the one hand, you can build in the liberal content (with possibility of misrepresentation) to the representation and the organism survives because it represents that which is conducive to survival. On the other hand, you could allow only the conservative content (with no possibility of misrepresentation) and the organism survives because survival happens to be in the direction of that which it infallibly represents. In the case of northern-hemispheric magnetosomes in their natural hemisphere, they represent the direction of magnetic north, and this happens to be the direction of oxygen-depleted deep water that they need to survive. On both readings of function and content, the organism's representation fulfills a biologically advantageous function, and, because of this, it is not the biological advantageousness that is behind choosing the liberal content (and the possibility of misrepresentation) to the conservative content (without the possibility of misrepresentation). So, in this supposedly biological and teleological account of content and misrepresentation, the teleology is not what provides the distinction between meaning-fixing and non-meaning-fixing functions. Actually it is not clear what provides that distinction, other than either an arbitrary decision or knowing the 'right answer' in advance and begging the question.

5. Neander: High Church versus Low Church Teleofunctionalism

Looked at very generally, the link between misrepresentation and functions is something like this: Content must be normative to allow misrepresentation, and, since teleological functions are normative, the normativity needed for misrepresentation can come from functions. This may initially sound right, but it works only if the normativity needed for misrepresentation is the *same* normativity that we get from teleological functions. It is not clear that we do get the *same* normativities on each side. And, when we look closely at functions, there are various different normativities they could yield. Karen Neander (1991, 1995) lays out some of the alternatives and recommends a different reading of function-based content than we see in Millikan. Neander argues that representational content based on functions gives us fewer misrepresentations than Millikan's view would have, but Neander still thinks functions can provide content assignments that give some misrepresentation.

Like Millikan, Neander defends an etiological theory of proper functions—that is, she defends the notion that biological functions are to be defined as effects for which traits were selected by natural selection. 'It is the/a proper function of an item (*X*) of an organism (*O*) to do that which items of *X's* type did to contribute to the inclusive fitness of *O's* ancestors, and which caused the genotype, of which *X* is the phenotypic expression, to be selected by natural selection' (Neander 1991: 174). The opposable thumb has the function of grasping objects, Neander explains, because opposable thumbs contributed to the fitness of our ancestors. She rejects the identification of function with actual causal power or role (as Cummins defines it) precisely because she wants to use function as a basis for malfunction and misrepresentation. 'A trait has a proper function if there is something that it is supposed to do. According to my etiological theory, a trait is supposed to do whatever it was selected for by natural selection' (Neander 1991: 183). We cannot identify proper function as actual causal role, statistically typical contribution to fitness, or disposition, because that would leave no room for dysfunctional traits. 'Dysfunctional traits do not actually have the disposition to perform their proper function', (Neander 1991: 183) and they do not actually play the causal role that a normally functioning trait does. This is what allows for the very possibility of a dysfunctional trait. We must have this gap between what a trait actually does and what it is supposed to do if we are ever to say it is dysfunctional. If we understand malfunctioning and misrepresentation as instances of (at least occasionally) dysfunctional traits, then this gap, made possible by appeal to functions, is essential in allowing for malfunctions and misrepresentation.

However, Neander differs from Millikan on the issue of what I have been calling 'liberal' versus 'conservative' readings of functions. Recall once again the frog (toad) story and the magnetosome story. When the frog snaps at a bee-bee, the liberal reading is that, based on its function, the frog's representation means *fly*, and he misrepresents the bee-bee as a fly. This is Millikan's position (1990)—it was snapping at flies that contributed to the proliferation of frog ancestors. Neander (1995: 126) refers to such insistence on the highest level interpretation of function (and thus of representational content) as 'High Church Teleology'. Of course, there is an intermediate level also, as in Griffiths and Goode's claim (1995) that the frog's thought content is really *frog-food*, since frogs were interested in flies only because they are nutritious. But, while less exacting than *fly*, the *frog-food* content still makes snapping at a bee-bee a mistake, and the identification of the bee-bee as *frog-food* a misrepresentation. So, while this may be a different denomination, but it is still roughly High Church Teleology. The High Church position on functions is specifically geared towards getting functions to allow malfunction and misrepresentation. But Neander favors what she calls 'Low Church Teleology', which reads the functions more conservatively.

The Low Church will interpret the frog as snapping at a *small, dark, moving thing*. The Low Church reads the content as more proximal than distal, yet Neander notes that this preference for the 'lower level of description' is not insistence on the lowest level, which would be down in the area of sub-cellular neuronal components. Is it 'the lowest level at which the trait in question is an unanalyzed component of the functional analysis', where such a component is used to analyze the system in which it is a part, but is itself not analyzed into the parts which compose it (Neander 1995: 129). On this account, we should be considering the content involved in the frog's detection system at the lowest level at which it is still an unanalyzed device. To call the content *fly* would be too high a level, inserting the frog into environmental and historical context, where to call the representation *activation of retinal cells 18–47* would be too low, already analyzing the detection system into components. So the content of the frog's representation content is *small, dark, moving thing*, because, Neander (1995: 130) says, 'it is *by* detecting small dark moving things that the frog detects frog-food and flies'.

The big question for the Low Church is whether it can allow for misrepresentation. Neander does not see the frog snapping at the bee-bee as a case of misrepresenting—the frog correctly represents the bee-bee as a *small dark moving thing*. Neander correctly diagnoses the risk here: without separating the causes that are properly part of the content of a representation R and the causes that are not, all the causes will be, eliminating all possible misrepresentation. Neander (1995: 131) writes:

> It doesn't follow that unless the frog misrepresents when it R-tokens at bee bee it cannot misrepresent. . . . On the account I favour, if the frog R-tokens at anything which reflects onto its retina a pattern that falls outside of the specified parameters, then it misrepresents. The images cast by snails, for example, will fall outside these parameters, even when the snails are at the smallest and most sprightly, so R-tokening in response to a snail is a misrepresentation. So the frog *can* still misrepresent.

Even if frogs do not actually make these mistakes, they could, and, since a snail or lily pad could make the frog token R, the frog can misrepresent. As Neander (1995: 131) explains, 'Damaging the frog's neurons, interfering with its embryological development, tinkering with its genes, or giving it a virus' could cause the frog to have dysfunctional detection mechanisms, and thus misrepresent. So, given the possibility of mechanical malfunction, Neander claims Low Church Teleology has room for misrepresentation. Or does it?

First consider the cases where misrepresentation is supposed to arise from malfunction. Where High Church Teleology like Millikan's seeks to interpret functions at a highly distal level (*fly* instead of *little black moving thing*, or *black retinal image*), Neander's Low Church goes proximal. There is no biological reason to insist on the most distal reading of the function, and, as we move to the proximal, we eliminate some of the misrepresentations that would occur on a more distal interpretation. But it is important to stop somewhere in the middle—the Low Church does not want to stoop so low that it eliminates all possible misrepresentation. If the function relates to *activation of retinal cells 18–47*, no frog snapping in response to such activation would be a malfunction, because a bee-bee will cause that retinal activation as well as a fly. This erosion from distal to proximal is a familiar problem—it is a move from the liberal to conservative reading of function. Critics of Dretske's causal/informational theory of content (1981) were often tempted simply to read the information as defined by such proximal features as to make it incapable of allowing for misrepresentation. As we saw, Dretske makes the same point himself in the magnetosome story (1986), reserving the content yielded by 'distal' readings only for very complex organisms. The problem is, if you are willing to go a little proximal, why not all the way?[9] You cannot have your only reason for stopping in the middle be that this still retains the possibility of malfunction, and thus misrepresentation. There should be an independent reason, and establishing it is no easy trick.

[9] Enç (this volume) calls this the 'landslide argument'.

Along the way there seems to be a confusion going on here between two distinctions: distal/proximal and normal/abnormal. It is important to remember that they are not the same. Reminiscent of Neander's description of altered frogs, one cognitive psychology experiment takes a ferret in early development, detaches its auditory nerves, and attaches its visual nerves to the auditory center. The ferret can then see, but not as well as 'normal'. But the ferret undergoes normal development in unusual conditions. Is the developmental system malfunctioning? No, it is functioning correctly in an unusual environment/circumstance/context. But, while correctly functioning, it is certainly not a normal ferret. But this abnormality does not arise specifically from any difference between proximal and distal stimuli.

So why should we stop at the level between distal and proximal that Neander recommends? Why give up making the frog's target *fly*, insist on stopping halfway at *little black moving thing*, yet refuse to go proximal all the way and read the content as *black retinal image*, or even *activation of retinal cells 18–47*? Neander's answer is that we should stop at the lowest level of description at which the component is still unanalyzed. The reason is that this level is supposed to be the most informative. But informative for/about what? Explanation of the past? Presence of the trait? Predominance of the trait? Functions may have the answers to some or all of these questions, but not necessarily all at the same level of analysis. The degree of informativeness is relative to context.

This choice of level of analysis is a particular problem for Neander and her example of the genetically or biologically altered frog. Once we mess with the frog's neurons, genes, or cells, we have jumped down to that level in our analysis. According to Neander's own view, that is the level at which our reading of function should be done. So we must describe the function of the detection systems and representations of frogs altered in any of these ways at the level of cells in the systems, or neurons, or genes. In doing so, we will not get the content Neander wants—a content that allows for misrepresentation. Specifically, if we were to reroute the neurons so that the detection system functioned in bizarre ways and the frog snapped at passing cars and large orange beach balls, the level we would describe its functioning would be neuronal. So the frog's representation content might be something like *neurons A6–D17 firing*, and the frog is now altered so that it snaps when those neurons fire. Who cares if, because of neuron rerouting, an orange beach ball now causes those firings—they are still firing. The frog correctly identifies those neurons as firing, and its snapping at the big orange beach ball is not a mistake or malfunction, since the organism is now hooked up so that it is supposed to snap at things that cause those neurons to fire. The promised misrepresentable content has disappeared.

There is also a problem of types and tokens lurking here in Neander's theory. Neander's view is that some misrepresentations are due to malfunctions. If we alter the frog's physiology in the ways Neander imagines (the kinds of things that are actually done by experimental psychologists), the organs and representations of the altered frog have their functions defined by what their functions are (or would be) in normal frogs. But what makes these tokens of the same type?[10] If we reroute nerves and brain circuitry and so on, it is not obvious that the organism will have developed the same functional organization. Is sight in the altered ferret really the *same trait*? It becomes unclear what right we have to expect the relevant counterfactuals to remain unchanged—the alteration may result in a different organ. The altered frog is still an animal, and still (genetically) a frog, but does it still have a frog brain? It is a brain, and it is in the frog, but that does not make it the same type of organ that a normal frog has. Would we be willing to put this frog with the others and call them all of a type? We certainly would not use it in a collection of frogs for research or taxonomy. Why would the weird detection mechanisms of a Frankenstein-frog be relevant to an analysis of the function of the detection systems of normal frogs? Or vice versa? For counterfactuals to count, they have to be counterfactuals about the same type of objects, and these altered organisms seem like a different type of beast.

So, Neander's explanation of how some misrepresentations will occur in virtue of malfunctions fails. Overall, she lacks a convincing (non *ad hoc*) rationale for stopping her descent into the proximal realm where she wants to stop. And, specifically, when we consider genetic or biological alterations of an organism, such alterations force us to go lower in level of analysis than where she had wanted it stopped, and they cast doubt on whether the resulting organs, mechanisms, and representational systems are even of the same type as those in normal organisms.

But now consider those misrepresentations that are not malfunctions, as in Fodor's example (1987) of mistaking a cow in dim light as a horse. Fodor wants his causal theory of asymmetric dependence to allow for this case of misrepresentation. But Neander (1995: 132) is happy to say that, in this kind of case, 'early visual processing does not represent the cow as a horse (or as a cow) but as something which *looks a certain way*—as having certain outline, texture, color, and so on. That is, according to conventional computational theories of perception, *initially there is a representation of the physical parameters of the environment as measured by the visual system*.' So how does Neander think error and misrepresentation could occur? Because, she says, perception involves inference in its later stages, and our beliefs and expectations and other background information could make our inferences go wrong. We might have expected the thing in the field to be a horse, and that expectation might

[10] One might be tempted to say the resulting organism is no longer a frog at all, but that seems too extreme; a human with an artificial limb, or blindness, or aphasia, is still a human.

override the perceptual data. We can also revise our perceptions in the light of further data and information. The frog is too simple to make such inferences, so it cannot make such sophisticated errors. But human beings can.

Does this solve the problem? Misrepresentations not due to malfunctions are due to bad inferences? The trouble is that this appeal to inference and background information depends crucially on determining the content of the representations involved in the inferences. But this is just what we are worrying about. So, if we can already have determined that background information means such-and-such, then a perception or detection can be a misrepresentation if it is due to inferential interference from these background representations of known content. But this begs the question. The theory will have us use functions to determine the contents of those representations as well. For each reading of those other representations that would make the detection or perceptual instance a misrepresentation, there is another function that makes it a correct representation. Round and round we go.[11]

Neander calls her 1995 paper 'the philosophical marriage of Fodor and Millikan. This is teleosemantics for those who love their Language of Thought and Computational Theory of the Mind' (1995: 137). Where she recognizes the union of these as improbable, we can see that this marriage is doomed to fail. We have already seen that the High Church has no good defense of the high ground, and now the Low Church fails to find a secure foothold in the middle. Grounding representational content on teleological functions erodes to a position I am tempted to call 'No-Church Teleology'. But this would imply an outright rejection of functions, for which I have not argued, and which I do not accept or desire. There are indeed functions, and we do have ways of describing the functions of traits and of organism components. But, unfortunately for fans of biological teleology, whatever virtues functions and teleology might have, functions do not provide a basis for describing representational content that admits of misrepresentation. So it is not quite No-Church

[11] Neander appeals to a two-factor theory at one point (1995: 134) in arguing that we should prefer Low to High Church Teleology. I think this is unwise, and have argued elsewhere against two-factor conceptual role theories (Perlman 1997). The acknowledged use theories, 'conceptual role' theories, in general add a causal or teleological factor to straight conceptual role, in hopes of having this give them a way to establish misrepresentation. The ironic thing is that causal and adaptational role theories are themselves just limited or restricted conceptual role theories, and adding a restricted conceptual role theory to a full-blown conceptual role theory is not adding much. For the added causal or teleological theories to be of any help, they would have to establish the possibility of misrepresentation on their own, and we have just seen that they fail to do this. Failing this, they would have to pick out a different pattern of use than full-blown conceptual role theories do, so that we have a misalignment of the two factors in the 'two-factor conceptual role theories', and then we would have to have independent grounds for assigning the content of the representation solely as that determined by the causal or teleological factor. But this is an arbitrary assignment at best, and at worst it is circular and begs the question. So two-factor theories also fail to establish the possibility of misrepresentation.

Teleology, but rather 'Pagan Teleology': we can still believe in functions, but not the way the organized churches, either High or Low, would have liked. Teleology does not buy us what the Churchgoers advertised.

6. The General Problem with Adaptational Role and Function

How are these adaptational theories relevant to the general complaint against use theories, that they cannot account for misrepresentation? We should recognize that adaptational/teleological theories are really just specific kinds of causal theories, which themselves are really specific kinds of conceptual role theories. The contents of the representations are fixed not by entire conceptual role, but (on the teleological accounts) only by the causes of the tokening of the representation. Even then the content is not fixed by all the causes, but only by those causes that historically have served a biologically adaptive function. Exactly which function it is said to serve varies among the adaptational theories. The function can be to indicate the presence of certain objects or events (in Dretske's view (1986)), or to help in the reproduction of the representational mechanism (in Millikan's view (1984, 1986)), or to satisfy the desires of the organism (in Papineau's view (1987)). These are all ultimately functions to assist in the survival of the organism. So the teleology is simply another way to cut apart the uses of a representation that fix its meaning from the uses that do not: the uses that count are the ones that increase either the chances of satisfaction of desire, survival of the organism, transmission of genes, or the reproduction of the cognitive mechanism involved in the representation.

The problem with teleological/adaptational theories is, not surprisingly, just the same as with the other causal theories: alternative formulations of the meaning-fixing property. Just as there is always both a specific and a more general cause to cite, there is a liberal and a conservative reading of every function. Where the specific version of cause and the liberal reading of the function can make room for misrepresentation, the general/disjunctive cause and the conservative reading of the function cannot. But both readings of the function are consistent with the goal of survival of the organism, reproduction of the cognitive mechanism, and passing on of its genes. Hence, if both readings of the function are equally consistent with biological advantageousness, it turns out that nothing about adaptational role tells us which reading of the function is the one that fixes meaning. Basically we choose on the basis of what we antecedently have decided is the content of the representation. Yet it is often the case that representations with the conservative function are more adaptive, since a less accurate but accurate enough representational mechanism will do better than a more accurate one that is slower and uses more cognitive resources. This consideration gives us reason to think not that the conservative and liberal readings of function are equal relative to adaptiveness, but that the system with the more conservative function is more adaptive. On the one hand, this makes us prefer the representations and representational mechanism that is always wrong but non-lethally wrong—always close enough to the right answer for the organism to survive. This makes it sound as if teleology does indeed allow for misrepresentation, and in fact that it prefers misrepresentation. But this description is slightly misleading, since even that assignment of content has a more conservative alternative reading, such that the system's function is much less ambitious, but more successful in meeting its ambition. Biological considerations of fitness and adaptation show that a cognitive system is most adaptive when it has the most conservative function that is successful enough to allow survival and passing on of genes. And in lowering the target to something so mundane that the system can always represent it correctly (like *little black moving thing*, instead of *fly*), we eliminate misrepresentation. Again, here the possibility of the alternative, conservative reading of function shows that teleology does not univocally cut uses of representations apart into the correct and incorrect ones. So all of the teleological talk is in the end irrelevant to establishing a means of distinguishing the meaning-fixing uses from the non-meaning-fixing uses.

7. Conclusion

The split between the meaning-fixing uses and non-meaning-fixing uses of representations frequently reappears within the criterion that is supposed to make the larger distinction clear. To separate the meaning-fixing from non-meaning-fixing uses we need a criterion that is univocal and not itself susceptible to being split apart. This general skepticism about teleology being capable of dividing the meaning-fixing uses of a mental representation from the non-meaning-fixing uses is in accord with Fodor's comments that teleological notions 'always have a problem about indeterminacy just where intentionality has its problem about disjunction' (1990*a*: 70). 'You get indeterminacy about the function of the mechanism wherever there is ambiguity about the content of the state' (Fodor 1990*a*: 71).[12]

If functions are to be the basis of misrepresentation, by dividing the uses of a representation into those that fix meaning and those that do not, we must

[12] I argue elsewhere (Perlman 2000) that a similar ambiguity plagues the kind of causal theory that Fodor advocates, in that you always get indeterminacy about the counterfactuals that are relevant to fixing mental content whenever there is ambiguity about the content itself. (The same thing holds for informational theories such as Dretske's, but I will not enter into the details of either argument here.)

then clearly specify the functions that are to have a role in fixing meaning. Only some of the functions (the liberal readings) can count in splitting the uses. Other functions (the conservative readings) do not make any split between meaning-fixing and non-meaning-fixing uses. But, within the factor that is supposed to divide uses apart, the basic distinction re-emerges. Functions all need to be divided just where the uses themselves needed to be divided. So these teleological theories all presuppose the very distinction they are being proposed to ground. Hence, we never get a non-arbitrary and non-question-begging distinction, and, thus, we never get a teleological account of how misrepresentation could be possible. We never end up grounding the distinction between meaning-fixing and non-meaning-fixing uses—all that these disguised use theories do is move the distinction to deeper levels. But at no level is it grounded by some criterion that itself does not also need to invoke that same distinction. So adaptational role theories of mental content are in the end really use theories after all, and they turn out to be use theories that cannot allow for misrepresentation.

References

Block, Ned (1986), 'Advertisement for a Semantics for Psychology', *Midwest Studies in Philosophy*, 10: 615–78.

——(1993), 'Holism, Hyper-Analyticity, and Hyper-Compositionality', *Mind and Language*, 8/1: 1–26.

Boghossian, Paul (1993), 'Does an Inferential Role Semantics Rest upon a Mistake?', *Mind and Language*, 8/1: 26–40.

Chomsky, Noam (1957), *Syntactic Structures* (The Hague: Mounton & Co.).

Churchland, Paul M. (1979), *Scientific Realism and the Plasticity of the Mind* (Cambridge: Cambridge University Press).

——(1981), 'Eliminative Materialism and the Propositional Attitudes', *Journal of Philosophy*, 78/2: 67–90.

——(1990), *A Neurocomputational Perspective: The Nature of Mind and the Structure of Science* (Cambridge, Mass.: MIT Press).

Cummins, Robert (1975), 'Functional Analysis', *Journal of Philosophy*, 72: 741–64.

——(1989), *Meaning and Mental Representation* (Cambridge, Mass.: MIT Press).

——(1996), *Representations, Targets, and Attitudes* (Cambridge, Mass.: Bradford/MIT Press).

Dennett, Daniel (1969), *Content and Consciousness* (London: Routledge & Kegan Paul).

Dretske, F. I. (1981), *Knowledge and the Flow of Information* (Oxford: Blackwell).

——(1986), 'Misrepresentation', in R. J. Bogdan (ed.), *Belief: Form, Content, and Function* (Oxford: Oxford University Press), 17–36.

Fodor, Jerry (1987), *Psychosemantics* (Cambridge, Mass.: MIT Press).

——(1990*a*), *A Theory of Content and Other Essays* (Cambridge, Mass.: MIT Press).

——(1990*b*), 'Information and Representation', in Philip P. Hanson (ed.), *Information, Language, and Cognition* (Vancouver: University of British Columbia Press).

Fodor, Jerry, and Lepore, Ernest (1992), *Holism: A Shopper's Guide* (Oxford: Blackwell).

——— ——— (1993*a*), 'Reply to Block and Boghossian', *Mind and Language*, 8/1: 41–8.

——— ——— (1993*b*), 'Replies', in J. Fodor and E. Lepore, *Holism: A Consumer Update* (Amsterdam: Editions Rodopi, 1993).

Goldman, Alvin (1976), 'Discrimination and Perceptual Knowledge', *Journal of Philosophy*, 73: 771–91.

——(1986), *Epistemology and Cognition* (Cambridge, Mass.: Harvard University Press).

Griffiths, Paul E., and Goode, Richard (1995), 'The Misuse of Sober's Selection for/Selection of Distinction', *Biology and Philosophy*, 10/1: 99–108.

Harman, Gilbert (1982), 'Conceptual Role Semantics', *Notre Dame Journal of Formal Logic*, 23: 242–56.

Israel, David (1987), *The Role of Propositional Objects of Belief in Action* (CSLI Monograph Report No. CSLI-87-72; Palo Alto, Calif.: Stanford University Press).

Millikan, Ruth Garrett (1984), *Language, Thought, and Other Biological Categories: New Foundations for Realism* (Cambridge, Mass.: MIT Press).

——(1986), 'Thoughts without Laws: Cognitive Science without Content', *Philosophical Review*, 95: 47–80.

——(1990), 'Truth-Rules, Hoverflies, and the Kripke–Wittgenstein Paradox', *Philosophical Review*, 99/3: 323–53; repr. in Millikan (1993), 211–39.

——(1993), *White Queen Psychology and Other Essays for Alice* (Cambridge, Mass.: MIT Press).

Neander, Karen (1991), 'Functions as Selected Effects: The Conceptual Analyst's Defense', *Philosophy of Science*, 58: 168–84.

——(1995), 'Misrepresenting and Malfunctioning', *Philosophical Studies*, 79: 109–41.

Papineau, David (1987), *Reality and Representation* (Oxford: Blackwell).

Perlman, Mark (1997), 'The Trouble with Two-Factor Conceptual Role Theories', *Minds and Machines*, 7: 495–513.

——(2000), *Conceptual Flux: Mental Representation, Misrepresentation, and Concept Change* (Studies in Cognitive Systems; Dordrecht: Kluwer Academic Publishers).

Stampe, Dennis (1977), 'Toward a Causal Theory of Linguistic Representation', in P. French *et al.* (eds.), *Midwest Studies in Philosophy*, vol. ii (Minneapolis: University of Minnesota Press).

——(1986), 'Verification and a Causal Account of Meaning', *Synthese*, 69: 107–37.

Sterelny, Kim (1990), *The Representational Theory of the Mind: An Introduction* (Oxford: Blackwell).

11. Indeterminacy of Function Attributions

Berent Enç

ABSTRACT

A popular thesis maintains that among the many effects brought about by a system, at most one can be identified with the function of that system. An examination of this thesis reveals that there are good reasons for denying it. It is then observed that the indeterminacy entailed by this denial will have consequences to teleosemantics—to our attempt at finding a unique determinate representational content for detection systems by looking to see what specific function they have been selected for. It is then suggested that the non-uniqueness involved is appropriate to the representational richness of these subdoxastic systems.

The thesis that minds are part of the natural scheme of things, and that intentionality of mental states is a manifestation of the causal relations and laws of the physical world, is a familiar one. A number of philosophers, myself being among them, are committed to it. And some of these philosophers have sought to show that a theory of content for mental states could be developed by appealing to the functions of our perceptual and mental organs. The rough thought was simply this: talk about functions is in its surface grammar teleological. Teleology is essentially intentional. So, if function assignments can be given a reductive analysis in terms of causal relations and causal laws, then we will have found a key to naturalizing intentionality, and this will be the first step towards understanding the source of the intentional nature of minds.

A familiar example might illustrate this thought. When we hear something, our auditory experience has a determinate content: we hear *the sound of the door bell.* On a causal account of perception, it is the sound that causes us to enter that perceptual state. But the causal account does not by itself yield the content in question, for the bell button's being depressed, as well as the vibrations of the ear drum, are equally active causes of our entering the state, and yet, we hear neither the button's being depressed nor our ear drums vibrate. However, if we could argue that, on some properly naturalistic analysis of

I am grateful for their help in improving earlier versions of this chapter to the following: André Ariew, Fred Dretske, Mohan Matthen, Chris Stevens, and Larry Shapiro.

functions, the function of the auditory system is to convey information about the sound waves within a certain frequency range in the ambient medium, then we could zero in on the sound of the bell as the proper content of our auditory state, and thus begin the task of deriving mental content out of a world of causes.

As a result of this line of thought, many philosophers have turned to biology to see if some suitable principle for assigning functions to organs, or traits of biological organisms, could be developed and then used in a theory of mental content. I have in the past personally believed in the fruitfulness of this method, and have joined the crowd both in looking for a suitable principle for function attributions and in applying the principle to show how content can arise in a simple system (see Enç 1979, 1982).

This chapter is a counsel of despair—I have now come to think that the method in question is ultimately doomed to failure.

Using a theory of functions in the attempt to naturalize mental content requires the accomplishment of two separate goals. (1) The account of functions that we want should explain how functions of organs and traits arise. And this account should yield a principled way of giving a *unique* answer to the question 'what is the function of this organ, trait, or system?' in a way that is in harmony with function assignments in textbooks of ethology or biological physiology. (2) The principle that yields the *unique* function assignment, when applied to the representation generating subsystems of organisms, should yield contents for these representations that do not violate our basic common-sense intuitions. It should not, for example, force on us the result that we hear our ear drums.

In the following pages, I propose to survey the ever-growing literature on functions, and try to show in some principled way why in general we cannot hope to satisfy both of these goals. I will then go on and suggest that, in spite of this negative result, function assignments still promise to make a significant contribution to understanding the intentional nature of certain types of mental states. This rather surprising result justifies the interest we, philosophers of mind, have in naturalized teleology.

Ever since Boorse (1976) offered his counter-examples to Wright's original analysis of functions (in Wright 1973), it has become clear that the assignment of a function to an individual object (gadget, trait, organ, subsystem of a system, and so on) needs to be evaluated relative to a kind to which the object belongs. Wright's original account, which, I tend to think, was the seed from which all the etiological accounts of function sprung, simply asserted that the following two conditions were necessary and sufficient for the truth of the schema 'The function of X is Z'.

(i) Z is the result of X's being there.
(ii) X is there because it results in Z.

I will refer to this analysis as (W).

Boorse gave examples designed to show the inadequacy of (W), among which the example of a hose designed to carry carbon monoxide to the outside is perhaps the clearest. The hose accidentally develops a tear, and the leaking gas causes the mechanic to become unconscious and incapable of fixing the leak. On (W), the tear ends up having assigned to it the function of keeping the mechanic unconscious.

But one can block counter-examples of this kind, of which dozens have found their way into print in response to Wright's original paper, by a simple addition to (W).[1] The addition consists of a suitable requirement that function assignments be made relative to kinds. And since, in the Boorse example, neither the tear that has these causal consequences, nor the whole system that encloses the hose, is a thing of a relevant kind, the example ceases to be a counter-example to (W).

Once kinds are introduced into the analysis of functions, and some suitable specification of what constitutes a kind is found, it might seem that we will be closer to satisfying the first of the two goals mentioned above—that is, that of giving a determinate answer to the question 'What is *the* function of this system?' Such a motive may have been what originally guided Millikan (1984) to propose her formula for the kinds relevant to function attributions. The formula introduces the technically defined notion of a *reproductively established family*. There are two types of reproductively established families: (1) some individual is a member of a *first-order* reproductively established family if it has some traits in common with other members as a result of reproduction, understood as a law-governed copying process that preserves the traits in question, or (2) some individual is a member of a *higher-order* reproductively established family if it has some traits in common with other members as a result of belonging to members of some first-order reproductive family. So cats form a first-order reproductively established family, and a cat's eye is a member of a higher-order family. (See Davies 1994 and Davies 1997 for detailed critical discussions of Millikan's analysis.)

This conception of kinds renders a whole class of counter-examples, like Boorse's leaky hose, harmless. They cease to be a threat to the analysis just because the tear in the hose is not a member of a (higher-order) reproductively established family.

On the other hand, the introduction of the notion of reproductively established families, or any other specification of kinds for that matter, ultimately fails to be of any help in determining what the unique function specification is among a number of plausible candidates.

[1] See Adams and Enç (1988) for a discussion of some of the counter-examples that exploit the uniqueness feature of the systems described.

To argue for this point, I will stipulate some barebones general etiological account of functions as a working hypothesis:

(E) The function of X, some item in system S, is Z if and only if:
  (i) X is a member of a kind of thing in a kind of system S, and X is such that because of its activity or its structure a sequence of events takes place in S, and these events result in Z;
  (ii) the history of Xs and of Ss contains episodes that explain why X's activity or structure, in terms of their capacity to give rise to Z, has been retained in Ss.[2]

In biological systems, Xs are traits that have been selected for.[3] And this selection for is due to the fact that Xs' capacity to produce some outcome, Z, has conferred some advantage to the ancestors of the present population of Ss.[4]

In order to show how the essential reference to kinds in (E) does not help us to assign a unique function to biological traits, it is sufficient to explore the relation between assigning a *function* to traits and deciding on the identity conditions for *kinds* of traits.

Let us begin with the mammal eye: it works like a camera; it has a lens, which casts an inverted image of the shapes in the mammal's field of vision on the retina, and a complex network of rods and cones detects gradations of light energy and generates information about shapes, colors, and movement that are partially contained in these gradations.

Now the account we have adopted—analysis (E)—promotes at least the following two function assignments to the eyes:

F_1 detecting shapes, colors, and movement;
F_2 detecting gradations of light energy.

If we start with F_1, then it will be easy to maintain that the compound eye of the insects have the same function, because that result is what those eyes, too, have been selected for. There would be no principled difficulty in maintaining the same function for the insect eye, because one of the platitudes of functional

[2] When I originally offered an account of functions (Enç 1979), I aimed at a more general condition that required the availability of an explanation of why X's ability to bring about Z in S is retained, and I required that this explanation be given in terms of the consequences of X's doing Z in things of kind S. This more general specification, also an element in more recent accounts of functions (see e.g. Averill 1998; Price 1998) is friendly both to etiological and to propensity approaches to functions. For an intriguing and persuasive defense of the relevance of propensity considerations, see Walsh (1996), and Walsh and Ariew (1996). However, I do not think that anything I say in this chapter is affected by these subtle issues. To keep matters simple, I will focus my remarks on just the etiological versions of (W).

[3] The notion of selection *for*, as opposed to selection *of*, introduced by Sober (1984), is commonly used to good effect in this context.

[4] We will see shortly that, by fine-tuning these conditions, we can arrive at different versions of the etiological account of teleology, where each account is clearly reductive.

properties is that they are multiply realizable. The fact that structurally there is very little that is in common between the compound eye and the camera eye, and the fact that developmentally they arise from different kinds of cellular structures, are of no relevance to the claim that they both have the same function. However, when we look at the so-called pineal eye of the lizard, or the eye of the mole rat, we will reach a different conclusion. During embryonic development, the pineal eye branches off from the cells that divide to produce the two lateral eyes of the lizard and protrudes from the top of the head. It detects ultraviolet light and the signals from it cause the darkening of the skin coloration. The mole rat's eyes are behind sealed eyelids and, by detecting levels of illumination, they help adjust the mole rats' circadian rhythms. Now, since neither the pineal eye nor the mole rat's eye is capable of detecting shapes, colors, and movement, they cannot have the function F_1. And, if the *eye* is an organ the identity of which is determined by its function (and it is reasonable to suppose it is, for why else should two organs like the insect eye and the mammal eye, which have no structural features in common, be *eyes*), then the mole rat eye and the pineal eye are not real eyes.

If, on the other hand, we decide that F_2 is the proper function of the eye—a decision that is also fully consistent with the etiological account I have adopted—then, on the same assumption that kinds are determined by function, we would have good reason to classify the mole rat's eye and the pineal eye as eyes. But this consideration makes it clear that we cannot, without circularity, appeal to some conception of *kinds* to choose between F_1 and F_2 because the kinds themselves are determined by the choice between F_1 and F_2.

		Choice of function assignments	
		F_1	F_2
Choice of identity criteria	Eyes are functional kinds	m and s are eyes. p and r are not real eyes.	m, s, p, and r are all real eyes.
	Eyes are structural kinds	m, s, p, and r are different kinds of organs. m and s have the same function, whereas p and r each has a different function.	m, s, p, and r are different kinds of organs. They all have the same function.

Fig. 11.1. Kinds of eyes and their function

To summarize, we have a matrix of four choices here: mammal eye (m), insect eye (s), pineal eye (p), and mole rat eye (r) (see Fig. 11.1). This shows that, whatever function assignment we choose, also choosing the option where eyes are defined functionally threatens to be circular: whatever we choose to count as a real eye validates the function assignment according to which the kind *eye* is defined. So the methodologically sound choice would be to start with purely structurally defined kinds, and to try to determine what a real eye is only after the proper function assignment to these different structures has been determined.

This is not noteworthy at all. The fact that methodology directs us towards initially choosing kinds that are structurally defined demonstrates that relativizing function assignments to kinds does not help us in the least in choosing between F_1 and F_2 as the proper function of the eyes. Only after a decision has been made as to whether the function of the structures m, s, p, and r is F_1 or F_2 on independent grounds can we decide which are real eyes.

This issue is stated very succinctly in a textbook on Animal Physiology. In the section on the auditory systems of animals, the author says: 'We know that sound consists of regular compression waves that we perceive with the ears; in turn, an ear can be defined as an organ sensitive to sound. However, compression waves can be transmitted in the air, in water, and in solids, and we cannot restrict the word hearing to perception in air only' (Schmidt-Nielsen 1990: 527). The author then goes on to discuss how snakes are very sensitive to head vibration, and fish to lateral water vibrations. Given that what constitutes an ear cannot be decided on purely structural grounds, the identity conditions for the kind of thing ears are have to look to the discovery of their function.[5] And, given the threat of circularity, their identity cannot help us in discovering their function.

A second example, one that Dretske (1986) uses to good effect, will help us make the same point more perspicuously.

Anaerobic water-dwelling bacteria that inhabit the northern hemisphere have small magnetic particles within their cells, called magnetosomes, which point the bacteria to the magnetic north, and thus they swim towards the bottom of the ocean away from oxygen-rich environments. What is the function of these magnetosomes? At least two possible answers readily present themselves:

[5] The link between identity conditions for organs and function assignments to them is not a controversial issue in the literature. Price (1998) points out that, if the indeterminacy for function assignments cannot be solved, it would create immense problems for the taxonomy of biological categories. In illustration of how many biological categories, although structurally similar, are differentiated, at least in part, on functional grounds, she cites wings and flippers, whiskers and bristles, and veins and arteries. Also see my attempt (Enç 1979) to describe how such links are empirically discovered.

F_1 detecting the direction of the magnetic north;
F_2 detecting the direction of an oxygen-free environment.

We should also remember that in the southern hemisphere we find bacteria with magnetic structures that do F_2 but not F_1. So we have the same pair of choices we had with the eye (see Fig. 11.1). It is again clear that, if we want to avoid circularity, the search for the function of these magnetosomes should start with the choice represented by the second row. But, once we do that, the general etiological account of functions does not yield a clear choice between F_1 and F_2. This supports Dretske's conclusion, for he maintains that there is no determinate answer to the question about the function of the magnetosomes.[6]

It might seem here that an appeal to the phylogenetic lineage of first-order reproductively established families should be a useful constraint in our choice between F_1 and F_2. That is, if we know that the mammal eye and the insect eye both evolved from a structure of an organism that was the ancestor of both the mammals and the insects, whereas the pineal eye did not, this factor may favor assigning F_1 (detecting shapes) over F_2 (detecting gradations of light) to both the camera eye and the compound eye. But having common ancestors is neither a necessary nor a sufficient condition for having the same function. It is

Choice of identity criteria		Choice of function assignments: F_1	F_2
	Magnetosomes are functional kinds	The two types of magnetosomes are different kinds of things.	Both types of magnetosomes are the same kind of thing.
	Magnetosomes are structural kinds	The two types of magnetosomes are different kinds of things and have different functions.	The two types of magnetosomes are different kinds of things but they have the same function.

Fig. 11.2. Kinds of magnetosomes and their function

[6] Fodor (1990: 106) echoes Dretske's pessimism for the prospect of getting help from naturalized teleology in the determination of content for representational systems: 'Teleology goes soft just when you need it most; you get indeterminacies of function in just the cases where you would like to appeal to function to resolve indeterminacies of content.'

not necessary because homoplasies, organs that have evolved independently of each other because of the same type of selection pressures, arguably have the same function. The wings of insects, of bats, and of birds form a good example. And it is not sufficient because, as Peter Godfrey-Smith (1994) has argued convincingly, in order to avoid having to assign functions to vestigial organs, we need to look at only the recent history and ignore what selection pressures may originally have shaped the ancestors of such non-functional organs.

Thus, in order to see if we can find a principle for the choice between F_1 and F_2, we need to look into the specific versions of the general etiological approach summarized by (E).

In an earlier effort of mine (Enç 1989), I had proposed a series of counterfactual questions to help locate the proper function of an organ:

Q1 Would the magnetosomes be pointing in the direction of the magnetic north if the environment of the northerly bacteria were rich in oxygen in the direction of the magnetic north?

Knowledge of the way natural selection works, I thought, would yield a negative answer to this question, and a negative answer would disqualify pointing in the direction of the magnetic north from being a candidate for the function of these magnetosomes. On the other hand,

Q2 Would the magnetosomes be pointing in the direction of oxygen-free water if the environment of the northerly bacteria were rich in oxygen in the direction of the magnetic north?

receives an affirmative answer if we appeal to the same understanding of the way forces of natural selection operate. And an affirmative answer was supposed to yield pointing in the direction of the oxygen-free water as the function of the magnetosomes. This, I had thought, was a good method for determining functions.

Edward Averill (1998) illustrates the idea that guides these counterfactual questions very clearly. He describes four different tanks. Tanks 1 and 4 have oxygen-rich water in the northerly direction and Tanks 2 and 3 have oxygen-rich water in the southerly direction. In Tanks 1 and 3 north is away from the bottom of the tank and in Tanks 2 and 4 north is toward the bottom of the tank. Now Averill points out that, when a group of bacteria containing both north-seekers and south-seekers is placed in each of the tanks, after three weeks, south- and downward-seekers predominate in Tank 1, north- and upward-seekers predominate in Tank 2, and so on. Averill's carefully drawn conclusion is that, if we want to discover the law that describes the way natural selection selects for traits across these four environments, the trait over which the law would generalize would be turning in the direction of oxygen-poor environments, not, say, turning in the northerly direction. And, according to

Averill, properties referred to in such laws point to the determinate function of the traits in question. This is an elegant exposition of the idea I was hoping to capture with my counterfactual questions.

The idea is basically in the same spirit as that defended by Millikan (1984, 1990) and also perhaps that proposed by Sterelny (1990). The identification of the proper function of the magnetosomes is grounded in the fact that the property of the environment in the relatively recent history of these organisms for which the structures were selected was the property of pointing in the direction of the oxygen-free environment, not the property of pointing in the direction of the magnetic north because the empirical laws that describe the selection process make reference to the former property. And we can determine which property the laws generalize over by looking to see what happens in possible worlds in which magnetic north is rich in oxygen, and also in those possible worlds where some other direction is that of oxygen-free environments, while keeping the fitness-conferring needs of the bacteria constant.

The magnetosomes of the northerly and the southerly bacteria thus end up having the same function. And this judgment is independent of whether they are members of the same reproductively established family or not.

Although Sterelny and Millikan would come up with the same assignment of functions to the magnetosomes that Averill and I do, their views do not yield the same judgment in other, more complex situations.

When we ask 'what is the function of the frog's eye?' at least three possible candidates present themselves: F_1: to detect dark moving spots; F_2: to detect flies; F_3: to detect food.

My counterfactual questions will yield F_3 as the answer, or so I thought, because, if the historical environment of the frogs' ancestors had only indigestible flies, the detector would be expected to evolve so that it would end up being able to discriminate between flies and food. Millikan arrives at the same answer by looking to see what the consumer systems consume, consumer systems being those that take the output of the detector as input and process it to produce an output that contributes to some survival-related need of the frog.[7]

Sterelny, on the other hand, requires that the counterfactual worlds track natural kinds. According to him, in those possible worlds in which flies are ground-dwellers, frogs would have detectors that can discriminate those ground-dwellers from other things—that is, if the environment of the *flies* were to contain anything that might impede fly detection, the eye would

[7] As I discuss below, in connection with a similar view employed by Price, it is not clear how we can determine what it is that the consumer systems (need to) consume. Is it just common sense that makes one balk at the idea that the snapping mechanism (or perhaps the stomach), which is the consumer relative to the frog's eye, consumes a black dot, when black dots are locally matched one-to-one to flies and to food? This kind of a step, away from extensional contexts to intentional idioms, must be backed by something (say, finding a determinate function for the consumer system) other than common sense!

develop so as to enable it to discriminate flies from other things in *that environment.* Thus F_2 (detecting flies) is assigned as the function of the frog's eye. So, if the frog's ancestral environment had been from the start bee-bee infested, natural selection would have tended to construct mechanisms that could discriminate flies from pellets (Sterelny 1990: 127). For Sterelny, those possible worlds in which the nourishment for the frog includes small flying creatures that are not flies are not relevant in figuring out the function of the frog's eye. This seems arbitrary.[8]

The third candidate, that of assigning the function of detecting small moving spots, is favored by Neander (1996). She argues that, in assigning functions, we should try to obtain a match between the malfunctioning of the organ and the organ's failure to do what it is its function to do. Her point is that, if detecting flies or detecting food were the function of the frog's eye, then, when the eye was not malfunctioning, it could easily fail to perform its function because of the uncooperativeness of the environment. On the other hand, if the function of the eye is the modest one of detecting moving dark spots, then, as long as the eye of the frog was physiologically intact, it would be performing its function even in the lab when the technicians were fooling the frog with bee-bee pellets.[9]

I personally do not find the need for such a match pressing. In addition, if one respected Neander's desideratum, one could point out that a scientist could elicit the tongue-flicking response by inducing some sequence of neuron firings on the frog's retina when no dark spot was moving in the frog's environment, and argue, consistently with her line of reasoning, that detecting moving dark spots is not the function of the frog's eye; the function is rather that of detecting a sequence of neuron firings on its retina.

For completeness's sake, I should add to the catalog of differing views that are designed to generate unique function assignments to biological organs one that has recently been proposed by Nicholas Agar (1993).

Agar's view has the net result of conjoining Millikan's and Neander's function assignments. Agar proposes that we look at the organ in question, see what kind of a structure it has, what kind of a variable it takes as input, and then ask what sort of selection pressures have given rise to *that structure.* So, in

[8] Elder (1998) gives a plausible sympathetic reading to Sterelny's suggestion. The reading involves viewing the respective developments of the frogs' detection systems and the flies' protection systems as an 'arms race'. And if the 'arms race' is between two natural kinds of systems, the frog and the fly, Sterelny's choice of relevant possible worlds makes sense. But, as Elder rightly concludes, if the metaphor of an arms race is applicable, it should be viewed as an arms race between the frog and its environment, in which case those possible worlds in which frog food is in the form of small flying creatures that are not flies are relevant to discovering the laws of natural selection that govern the development of the frog's eye.

[9] Elder (1998), in criticizing Neander at this point, seems to imply that Neander ignores the distinction between malfunctioning and failing to perform a function. As I read Neander, she gives reasons as to why the distinction is not very helpful.

looking at the possible worlds, we keep the structure of the frog's eye constant, and see what property in the different worlds is responsible for selecting such a structure. Agar's conclusion is that, in the possible worlds we are using to identify selection pressures, we would be allowed to vary the fly's genetic make-up, its phylogenetic relatedness to bees, for example, but we would need to maintain the fly's mobility, its dark color, and its protein content. So the function of the frog's eye on Agar's view is that of detecting fast-moving, dark, non-dangerous food.[10]

Agar contrasts frogs with the imaginary frugs. Frugs have eyes that are structurally identical to those of frogs. But, after the frug has swallowed a sufficient number of flies, it uses them as ballast, sinks to the bottom of the pond, hunts there for the fish it feeds on, and spits out the flies to resurface. Agar points out that both Sterelny and Neander would assign the same function to the frug's eye that they assigned to the frog's eye. By Agar's lights, the function of the frug's eye is to detect fast-moving, dark, non-dangerous ballast. I tend to think it is to Millikan's credit that her appeal to consumer systems yields the assignment of the function of detecting food to the frog's eye and detecting ballasts to the frug's eye. But what is the basis of the raw intuition that makes us cheer this result?

The plethora of views briefly catalogued here makes one worry that Dretske may have been right in his claim about indeterminacy. Furthermore, there is an independent consideration that supports Dretske's pessimism, and makes one doubt the reliability of the intuition that gravitates us towards Millikan's conclusion.

To develop this consideration, let us suppose that wings on some birds enable them to fly; and wings on some other kinds of birds (like domestic ducks, say) enable them to run faster than they would without them. Suppose further that the two types of wings have been selected for their respective abilities. The predators of the earthbound birds were relatively slow creatures, so that outrunning them by flapping their wings was sufficient for escaping predation for these birds. So the force of selection that presents itself in the respective evolutionary histories of these birds is escaping predation. This is at least as good a candidate for assigning function, if not better, than maintaining that the wings of the earthbound birds have the function of making them run faster, and the wings of the other birds have the function of making them fly. To opt for the

[10] Actually the idea that, when we are in search of the function of the *frog's eye*, we should keep the structure constant across possible worlds is an attractive one. Fodor (1990), who proposes to dispense with function talk in preference for his asymmetric dependencies, also keeps frogs the way they are in the actual world in considering different possible worlds. In showing why the frog's snaps at flies are asymmetrically dependent on its snaps at black dots, he says, 'frogs continue to snap at dots in worlds where there are dots but no flies; but they don't snap at flies in worlds where there are flies but no dots' (1990: 107). Fodor's conclusion is that the frog's eye is a moving-black-dot detector.

function of escaping predation is in the same spirit as choosing the function of pointing in the direction of oxygen-free environments for the magnetosomes, and the function of helping obtain food in the case of the frogs.

This is in fact what my counterfactual questions would yield:

Q1 Would the wings of the earthbound birds enable them to run fast if they could not escape predation by running fast? No.

Q2 Would the wings of the earthbound birds be helping them escape predation if the wings could not help the birds to escape predation by running fast? Yes.[11]

This yields the function assignment of helping escape predation to the wings of both types of birds. I think the same result would be forthcoming from Millikan's appeal to consumer systems.

But, if we find the line of reasoning that leads to this function assignment harmless, it is easy to show that, by applying the reasoning recursively, we will find no principled way of stopping short of assigning the function of survival, or perhaps more accurately, the function of conferring fitness to all organs of organisms. I will call this 'The Landslide Argument'.

Here is the general form of the Landslide: we find in some system S, a structure, X. X actually tracks several properties $P_1, P_2, \ldots P_n$, which are coextensional in S's actual and historical environments. Let us suppose that the series of properties is constructed so that P_m is more *general* than P_n if m is smaller than n. (In the example of the northerly bacteria, X would be the magnetosomes, P_1 might be the direction of survival, P_2 the direction of oxygen-free environment, and P_3 the direction of the magnetic north.) We now consider an alternative actual environment in which structures similar to X have been selected for in organisms similar to Ss. Let us suppose that these structures in the alternative environment track a bunch of properties coextensional in that environment, and that these properties overlap but are not identical with properties $P_1, P_2, \ldots P_n$. Let us label this second set of properties $Q_1, Q_2 \ldots Q_m$. Again we will suppose that Qs are ordered from the most general to the most specific. By hypothesis the two sets of properties overlap in some subset, that is, $P_1 = Q_1, P_2 = Q_2, \ldots P_k = Q_k$; and they diverge after P_k so that $\{P_{k+1}, \ldots P_n\}$ have different members from the set $\{Q_{k+1}, \ldots, Q_m\}$. Pursuing the example of the magnetotactic bacteria might make this clearer. The alternative environment is the southern hemisphere. Here the coextensional property set

[11] These answers are contingent on a whole series of assumptions about the environments and about what traits natural selection would favor in the relevant possible worlds that need to be considered in evaluating the counterfactuals. (For example, for Q1, we assume that running fast would not be making a contribution to the fitness of these birds in the possible world in which it fails to help them escape predation; and, for Q2, we are assuming that the conditions are there for the ability to fly to evolve, etc.) Let us suppose, for the sake of the argument here, that these assumptions are defensible.

consists of three properties: Q_1: survival, Q_2: direction of oxygen-free environment, and Q_3: direction of the magnetic south. The two sets overlap as far as P_1, P_2, Q_1, and Q_2 are concerned and diverge at the third property. When we ask for the function of Xs, the reasoning that we have been finding attractive identifies the most specific property that is common to the two property sets, and uses that property in specifying the function of the X. It is this recipe that gets us to the function of pointing in the direction of oxygen-free environment for the magnetosomes of the bacteria. And this recipe is in the spirit of Millikan's appeal to the consumer subsystems.

But this line of thought is the beginning of a slide into choosing the less and less specific members of the overlapping property sets. The reasoning that was used in assigning a function to the wings of birds involved reference to *actual* alternative situations. But whether there are such alternative birds or not is not a relevant fact. What is relevant is what would happen in certain counterfactual situations. It is those situations that will help determine for us the unique property for which a trait was selected, if such a property exists. The reason for the appeal to the counterfactual environments is this: function ascriptions are opaque to substitutions of coextensional predicates. If the source of opacity is not going to come from original teleology (i.e. God's design), then it must come from some modal consideration. And modality cannot be explicated by what is only contingently true.[12] However, including in our reasoning such counterfactual environments forces us to many different sets of coextensional sets of properties so that the only property that is guaranteed to remain common to all sets could easily end up being that of conferring fitness. Since there is no principled way of limiting the range of possible worlds and thus constraining the divergence of coextensional properties across these worlds, there may remain no principled way of stopping short of the most general property in the specification of the function of Xs.

Evaluation of counterfactuals is like a minefield. In my use of the counterfactual questions, I was blithely assuming that I was treading on safe ground.

The argument I have just rehearsed seems to me to be powerful enough to make me think that, whatever principle we come up with, that principle will face the horns of a dilemma: either it will yield, at least some of the time, counter-intuitive answers (here Sterelny's and Neander's principles are cases in point) or else the principle will be flexible enough so that its application will be partly directed by the answers our intuitions demand (here Millikan's appeal to consumer systems and my counterfactual questions are two examples).

At this point several recent essays in which the authors anticipate the Landslide, and propose ways of stopping it, deserve discussion.

[12] This is basically the reasoning that is used in discovering the correct *law* that describes the selection process (Averill 1998). Nomic necessity provides the source of modality for counterfactual considerations.

Averill's argument for excluding the most general property is to show that a generalization that makes reference to a property like survival becomes a tautology and is therefore not a proper candidate for an empirical law.[13] Elegant as his suggestion is in blocking the last stop of the Landslide, it still allows the Landslide to yield counter-intuitive generalities. For example, using possible environment considerations, we could show that the eyes of ungulates, as well as their legs and ears, have the function of helping them escape predation. This result is no more palatable than everything being assigned the function of surviving better.

Price (1998) proposes to block the Landslide by imposing a condition (she calls it the Independence Condition) on function ascriptions. This condition requires that the description of the function of a device capture an activity or an effect that the device is capable of achieving on its own, without the cooperation of other devices with which it is linked. She quite reasonably points out that, if functions arise when the persistence of some device in some system is explained by its effect, the effects selected must be those that are peculiar to just that device, in contrast with those that require contributions from other components in the same system. This condition has the net effect of pushing the function ascription to the more specific descriptions of the effects of the device. Thus imposing this condition eliminates helping survive, providing nourishment from among the candidates for the function of the frog's eye. It might seem that this condition would have the net result of assigning the most proximal—that is, the most specific—description of the eye's achievement—that is, detecting a small moving dark object, and thereby agreeing with Neander's thesis. However, Price has a second condition, which she calls the Abstractness Condition, which blocks this result.[14]

The abstractness condition requires that, in specifying the function of a device, we must (i) abstract from the design of the device, and also (ii) abstract from the design of the components with which it cooperates (those that Millikan has named 'the consumer subsystems'). The first move towards abstraction blocks assigning detecting small, dark, moving objects as the function of the frog's eye—because this specification describes how the device

[13] In Averill (1998) roughly, the form of the law that is envisaged is this: 'There is a large population of R-type organisms; T is a trait which some Rs have and some Rs lack; T and not-T are heritable; Rs that have T are better able to survive and reproduce.' The requirement that this be a law excludes a trait like 'provides better survival' because substitution of that for T yields a statement that is analytically true.

[14] Price (1998) also finds the function assignments generated by Neander's view (e.g. assigning to the heart the function of making squeezing motions) counter-intuitive. She says, 'Neander's reason for taking this line is . . . that the lower the level of description, the more informative the claim that the device is failing to perform its function. But why should this be the relevant consideration? After all, the more information a function statement gives us about how a device may fail, the less information it provides about what the device achieves when it succeeds performing its function' (1998: 65).

achieves its result *given* its design.[15] Price defends this move by first endorsing Millikan's insight that the function of a device is partly determined by the demands of its consumer systems, and arguing that, since the production of an effect constitutes the device's contribution to its consumers, we must describe the function of the device in terms of the causally effective property of the device's activity in the production of this effect. Her helpful analogy is one's action of moving the finger, flipping the switch, turning on the light, and frightening the prowler. If the prowler's being frightened is the result, then the *causally effective property* of what one did was turning on the light, not moving the finger. In the same vein, if the consumer system for the frog's eye is the snapping mechanism, then we need to describe the function of the eye in terms of the property of the eye's activity that will causally explain how it contributes to the function of the snapping mechanism, which presumably is that of presenting something nutritious to the stomach. This consideration favors describing the function of the eye in terms of detecting a fly, or, more accurately, detecting some nutritious object.

The second move towards abstraction, that the function specification be made in abstraction from the design of its consumer systems, blocks assigning the function of detecting small, slow, slimeless things to the frog's eye.

Price's account is the best-defended and clearest attempt to date that I know at developing a defensible set of conditions that will generate determinate function assignments. And, just like Averill's, it blocks the Landslide from reaching totally vacuous function ascriptions, like the function of contributing to survival. But, if I understand it correctly, it still allows assignments that are too general to be acceptable. For example, it seems that her condition would have the result of assigning to both the eye and the ear of the antelope the function of detecting predators. And, given that she herself is committed to functional categorization of organs, I would not think she would welcome this result.[16] But, more generally, according to her account, in order to decide how to apply the first leg of the abstraction condition in figuring out the function of one device, one needs to have figured out the functions of the devices that cooperate with that device. Hence it seems that one can exploit the potential circularity involved and show how several function assignments to one and the same device are consistent with her conditions, provided that one chooses the functions of the cooperating devices in different ways. It is not clear why the function of the snapping mechanism is that of presenting some nutritious thing to the stomach, rather than, say, presenting a fly or even presenting a source of survival. That question would have to be decided in turn by discovering the function of the stomach, and so on; and

[15] This approach is diametrically opposite to that of Agar's discussed above.

[16] She explicitly disapproves of an account that assigns *different* functions to a lion's heart and an antelope's heart. I expect she would not want an account that assigned the *same* function to the antelope's ear and the antelope's eye either.

it is not clear to me how one would determine the ultimate consumer system that would end this series, and how one would assign a function to *it*.

Crawford Elder (1998) has objectives similar to Price's. However, he argues that the representational content of a detector should be distinguished from the function of the detector. Its function, according to him, is to orient the representation to consumer devices so that they can perform *their* proper functions. On the other hand, 'what the representation represents is a state of affairs which causally explains why its steering was effective on the occasions when it was effective—why the consumer device, acting as steered, succeeded on those occasions' (1998: 352). So it seems that Elder has the same explanatory concerns that Price articulates. He also achieves the same result that Price's abstraction condition is designed to achieve by distinguishing between *what* a representation-producing device is supposed to do and *how* it is supposed to do it, as well as between *what* affairs in the outside world the device is supposed to track and *how* it is supposed to track them. He argues that function and representational content assignments would be in error if they looked for answers in the *how* category.

In so far as Elder's discussion parallels Price's concerns, it would seem that Elder, too, would admit that the antelope's visual and auditory systems produce states that have the same representational content.[17] And this seems to violate our folk-psychological intuitions, because we would expect the kinds of properties represented by the states of the visual system (for example, shapes and distances) to be different from those represented by the states of the auditory system (for example, sounds). In addition, since Elder has nothing that matches Price's independence condition, the Landslide argument seems not to be blocked by his argument.[18]

A totally novel and different approach is advocated by Walsh (1996).[19] Walsh argues that, when we look to the function of a device, we should confine our search to the 'selective regime' in which the device confers fitness to its organism. This argument has the net result of rendering function assignments relative to the actual environment in which selective forces are operative. On this view, I would assume that, since the selective regime of the northerly bacteria favored bacteria that swam in the direction of the magnetic north, the function of the magnetosomes would be that of pointing in the direction of the

[17] Since Elder does not give a clear set of conditions for determining representational content, I cannot be sure that my diagnosis is correct.

[18] In fairness, I should add that Elder is not interested in addressing the indeterminacy problem. For all I know, he might be in sympathy with the main thrust of this chapter.

[19] See also Walsh and Ariew (1996), which studies the relation between an approach that focuses on the functional organization of the system (Cummins-function), and one that captures the teleological aspects of function (Evolutionary-function). The authors argue that, if the latter is understood as a relational concept, the interplay between these two approaches can be made transparent.

magnetic north. But this relativization to the selective regime would not exclude the assignment of detecting the direction of oxygen-free environments, because that, too, is a factor that contributes to the fitness of the bacteria in the northern hemisphere. Admirable as this move may be for understanding biological functions, it would still permit the Landslide argument and thus would not help us with the indeterminacy problem.[20]

In summary, the foregoing has been a roundabout way to the conclusion Dretske had reached in 1986: function assignments to products of natural selection cannot satisfy the uniqueness requirement. In that same paper, Dretske goes on to develop a thesis that maintains that unique content or unique function can be assigned in cases where the organism undergoes a learning process during its individual life time.

His thesis can be illustrated by Fig. 11.3. Here F represents some property, like food, or toxic substance, and each of the Ss are stimuli that have become conditioned stimuli during some learning history of the organism, like a bell's buzzing, a light's going on, or some taste or some odor. R is a type of state that is induced by any of the Ss and is the trigger for some type of output, say, eating behavior, or avoidance behavior. Dretske argues that the function of R cannot be that of indicating the presence of any of the Ss, nor indicating the disjunction of all the Ss because which stimuli have become conditioned stimuli for the behavior in question will depend on the individual's learning history.

> In terms of the S_i that produce R, R can have no time-invariant meaning. Of course throughout this process, R continues to indicate the presence of F. It does so because, by hypothesis, any new S_i to which R becomes conditioned is a natural sign of F. . . . Therefore, if we are to think of these cognitive mechanisms as having a time-invariant function at all (something that is implied by their continued—indeed as a result of learning, more efficient—servicing of the associated need), then we must think of their function not as indicating the nature of the proximal (or even distal) conditions that

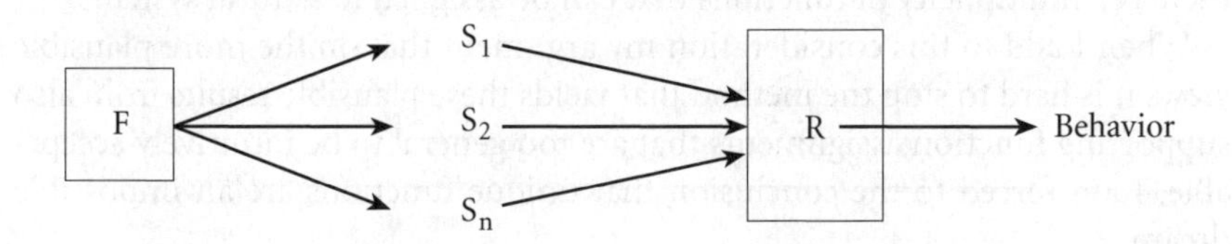

Fig. 11.3. The representational content of an indicator state

[20] Again I should note that, since Walsh is not concerned with the indeterminacy problem, this is not a criticism of his view.

trigger positive responses (the S_i . . .) but as indicating the condition (F) for which these diverse stimuli are signs. (Dretske 1986: 36)

I used to think at the time that, in the learning situation, my counterfactual questions would yield the same result, and the convergence of two different approaches to the same question on the same answer was good evidence that the answer was right.

Q1 Would S_i be producing R (the cause of the behavior in question) if S_i weren't a sign for F (for any S_i across time)? No.

Q2 Would F be producing R if S_i weren't a sign for F? Yes.

Hence indicating F is the function of R, not indicating any of the Si.

But now that I have seen how the counterfactual questions fail to stop the slide into generalities, I have become skeptical of Dretske's argument, too. Here is the source of my skepticism: it makes sense to think that certain odors, colors, and shapes could be the stimuli that a dog gets conditioned on for the presence of dog chow. So the state of the dog that initiates the eating response has the function of indicating not the odor the dog is smelling, but the presence of the dog chow. But the dog has also learned, through conditioning, that the smell of dog chow, the smell of steak bone, the smell of cooked duck carcass are signs of food. There seems to be no principled reason to stop us from applying Dretske's argument in the quoted passage to get us to assign to R the function of indicating food, and then repeat the exercise until we reach some dreaded generality.

Let us take stock. I have argued that it does not seem likely that we will come up with a principle that will give us unique function assignments to natural organs, traits, or detecting devices, regardless of whether they have been shaped by natural selection or by learning. This conclusion was partly based on the existing diversity of opinion in the literature—the fact that there are so many thinkers who arrive at such different unique function assignments makes it unlikely that all of them, with the possible exception of one, are fools. The more reasonable explanation is that most of them are correct, but that there is a multiplicity of functions that can be assigned to natural systems.

When I add to this consideration my argument that on the more plausible views it is hard to stop the method that yields these plausible results from also supporting function assignments that are too general to be intuitively acceptable, I am forced to the conclusion that unique functions are an impossible dream.

Does this mean, then, that a person committed to naturalizing the mind will have to give up any hope of getting help from naturalized teleology? I do not think so. It means only that the help from function talk will be a lot more modest than many of us had thought. In the remaining pages, I will try to develop this claim.

First let me make a distinction explicit. This is a distinction that has become familiar in the literature, and it is one that I have been presupposing in my negative arguments above.[21]

Let us consider a chain of cause and effect relations. I will assume here, for the sake of ease of presentation, that a cause in such a chain is constituted by an object and a set of properties (see Fig. 11.4). Following Stampe (1977), I will call all the properties of any particular object, o, in such a chain the set of *synchronic* properties of some causal antecedent of a chosen effect. In the example of the frog's eye, we have a causal chain that includes the fly (o_1), the light array that reaches the frog's eye (o_2), the pattern of neuron stimulation on the frog's retina (o_3), and the perceptual state, R, induced in the frog (o_n). According to the definition of synchronic properties, all the properties of the fly—for example, its being a moving dark spot (F_1), its size (F_2), its gender (F_3), and so on—make up a set of synchronic properties of one of the causal antecedents (that is, o_1, the fly) of R. Some members of this set (for example, the fly's being a moving dark spot) are *causally relevant* in the production of the kind of perceptual state that R is; others (for example, the fly's being a female) are not. In contrast, any set of properties that is constituted by properties of *different* objects in the causal chain will be called a set of *diachronic* properties of the causal antecedents of a chosen effect. So the set composed of some property of the neurons on the frog's retina, some property of the light array that reaches the frog's eye, and some property of the fly is a set of diachronic properties of the causal antecedents of R.

When we look for a principle that yields function assignments to a trait, we try to identify the task for which that trait's features or activities were selected in the recent history of the organism. This helps us eliminate a large subset of the set of synchronic properties of the objects in the environment in which selection was taking place. Since the property of having a compound eye (one of the fly's synchronic properties) is irrelevant to the selection story, that property gets eliminated from those pertinent to the function of the frog's eye:

o_1	o_2	o_3	...	o_n
F_1	G_1	H_1	...	P_1
F_2	G_2	H_2	...	P_2
...	...	...	...	...
F_k	G_m	H_n	...	P_r

Fig. 11.4. Synchronic and diachronic properties

[21] I learned this distinction first from Dennis Stampe's seminal paper (1977). Most recently Price (1998) makes the same distinction in different terms. She calls the diachronic problem 'the distality problem', and the synchronic problem, 'the description problem'.

although the frog's eye does detect things that have compound eyes, it does not have the *function of detecting them.* The same holds for the property of the fly's being a female. In the light of this observation I should state the conclusion of the negative argument I gave above more accurately. The conclusion will now read as follows: although a large number of the synchronic properties of the fly is disqualified from being relevant to the function assignment, there always remain more than one property in the synchronic set (like the fly's being a moving dark spot, its being food for the frog, its being a source of survival for the frog) that are so relevant. And we have no principled way of selecting a unique property from among the remaining ones.

The first point I want to emphasize here is that nothing in the arguments I produced addressed diachronic properties. And it is not that hard to show that, whichever of the various views we take in assigning function to the frog's eye, they will all converge on one link in the causal chain (that is, the object that has the property of being a fly, being food, being a dark moving spot, being non-harmful, being a source of survival and fitness, and so on), and none will yield the light array that reaches the frog's eye, or the actual firing of the neurons on the frog's retina. In other words, these views will always eliminate all members of the diachronic sets, except those that are among the synchronic properties of the fly. So the betrayal of the uniqueness requirement in natural teleology does not result in a free-for-all. More specifically, the legitimacy of using any one of, or all of, a subset of synchronic properties in function assignments is fully consistent with insisting that at most *only one* of several diachronic properties can be relevant to functions. Not only is this consistent; I would maintain that this is what our original working hypothesis (E) yields for us.

Now, what does this have to do with the intentionality of mental states? Here I suggest we divide the question. On the one hand, we have states of the organism that are the outputs of their perceptual systems—these states are constituted by percepts. On the other hand, we have states of the system that involve some cognitive element—states that are comprised of the so-called propositional attitudes.

If a frog believes that there is a fly out there, the frog is in a state distinct from the state it would be in if it believed that there is a moving dark spot out there. That is the nature of the intentionality that pertains to beliefs and to many other propositional attitudes. The negative result I have argued for above entails that function talk does not yield the level of intentionality that belief states (and other propositional attitudes) require. So the principle that could generate function attributions is not going to be a useful tool in our efforts to identify the source of intentionality involved in these mental states. We need to look beyond teleology to understand this source. The speculative thought I have here is that some naturalistic theory of concept formation will be needed before the mystery of cognitive mental states can be unraveled.

However, it is possible to argue that an important difference exists between propositional attitudes and perceptual states. It is uncontroversial that perceptual states embody representations. The difference between propositional attitudes and percepts derives from the fact that the content of perceptual representations is of a different kind from the kind of content that propositional attitudes have. The argument for this claim is easiest to formulate in the framework of the best bet we have for naturalizing perceptual states—the causal theory of perception. The theory says roughly, the perceptual state, R, that an organism, S, is in is a state that represents o's being F, if o's being F is what causes S to be in R. As it stands, this rough formula allows too many Fs to yield an intuitive answer to the question, 'what is the object of the representation and what does it represent the object as being?' All the diachronic properties that are causally relevant to the production of R become candidates for F. So how do we eliminate the bell button or the ear drum from among the objects represented by the auditory state? It is here that functions come to the aid of the rough formulation; we add, o is the type of object carrying information about which is the function of the perceptual system, and F is just one of the several synchronic properties that are proper to the specification of the function of the system. So the capacity that auditory systems have of carrying information about the state of the ear drum is not what the auditory systems were selected for; the fact that the vibrations of the ear drum are part of the cause of S's entering state R becomes irrelevant to the task of locating the object represented by R. On the other hand, we still have more than one property in the synchronic subset that recommends itself as the content of the representation: when the moth hears the high-pitched sound that the echo-locators of the bat produce, is its auditory state representing a sound wave of 50 Khz frequency, the sound of a bat, or the imminent danger of being eaten? I have argued that function assignment to the auditory system of the moth is not going to help us select one from among these candidates for the representational content of the moth's auditory state. Now I want to suggest that this is indeed the correct result. The state does represent each and every one of these properties. I will not be able to support this suggestion here by an argument. All I have is a conjecture. And here is the conjecture.

Perceptual states are, in their essence, subdoxastic. When they first emerge in the perceptual systems, they are non-epistemic. They contain information that can be extracted by a cognitive system that employs some conceptual apparatus. But when we ask for the content of a percept, we are asking for the specification of those properties that are represented in the percept. And that content better be richer than what the specification of just one property could provide; for that percept is expected to be such as to form the grounds for more than one non-inferential belief. So the employment of natural teleology to explicate the intentionality of subdoxastic perceptual systems gives us perfect results. It yields just the right degree of intentionality. The unavailability of unique function assignments mirrors perfectly the representational richness of such states.[22]

References

Adams, F., and Enç, B. (1988), 'Not Quite by Accident', *Dialogue*, 27: 287–97.

Agar, N. (1993), 'What Frogs Really Believe?', *Australasian Journal of Philosophy*, 71: 1–12.

Averill, A. W. (1998), 'Natural Representation and Laws of Natural Selection'.

Boorse, C. (1976), 'Wright on Functions', *Philosophical Review*, 85: 70–86.

Davies, P. (1994), 'Troubles for Direct Proper Functions', *Noûs*, 28: 363–81.

——(1997), '*Norms of Nature: Naturalism and the Nature of Functions*'.

Dretske, F. (1986), 'Misrepresentation', in R. J. Bogdan (ed.), *Belief: Form, Content, and Function* (Oxford: Oxford University Press), 17–36.

Elder, C. L. (1998), 'What versus How in Naturally Selected Representations', *Mind*, 107: 349–63.

Enç, B. (1979), 'Function Attributions and Functional Explanations', *Philosophy of Science*, 46: 343–65.

——(1982), 'Intentional States of Mechanical Devices', *Mind*, 91: 161–82.

——(1989), 'Causal Theories and Unusual Causal Pathways', *Philosophical Studies*, 55: 231–61.

Fodor, J. (1990), 'A Theory of Content II', in J. Fodor, *A Theory of Content and Other Essays* (Cambridge, Mass.: MIT Press), ch. 4.

Godfrey-Smith, P. (1994), 'A Modern History Theory of Functions', *Noûs*, 28: 344–62.

Millikan, Ruth Garrett (1984), *Language, Thought, and Other Biological Categories: New Foundations for Realism* (Cambridge, Mass.: MIT Press).

——(1990), 'Biosemantics', *Journal of Philosophy*, 86: 281–97.

Neander, K. (1991), 'The Teleological Notion of "Function"', *Australasian Journal of Philosophy*, 69: 454–68.

——(1995), 'Misrepresenting and Malfunctioning', *Philosophical Studies*, 79: 109–41.

——(1996), 'Dretske's Innate Modesty', *Australasian Journal of Philosophy*, 74: 258–74.

Price, C. (1998), 'Determinate Functions', *Noûs*, 32: 54–75.

Schmidt-Nielsen, K. (1990), *Animal Physiology, Adaptation and Environment*, 4th edn. (Cambridge: Cambridge University Press).

Sober, E. (1984), *The Nature of Selection: Evolutionary Theory in Philosophical Focus* (Cambridge, Mass.: MIT Press).

Stampe, D. (1977), 'Towards a Causal Theory of Linguistic Representation', in P. French, T. Uehling, and H. Wettstein (eds.), *Midwest Studies in Philosophy*, ii. *Studies in Semantics* (Minneapolis: University of Minnesota Press).

[22] This conjecture is Gibsonian is spirit, without necessarily forcing us to say that the frog *sees* edibility, or the moth *hears* danger. All I am suggesting is that the content of the perceptual state that the moth enters contains more properties than the property we refer to when we identify what it *hears*.

Sterelny, K. (1990), *The Representational Theory of Mind: An Introduction* (Oxford: Blackwell).

Walsh, D. M. (1996), 'Fitness and Function', *British Journal for the Philosophy of Science*, 47: 553–74.

——and Ariew, A. (1996) 'A Taxonomy of Functions', *Canadian Journal of Philosophy*, 26/4: 493–514.

Wright, L. (1973), 'Functions', *Philosophical Review*, 82: 139–68.

12. Brentano's Chestnuts

D. M. Walsh

> It is, I think, most unlikely, even on empirical grounds, that Darwin is going to pull Brentano's chestnuts out of the fire.
>
> (Fodor 1990: 70)

ABSTRACT

The teleosemantics programme is a reductive approach to naturalizing intentionality. It comprises three phases: (i) the reduction of intentional content to intentional function, (ii) the reduction of intentional function to evolutionary function, and (iii) the reduction of evolutionary function to selectional history. I argue that phases (ii) and (iii) fail. Phase (ii) fails because any notion of intentional function rich enough to serve the purposes of phase (i) is too rich to be reduced to evolutionary function. Phase (iii) fails because it misconstrues the relation between function and selectional history. I argue that phase (i) alone constitutes an adequately naturalistic account of intentionality. The teleology inherent in the notion of intentional function needs no reduction.

Consider two puzzles for philosophical naturalism: *Paley's problem* and *Brentano's problem.* The first is that of explaining the biological phenomenon of adaptedness: how can a trait of an organism be an adaptation for dealing with some state of the environment. The second is that of explaining the psychological phenomenon of intentionality: how can a state of an agent be about some (putative) feature of the environment. There are evident similarities: each of these phenomena, adaptedness and aboutness, involves a norm-governed relation between an individual—organism or agent—and its environment. Indeed, there is growing interest in the view that these are not so much two distinct problems as one and as such they have the same solution. Darwin's theory of natural selection yields a perfectly adequate naturalistic response to Paley's problem and it is thought by many that a simple extension of that theory can do the same for Brentano's problem. The programme to co-opt evolutionary theory in this way—the teleosemantics programme—has been pursued with considerable vigour of late. While I am broadly in sympathy with its motivations, I believe that in its current guise it is destined for failure. Teleosemantics, I shall claim, commits two serious errors. The first is that it misconstrues Darwin's solution to Paley's problem. The second is that it mistakes what is required of a naturalistic solution to Brentano's problem. The errors are related and it is the relation among them that interests me here. I claim that once it is understood just what evolutionary theory *does* say about adaptation it becomes easier to see what *would* constitute an acceptable naturalistic account of intentionality.

There is something approaching a consensus nowadays that any acceptably naturalistic account of intentionality will have to be a reductive one, that is one in which intentional states do not constitute a *sui generis* category. The point is made with characteristic elegance by Fodor (1987: 97):

> It is hard to see . . . how one can be a Realist about intentionality without also being, to some extent, a Reductionist. If the semantic and the intentional are real properties of things, it must be in virtue of their identity with properties that are themselves *neither* intentional *nor* semantic. If aboutness is real, it must be really something else. (emphasis in the original)

The teleosemantics programme is one among many reductive strategies currently on the market, one that appears to be increasing its share. As I understand it, it comprises three phases. A quick overview should give the general flavour.[1]

Phase 1: Intentional content to intentional function. Intentionality has a normative dimension. The content of an intentional state is individuated by the norm-governed, functional role it plays in an agent's cognitive economy. It would appear quite natural, then, to suppose that intentional states are teleological categories individuated by their teleological (intentional) functions. Simply, the function of the belief that *p* is to represent that *p* and the function of the desire that *q* is to bring it about that *q*.

Phase 2: Intentional function to evolutionary function. Evolutionary biology and its cognate fields regularly trade in *prima facie* teleological, normative notions like function and adaptation. In fact, in these sciences function and adaptation play a particularly important explanatory role; the presence of a trait, behaviour, or structure is explained by its function or by what it is an adaptation for. There is every reason to believe that the concepts of evolutionary function and adaptation are naturalistically acceptable (see Phase 3). So, if intentional function can be reduced to evolutionary function it may well lay claim to the same naturalist credentials.

Phase 3: Evolutionary function to causal history. Function and adaptation in biology can be construed in a wholly non-teleological way. Adaptations and functions are just the causal consequences of natural selection.

[1] I take this to be a fairly uncontroversial breakdown of the programme. It accords reasonably well with the account given by some of its prominent proponents (Papineau 1994; Dennett 1996) and at least one of its prominent opponents (Fodor 1996).

We might interpret the first two phases as an attempt to assimilate Brentano's problem to Paley's and the third as an attempt to apply Darwin's solution to both.

The position I wish to outline here is as follows. I am happy to acknowledge the feasibility of Phase 1, so I shall take it as read. Beyond that the prospects for teleosemantics are meagre. I believe that the programme fails at Phases 2 and 3—the heavily reductive stages. Contrary to the aims of Phase 2, the functions of intentional states cannot be reduced to the sort of functions encountered in evolutionary biology. And, contrary to Phase 3, evolutionary theory does not yield a reductive account of biological teleology in terms of the historical causal consequences of selection. That is all very negative, but, on a less polemical note, the failure of Phases 2 and 3 should not occasion widespread despondency among those, myself included, who are committed to the view that Darwin's theory illuminates Brentano's problem. On the contrary, what evolutionary biology *does* tell us about teleology in the natural sciences ought to make us more sanguine about accepting Phase 1 alone as the basis of a complete naturalistic account of intentionality.

1. Intentional Function as Evolutionary Function

It is by now a piece of well-entrenched philosophical lore that evolutionary theory gives us a naturalized, indeed reductive, account of the teleology inherent in the notions of biological adaptation and function. Pre-theoretically, one thinks of an adaptation as a trait with a biological function, a trait that subserves some biological *goal*. The theoretical treatment of this intuition appeals to the historical effects of natural selection: an adaptation is a trait that has been selected in the past because of its capacity to play a positive role in survival and reproduction. The effect it has been selected for is its function. The core of the theory of evolutionary functions is as follows.

> E: The function of trait *x* is to *F* iff tokens of *x*'s type are present (or prevalent) in a population because they have contributed to survival and or reproduction of those individuals possessing tokens of *x*'s type *in the past* by doing *F*.[2]

Among the many presumed virtues of E is the fact that it makes no unreduced appeal to biological goals or goal-directed processes. What a trait is *for* is simply a consequence of natural selection in the past, and natural selection is strictly a causal/mechanical process. Furthermore, E gives us a translation

[2] The aetiological theory and its variants are discussed in numerous places (Millikan 1984, 1989; Neander 1991; Griffiths 1993; Godfrey-Smith 1994). Buller's collection (1999) is perhaps the definitive source. I also recommend Buller's introductory essay to that volume.

schema for the functional (or adaptive) explanations commonly encountered in evolutionary biology. To say, for example, that the purpose or function of the mammalian stapes is to transduce airborne vibrations is simply to say that mammals now have the sort of stapes they have because this arrangement contributed significantly to survival and reproduction of mammalian ancestors by transducing airborne vibrations. Again no mention of goals or purposes is required.

If, as many teleosemanticists believe, *intentional* function is merely a species of biological or *evolutionary* function, then by extension, the reductive, aetiological theory of evolutionary functions should apply equally to intentional states. If evolutionary function, construed along the lines of E, is to act as a criterion of individuation for intentional states, it must meet two quite stringent requirements. I will call them the 'Determinacy Constraint' and the 'Adaptiveness Constraint'. I take these in turn.

1.1. The Determinacy Constraint

One lesson to be learned from the famous intensionality of intentional contexts is that the functional roles played by intentional states are extremely finely circumscribed. It may well be that the belief that *p*, when combined with the desire that *q*, motivates and hence explains action *A*, whereas the belief that *p*′ does not, even though '*p*' and '*p*′' are necessarily coextensive. Your desire for a record by Muddy Waters combined with your belief that *this* record is by Muddy Waters motivates and explains your record-buying action, whereas your belief that this record is by McKinley Morganfield combined with the same desire may motivate no action at all. This despite the fact that 'this record is by McKinley Morganfield' and 'this record is by Muddy Waters' are necessarily coextensive. If the intentional content of a state is determined by its *evolutionary function*, then evolutionary functions must be at least as finely demarcated as intentional states. The most common objection to teleosemantics is that they are not.[3]

The numbingly familiar frog–fly problem illustrates the point. Frogs catch flies by responding to visual cues. A fly crossing a frog's visual field causes a retinal image that, in its turn, guides the frog's glossal apparatus in the capture of the fly. Clearly it is the biological function of the retinal image to represent something, but what? The obvious answer is something like 'the retinal image represents the presence of a fly' and indeed this conforms well with E, given that the frog's visual apparatus has most likely been selected for its capacity to detect flies. But there are other, equally reasonable responses sanctioned by the aetiological theory of functions. Frogs' visual apparatus has been

[3] See particularly Fodor (1990: ch. 3, 1996), and Dennett (1996) for discussions.

selected for its capacity to respond to small, dark, moving objects (Neander 1995), catchable things (Price 1998), food (Shapiro 1992). There is no principled way to choose between these function attributions, at least as far as E commands. So, there is no univocal answer to the question 'what does the frog's retinal image represent?' As Dennett (1987*a*: 300) opines: 'When the "fact of the matter" about proper function is controversial—when more than one interpretation is supported—there is no fact of the matter.' Nor is this indeterminacy restricted to those biological states with representational functions. Similar arguments suggest that the indeterminacy is endemic to biological functions, at least where these are construed along the lines of E (Dretske 1986; Millikan 1989, 1990; Sterelny 1990; Agar 1993).[4]

Faced with the disparity between the looseness of evolutionary functions and the stringent requirements on any account of intentional functions, advocates of the teleosemantics programme tend to adopt either of two strategies. The first, exemplified by Neander (1995) and Price (1998), among others, seeks to amend the theory of functions itself or to refine our function-attributing practices. The other, followed by Millikan (1989, 1990, 1995) and Papineau (1998), bypasses the indeterminacy inherent in biological functions proper (so to speak). Instead it maintains that, however indeterminate biological functions are generally, the functions of intentional states are highly determinate on account of the peculiarities of the system of belief–desire psychology in which they ply their trade.[5]

As an example of the first strategy, Price's attempt (1998) to eliminate the indeterminacy inherent in biological function ascriptions is instructive. She supplements the standard aetiological theory, E, with two further constraints.

1. *Independence Condition.* The determinate function of a trait x is that effect that x has independently of the various consumer devices that depend upon it.
2. *Abstractness Condition.* The function of a trait is to be specified in ways that abstract from (i) the specific structure of the trait itself, and (ii) the structure of the consumer devices that depend upon it.

Thus amended, Price claims, the aetiological theory assigns a perfectly determinate function to the frog's retinal image in the intuitively right way. The representational content of the retinal image is determinately *fly* rather than, say, *small, dark, moving object.* The conditions under which the retinal images have contributed to fitness in the past are those in which they have contributed to the catching of *flies.* A frog's visual apparatus may detect a fly *as* a small, dark, moving object, but that fact is dependent upon the structure of the visual

[4] Berent Enç (this volume) suggests that there is no way effectively to make determinate or unequivocal ascriptions of biological function given E or any form of E.

[5] Rowlands (1997) may well also fall into this category.

apparatus. 'In the frog's environment there is a correlation between snapping at flies and snapping at small, dark moving things. But it is the flies' nutritional qualities, not their appearance or size, which helps explain how directing snaps at flies helps the mechanisms to survive' (Price 1998: 70).

I admit to some perplexity here. I do not see why, given Price's argument, the visual image represents *fly* rather than *food* (as Shapiro 1992 suggests), or maybe *fly or food.* Price's discussion of this issue does little to dispel my confusion. She argues that the retinal image 'signals the presence of an item having certain biochemical properties that make flies nutritious to frogs' (Price 1998: 71). Fair enough, but this seems to invite the wholesale indeterminacy of Fodor's original objections (1990) to teleosemantics. Frog chow pellets are formulated specifically to have these same biochemical properties. So the visual image would seem to be indeterminate between the representation of *flies or frog chow pellets.* I used to feed my pet frogs pieces of beef liver because, I was told, they too had precisely the right nutritional properties. So perhaps the retinal image represents *fly or frog chow pellet or piece of beef liver.* Price's proposal leaves us with a fairly virulent form of the disjunction problem.

The principal problem with Price's emendation of E, though, is not that it fails to wring out the indeterminacy inherent in aetiological functions. It is that it imposes undue constraints upon the function-ascribing practices of biologists. If the objective of Phase 2 is to present intentional functions as merely of the same kind as the functions encountered in evolutionary biology, then the theory by which one individuates intentional functions must be equally applicable to the determination of evolutionary functions in general. This is where Price's account fails. The Independence Condition alone would close up shop almost entirely on quantitative and developmental genetics. In these sciences genes are assigned functions on the basis of their phenotypic effects. The expression of these functions relies upon the co-operation of an enormous number of devices downstream of the genes themselves. In fact, the Independence Condition threatens to negate any sort of function ascription to genes; as Ariew (1996) reminds us, *genes have no independent effects.* Yet, manifestly, some of them have functions. Nor is the Abstractness Condition well motivated. Physiologists and biochemists often ascribe to chemicals functions that depend upon their specific structure or on the structure of the systems in which they operate. Vivid examples are to be found in the functions of both allosteric enzymes and their effectors. Allosteric enzymes are capable of adopting a variety of conformations (shapes). On an occasion, the particular effect—the function—of an allosteric enzyme depends upon its precise conformation. This is a violation of clause (i) of the Abstractness Condition. The conformation adopted by an allosteric enzyme is determined by allosteric effectors. Effectors induce highly specific changes in the activities of an enzyme by forming non-covalent bonds at sites that are exposed by the enzyme's particular conformation. An effector may have different functions in different contexts depending upon which enzymes are around and what conformations they are in. So, the function of an effector depends upon the highly specific structure adopted by the enzymes upon which it works, a violation of clause (ii) of the same condition.[6] These sorts of functions are not to be dismissed lightly. The discovery of enzyme function has been one of the most significant advancements in understanding the nature and evolution of metabolism (Monod 1967). Yet these functions are ruled out by fiat on Price's supplemented version of E.

As is obvious from the frog–fly problem, the functional indeterminacy of biological traits is a consequence of the fact that most functional traits have a vast number of effects each of which makes some contribution—small or large—to survival and reproduction, each of which lays some *prima facie* claim to being a function. The advocate of theory E might well avoid the problem of indeterminacy by allowing that these are all genuine, determinate functions. One way to go about this is to incorporate one of the many important insights of Cummins's theory of functions (1975), specifically that function is determined only relative to a containing system.[7] This system-relative approach is clearly in play in Shapiro's discussion (1992) of determinacy. In addressing the frog–fly problem, he contends that the visual apparatus is part of a system selected for its capacity to detect and procure food: so the representational content of the retinal image is (determinately) *food*. This strikes me as entirely correct. But, while this approach may yield a *determinate* function for the frog's retinal image, it does not yield a *univocal* one. The frog's retinal image is part of an inordinately large number of biological systems. Accordingly, the system-relative approach should yield a panoply of reasonably determinate function attributions for the frog's retinal image. As part of the visual apparatus, it has the function of representing whatever is in the visual field; as part of the glossal apparatus, it has the function of representing something catchable; as part of the food-procuring system, it represents the presence of food; as part of the protein-assimilating system, it represents the presence of a source of proteins; as part of the sugar-metabolizing system, it represents a source of sugar. The list goes on and on.

At first blush multiplicity of function looks to be as much an impediment to the teleosemantics programme as is indeterminacy. Neander (1995) offers a novel suggestion designed to alleviate the problem. She claims that, among a trait's array of functions, one—the most specific—is explanatorily privileged.

[6] I take the discussion of enzyme function from G. Zubay (1993: ch. 10.)

[7] To be sure, the notion of function being invoked here is not entirely in keeping with the spirit of Cummins (1975). That conception of function is expressly non-teleological. But see Walsh and Ariew (1996) for a discussion of the relation of Cummins function to evolutionary (teleological) functions.

A trait's most specific function is the effect it has *irrespective* of the systems with which it cooperates. Neander's rationale is twofold. First, privileging the most specific function 'makes talk of malfunction more informative' (Neander 1995: 120). Secondly, a trait's most specific function identifies the effect that accounts for its presence in a way that less specific functions do not. Applying the suggestion to the frog–fly problem produces yet another determinate function ascription, but not an implausible one. The frog's visual apparatus is sensitive to small, dark, moving things; its most specific effect is to fire in their presence. It malfunctions, not when it fails to detect flies, but when it fails to detect small, dark moving things. After all, the visual apparatus is not malfunctioning when it induces the frog to snap at a bee-bee. So, Neander maintains, the determinate, privileged function of the retinal image is to represent *small, dark, moving objects.*[8]

I think this is a perfectly plausible way of marking out a distinctive kind of function. If a trait has a number of functions, one of them may well be the most specific. If specificity is to arbitrate between competing function ascriptions, however, the most specific functions must really enjoy the privilege that Neander claims for them. Specific functions must be more informative than less specific ones. They must also play a more basic explanatory role in evolutionary biology than less specific functions.

How do more specific functions make talk of malfunction more informative than less specific functions? Suppose some mechanism x contributes to the activities (the function) f of some containing system s by doing e. If we observe that s fails to perform f, it is more informative to be told that this is because x cannot do e, than to be told that the entire system s malfunctions. This much is common sense, but I should not think this makes a case that specific function attributions are more informative *per se.* Relative informativeness should be set against a background of what is to be explained and what is already known. It is certainly true that we may learn something more specific about the nature of the particular malfunction of the whole system s when we are told of the breakdown of some specific part x. But suppose that we know that x is incapable of doing e but do not know what the consequences might be; it certainly adds information to be told that because of x's malfunction, s will fail to do f and s' of which it is also a part will fail to do f', and so on. If we allow that it is among the many functions of x to contribute to the production of f in s and f' in s', then applying *less* specific functions to x makes talk of its malfunctioning *more* informative. I see little need, on grounds of informativeness, for privileging any particular function ascription in the way that Neander proposes.

[8] Mark Rowlands (1997) also claims that this is the correct *algorithmic* proper function of the frog's fly-catching mechanism.

Nor do functions at the highest level of specificity have any claim to explanatory privilege. They certainly do not seem exclusively to capture the function ascribing practices of biologists. Biologists often call upon less specific effects in giving a functional explanation for a trait. Consider the functions of homeobox genes. Homeobox genes contain very highly stereotyped sequences of base pairs. Virtually identical genes may be found across a wide range of eukaryotes, yet these same sequences code for wildly different morphological structures in different groups. For instance, the *Dfd* gene of *Drosophila* and the *Hox 4.2* gene of humans have the same sequence of base pairs. However, the *Drosophila* gene controls the development of the mouth parts (structures not found in mammals), whereas the mammalian gene controls the development of cervical vertebrae (structures not found in insects). Despite the disparate morphological effects, the first products of transcription—the most specific functions—are the same for both *Dfd* and *Hox 4.2* genes (McGinnis *et al.* 1990; McGinnis 1994). Developmental biologists ascribe *different* functions to *Dfd* and *Hox 4.2* genes in virtue of the fact they explain the occurrence, development, and evolution of *different* morphological regularities. And they ought to be free to do so. Neander's proposal, like Price's, appears to place unwarranted constraints upon the explanatory practices of biologists.

I do not wish to leave the impression that attempts to render aetiological function ascriptions determinate, such as the two I have just canvassed, err merely on matters of detail. I think they are motivated by a misunderstanding of the role of function ascriptions in evolutionary biology. Aetiological theorists tend to attribute to functions both a distinctive explanatory role and a particular classificatory role. The explanatory role, I believe, is mostly an artefact of the aetiological theory. The classificatory role, I believe, is largely a piece of wishful thinking stirred up by Phase 2 of the teleosemantics programme.

It is often supposed that the attribution of a function to a trait implies a more or less complete explanation of that trait's persistence under natural selection.[9] After all, the aetiological theory holds that teleological explanation of a biological trait is really just an aetiological explanation of how it got there. But, as Amundson and Lauder (1994) point out, this is not the sole explanatory role of biological functions. More often than not, biologists not interested in historical explanation simply ascribe functions to traits as part of an explanation of how their effects contribute to the adaptively significant activities of the systems of which they are a part. In physiology, functional anatomy, endocrinology, immunology, there may be many functions for a given mechanism, none of which taken alone constitutes anything like a complete causal explanation of its prevalence. Given the explanatory roles of function ascriptions in biology, there is no general requirement that they be either univocal or determinate.

[9] See Neander (1991, 1995); Kitcher (1993); Godfrey-Smith (1994); Price (1998).

As for the presumed classificatory role, one frequently encounters the claim that biological categories are, for the most part, individuated by their functions (e.g. Millikan 1986; Dennett 1987*b*). If functions are by their nature indeterminate or multifarious, it is argued, they cannot be called upon to yield determinate, unambiguous classifications of trait types. As far as I can tell this is simply a piece of philosophers' myth-making. To the extent that the concept of a trait-*type* is useful in comparative biology, it is based on the principle of homology and not on function. Developmental and phylogenetic considerations are paramount in determining homology; function is not. Equivalence of function may, on occasion, count as an auxiliary consideration, or as evidence for sameness of trait-type. But it is never *by itself* sufficient to guarantee sameness of trait-type (Lauder 1994). Bats' wings and insects' wings have (some of) the same functions, but as they are non-homologous they do not count as traits of the same type. Conversely, traits of the same type may have radically different functions. The hyoid structure of vertebrates is a prime case in point. Depending on what organism a hyoid element is in, it may support the gills, evert the tongue, or help produce speech or even mating calls (Carroll 1987).

There is no compelling reason, then, to suppose that the functions of biological traits must be determinate or univocal, short of the requirements of the teleosemantics programme. The uses to which biological function attributions are actually put suggest that indeterminacy and multiplicity are endemic to evolutionary function. That said, it does not follow from the fact that evolutionary functions in general tend to be inappropriate for the purposes of teleosemantics that *all* are. If the intentional functions of Phase 1 are indeed evolutionary functions, it does not follow that *they* are not determinate and univocal in the required way. So how could the evolutionary functions of intentional states be determinate and univocal if evolutionary functions in general are not?

Again, the Cummins insight that function attributions are system relative becomes important. Just as biological functions are ascribed relative to the systems to which they contribute, the functions of intentional states are determined relative to the system in which *they* operate—namely, the system of belief–desire psychology. It may be difficult to pin down *the* function—and hence the content—of a frog's retinal image, but once one considers the role of an intentional state in an agent possessing full-blown belief–desire psychology, there is no problem whatsoever (Papineau 1993, 1998). Belief–desire psychology is an extremely complex, intricate, highly plastic and powerful piece of biological machinery. It seems plausible that there has been heavy selection for this biological apparatus. Within this system an intentional state plays a highly specific, determinate functional role, one that is easily picked out. The function of a belief is to represent that its truth conditions hold. The function of a desire is to bring about the fulfilment of its satisfaction conditions. In turn, the truth condition of a belief is that state that guarantees the satisfaction of any of a wide array of desires with which the belief works. The satisfaction condition of a desire is that condition that it brings about in concert with any of a wide array of beliefs. This is a time-honoured, venerable approach to the individuation of intentional content.[10] It serves equally well as the determinant of intentional function.

This is the approach to the determinacy problem pursued explicitly by Papineau and, I believe, implicitly by Millikan.[11] It certainly seems to meet the requirements of the Determinacy Constraint. Ascribing to states intentional functions (hence contents) on the basis of the norm-governed roles in the motivation of action preserves the intensionality of intentional contexts. Beliefs or desires with extensionally equivalent—even *necessarily* equivalent—contents may play *different* roles in the motivation and explanation of actions. If psychological states are individuated by these finely demarcated causal/explanatory roles, then they will naturally, almost trivially, support the intensionality of intentional contents.

In effect the attempts to render *all* evolutionary functions determinate miss the point. Phase 2 of the teleosemantics programme does not need them to be; it simply requires that *intentional* functions are determinate and it is obvious that they are. Beyond that, success in Phase 2 requires simply that the determinate functions of intentional states are shown to be *evolutionary* functions. It is here that teleosemantics confronts its second challenge, the Adaptiveness Constraint.

1.2. The Adaptiveness Constraint

If theory E is the full story on evolutionary function, then for *each* intentional content there must be a plausible story to be told about how *it* was promoted by natural selection in the past. It is clear that in rudimentary cases, like the desire for water, the desire for a mate, some such story will be available. But for many beliefs and desires there will be none. In particular, any account that ties the content of an intentional state to its evolutionarily significant past effects will face acute difficulties accounting for the contents of novel, one-off intentional states and those whose fulfilment conditions are adaptively *non*-advantageous.[12]

This problem, of course, has been anticipated and addressed. Millikan in particular proposes an intriguing solution (1986, 1989, 1990). Millikan's theory recognizes a special category of proper functions specifically for these

[10] For examples, see Mellor (1988); Whyte (1990); Papineau (1993, 1994).

[11] See Papineau (1993, 1998) and Millikan (1989, 1990, 1995).

[12] This is a less generalized version of what Peacocke (1995) calls the 'problem of reduced content'.

sorts of cases, Derived Proper Functions. These are 'functions derived from the functions of devices which produce them' (Millikan 1993*a*: 14). They need no historical antecedents. The idea is easily illustrated. The direct proper function of my stomach (anybody's stomach) is to digest items that are introduced into it. As a consequence of this generalized function, on an occasion of my eating some food item, my stomach goes into a state whose derived proper function is to digest what I just ate. Now suppose I eat something of a sort that has never been ingested before, or whose ingestion has no positive or negative selective consequences—let us say a piece of styrofoam. Clearly there is no evolutionary story to be told about the benefits of having stomachs that invokes the selective advantages of digesting pieces of styrofoam in the past; *ex hypothesi* there are not any. Nevertheless, it is a perfectly legitimate *derived* proper function of the current state of my stomach to ingest a piece of styrofoam.

The same strategy can be extended to the functions of novel or selectively non-advantageous representational states. Millikan offers the compelling example of the bee's waggle dance. The general function of the waggle dance is to signal the direction, distance, and abundance of a patch of nectar-yielding flowers. A particular waggle dance may (let us suppose) have the function of representing the presence of a substantial amount of nectar at a distance of 500 metres in a direction due north from the hive, even though no dance with precisely this representational content has been performed before. The novel dance derives its representational function from the function of waggle dances in general.

According to Millikan (1995), the contents of novel, non-adaptive beliefs and desires can be treated in the same way. Take some novel or non-adaptive intentional state, perhaps (let us pretend) the desire for some never before thought of element beyond the currently known Lanthinide or Actinide series; let us say you are the first person ever to harbour the desire for some Gargantium. Applying the machinery of derived proper functions yields a determinate content to this thought. The desire apparatus is one whose proper function is to produce actions that bring about those states of affairs that the agent desires; when a state with representational content *p* is introduced into it, the agent is in a state of desiring *p*—a state whose derived proper function is to bring about *p*. So, even though no one has ever before had the desire for Gargantium—or entertained any thought about the stuff—your desire has the determinate derived proper function of bringing about your acquisition of Gargantium. That is what makes it the desire for Gargantium. By applying the apparatus of derived proper functions, then, an advocate of E can accommodate the idea that novel, non-adaptive thoughts have determinate evolutionary functions.

The proposal is ingenious but flawed. There is a crucial disanology between the derived proper functions of *digesting* styrofoam and *desiring* Gargantium.

The disanalogy derives precisely from the fact that the latter function but not the former involves representation. Consider what factors determine that the state of my stomach has the function of digesting styrofoam. There are two: (i) that it is the general function of my stomach to digest what is in it and (ii) what is in it is styrofoam. Similarly, that your desire has the function of bringing about your acquisition of Gargantium is determined by two factors: (i) that it is the general function of your desire apparatus to bring about the state of affairs represented by what is in it and (ii) what is in it represents Gargantium. Compare the second factor in each case. What makes it the case that what is in my stomach is styrofoam?—clearly, its internal constitution. But what makes it the case that what is in your desire apparatus represents Gargantium?—clearly, its *representational content.* The specification of the latter derived proper function requires an appeal to representational content. But the teleosemantics programme is built upon the notion that *all* representational contents can be cast as functions. Unless this appeal to representational contents can itself be reduced, derived proper functions offer no assistance in meeting the adaptiveness constraint.

I suspect that this sounds unconvincing. After all, novel waggle dances have derived proper representational functions, so why cannot novel Gargantium thoughts? There are two crucial disanalogies between novel waggle dances and Gargantium thoughts: one concerns their respective histories, the other concerns their respective consumer systems.

Novel waggle dances derive their representational contents from general mapping rules that hold over a range of dances. These *mapping rules* have been selected for because instances of their application have conferred survival value sufficiently often in the past. Concomitantly, there has been selection for the capacity of bees to interpret novel dances under those rules. 'The bee dance is a representation because there is, we take it, a univocal relational explanation that is invariant over the bulk of cases in which ancestors of our bees found nectar as a result of reacting as they do to bee dances' (Millikan 1995: 287–8). Even though there has been no previous selection for the content of a given waggle dance, there has been selection for the *rules* that confer on it a representational content. Things are different with novel thoughts involving Gargantium. There has been no selection for either *instances* of mapping the presence of Gargantium onto success-guaranteeing behaviours or for *rules* that do so. There has been no selection of instances of Gargantium behaviours because there have been no such instances. If there has been no selection for instances of the Gargantium rule, there has been no selection for the rule either.

There is an obvious rejoinder to be made on behalf of teleosemantics: perhaps Gargantium-representing rules are themselves derived functions. There may have been no selection for specific Gargantium-related rules, but there may well have been selection for general representational rules, of which the Gargantium rules are instances. We might, for example, have an apparatus that, given any environmental condition *p*, produces a representation whose content is *p*. This representation, in turn, is apt to form the content of a belief that *p* or a desire that *p*. So, in the presence of Gargantium, the apparatus produces a representation of Gargantium that may become the content of either a belief about Gargantium or a desire for Gargantium.[13] Here is where the second disanalogy between waggle dances and thoughts becomes important. Waggle dances and thoughts both have representational functions. As we have discussed, the function of an entity is determined by the way it is used by the system to which it contributes—its consumer system. Waggle dance representations are used by the nectar-gathering system of bees. The relevant system in the case of thoughts is the whole of belief–desire psychology. The difference in complexity of these consumer systems makes all the difference concerning whether these representational rules can be the subject of natural selection.

Consider what, according to the account on offer, it is to represent some environmental condition *p*; it is to instantiate a rule that, given *p*, directs behaviour appropriate to the fulfilment of an individual's purposes.[14] If these rules are to be subject to natural selection, certain conditions must be met. The rules must be determinate, projectible, and straight. This, I take it, is what Millikan is latching onto by saying that there is 'a univocal relational explanation that is invariant over the bulk of cases . . .' in the past. Presumably, when a mapping rule meets these conditions, it is easy to make sense of the notion of a novel instance. Where the consumer system and the purposes are simple, these conditions are relatively easily met. The rules for the waggle dance, for example, map a set of properties of nectar patches, *p*, onto behaviours that, given *p*, guarantee success in nectar gathering. The relevant features of the behaviours vary as simple arithmetic functions of the direction, richness, and distance of the nectar patch. The mapping rule is simple, selectable, and projectible.

Compare this to the case of representations to be used by the entire system of belief–desire psychology. A general rule for representing environmental conditions, *p*, would need to map *p* onto behaviours that conduce to the fulfilment of the agent's purposes given *any* combination of beliefs and desires. For this rule to be selectable, it would have to be determinate, straight, and projectible. But it is difficult to imagine what such a rule could be. What straight, determinate, behavioural rule could one specify that, given any condition of the environment, *p*, guarantees successful fulfilment of *any of* one's psychological purposes? Such a rule would have to be fantastically general.

[13] Peter Carruthers, Stephen Laurence, and Walter Sinnott-Armstrong have all suggested this strategy to me.

[14] My account of mapping rules draws heavily upon Millikan's (1993*b*) discussion of these matters.

And the more general a rule is, the less straight, projectible, or selectable it is. 'Do the right thing under the circumstances' may be a determinate rule, but it is not projectible or straight. It doesn't specify a specific behaviour for any of the circumstances. Hence it is not selectable. The capacity of organisms to 'do the right thing' in the past may have determined their success in the past, but this capacity does not determine what 'doing the right thing' consists of in future, novel circumstances. In short, it seems that there could be no straight, determinate rule of the form 'in the presence of *p* bring about behaviours conducive to success given any of a range of beliefs and desires' that could be the subject of natural selection. In that event, there could be no selection for an apparatus whose proper function is to conform to such a general rule. If that is so, there could be no state whose proper function, as *derived* from this general function, is to produce a novel representation of Gargantium, or for that matter, a novel representation of anything.

I suspect that the Adaptiveness Constraint cannot be met by an appeal to the machinery of derived proper functions. Unless a case can be made that all novel and non-adaptive contents are constructs out of those with genuine historical proper functions, the teleosemantics project, at least in so far as it is wedded to the aetiological theory of functions, cannot provide a reductive account of the teleological functions of intentional states.

So what are the prospects for Phase 2? It seems that Determinacy and Adaptiveness pose antagonistic constraints upon the reduction of intentional function to evolutionary function. The Determinacy Constraint is easily met if we allow that intentional functions are fixed by the functional roles of thoughts in an agent's belief–desire psychology. But these are not the adaptive, evolutionary functions served up by the aetiological theory, E. For many or most of the teleological functions that determine intentional content of a psychological state, there is no adaptive story to be told about how natural selection in the past put them there. In effect, any account of intentional function that is rich enough to serve the purposes of Phase 1 is too rich to be reduced to evolutionary function in Phase 2. This ought to be unacceptable to supporters of the teleosemantics programme. It forecloses on what teleosemanticists see as the great promise for a naturalized theory of intentionality. Phase 2 of the programme is, after all, a mere way station on the road to Phase 3 in which the teleology inherent in the talk of functions is given its naturalistic licence. If Phase 2 fails, then no matter how successful Phase 3 may be it is inapplicable to the account of intentional content. But take heart: if this is reason to abandon teleosemantics, it is no reason to forsake the cause of applying teleological function to the naturalization of intentionality. A perfectly good account can survive the failure of Phase 2. One reason to believe this, oddly enough, resides in the failure of Phase 3.

2. Evolutionary Function to Causal History

The whole edifice of teleosemantics rests on the supposition that evolutionary teleology can be traded in for aetiology—these are shaky foundations. As we have seen, the aetiological theory, E, holds that all talk of adaptation and function is merely disguised talk about the causal history of natural selection. In order to explain what a particular trait is *for*, we do not need to advert to the goals of the organism; we simply need to invoke the effects of selection in the past. The teleological language of biology carries no real teleological commitments. I think this is a mistake fostered in part by a misconstrual of Darwin's solution to Paley's problem.

2.1. Paley's Problem

The story of Darwin's solution to Paley's problem is now familiar. Darwin agreed with Paley that any theory of biological form must account for the nature of adaptations but, unlike Paley, required that it do so *without* invoking the intentions of a designer. In *The Origin of Species* Darwin asks: 'How have all these marvellous adaptations of one part to another and of one organism to another arisen?' (1996: 51). The story continues that his immediate answer is 'natural selection'. In fact, Darwin tells us something quite different: 'All these results', he says 'follow from the struggle for life' (1996: 52) (alternatively the 'struggle for existence').

> Owing to this struggle for life, any variation, however slight, and from whatever cause proceeding, if it be in any degree profitable to an individual of any species, in its infinitely complex relations to other organisms and to external nature, will tend to the preservation of that individual and will generally be inherited by its offspring. I have called this principle, by which each slight variation, if useful, is preserved, by the term Natural Selection . . . (1996: 52)

The 'struggle for existence' and 'natural selection' are not merely synonyms. For Darwin, the struggle for existence is a process played out *within* individuals. Natural selection, in contrast, is a population-level phenomenon—the differential preservation of useful traits within a population. Selection requires variation, indeed competition, between individuals; the struggle for life does not. 'Two canine animals in a time of dearth, can be said to struggle with each other which shall get food and live. But a plant on the edge of a desert is said to struggle for life against the drought . . .' (1996: 33). It appears that the struggle for life and natural selection, according to Darwin, are related as cause and effect. It is the former of these and not the latter that he proposes as the cause of adaptations. There is, to be sure, no account in Darwin of *how* the struggle for existence brings about adaptive novelties. The important point for my purpose is that Darwin's theory of natural selection appeals to an unreduced teleological phenomenon, the *struggle for existence.*

In what sense, then, does Darwin's theory of natural selection banish the teleology inherent in Paley's Argument from Design? There is an important, but admittedly rough and ready distinction to be observed here between what might be called 'higher-order' and 'first-order' teleological explanations. In a higher-order teleological explanation, the nature, behaviour, or function of an entity of some type is explained by appeal to the goals of some *other* agent or controlling system. The explanation of the function of artefacts is clearly of this sort and so is Paley's account of biological adaptation. In first-order teleological explanation, the properties of an individual are explained by appeal to its pursuit of its own goals. Action explanations are typically of this sort, and so is Darwin's account of adaptation.[15] Certainly Darwin eliminated Paley's appeal to higher-order teleology, but he did so by invoking first-order teleology. The intentions of the Designer are supplanted as explanans by the struggle for existence *within* individual organisms. This struggle for existence appears in *The Origin* as a completely unreduced first-order, teleological notion. Like it or not, teleology is written into the very foundations of the modern evolutionary theory.[16]

The obvious response is that, thanks to the growth of evolutionary biology since that time, the residual teleology inherent in Darwin's theory can now be expunged. In Darwin's theory, the 'struggle for existence' acts as a placeholder for those processes going on within individuals that account for the generation of novel variants, their inheritance, and the development of phenotypes. Darwin knew very little about these processes, but now that we know more about mutation, heredity and ontogeny it is obvious that they are one and all causal/mechanical processes. So, no unreduced appeal to teleology ought to be required. It is certainly true that mutation, heredity, and development are causal process. But, even accepting that they are the causes of adaptation, it does not follow that adaptation is not a teleological phenomenon. In fact I think that truly modern biology appears to tell us quite the opposite: adaptation and function are indeed genuinely and irreducibly teleological.

[15] André Ariew (this volume) glosses my distinction between higher-order and first-order teleology as the distinction between the Platonic and Aristotelian conceptions of teleology. This distinction is borne out by Hankinson (1998: ch IV). Paley is Plato to Darwin's Aristotle.

[16] In various places in his correspondence Darwin is perfectly sanguine about the inherently teleological cast of his theory. He famously endorses Asa Gray's comment that Darwin has done a 'great service to natural science in bringing it back to Teleology; so that instead of Morphology *versus* Teleology we shall have Morphology wedded to Teleology' (quoted in Amundson 1996: 28).

2.2. Goals, Development, and Adaptation

Organisms are the paradigm cases of goal-directed entities (Nagel 1977). They are highly complex, self-organizing systems, capable of attaining and maintaining finely balanced, persistent equilibrium states. Uniquely, organisms achieve and persist in these end states by mounting highly plastic adaptive responses to perturbations occurring both within their developmental programmes and within their environments. In fact, on one prominent account of goal directedness, what it *is* to be a goal-directed entity is just to manifest these properties (Nagel 1977; Bedau 1992). Of course, a goal-directed process is just a teleological process. So the question for our purposes is: 'what is the significance of this goal directedness for the phenomena of function and adaptation?'

Recent work in the dynamics of complex systems suggests that only populations comprising individuals capable of mounting such adaptive responses are capable of undergoing adaptive evolution. The reasons are fairly simple: (i) only those organisms capable of maintaining metabolic stability in the face of unpredictable environmental perturbations are capable of survival, (ii) only those systems capable of reproducing and developing with reasonable fidelity *despite* perturbations (mutations) in their developmental systems are capable of development and reproduction, and (iii) only those systems capable of mounting supple, adaptive responses to mutations in their developmental programmes are capable of generating the sort of novelties on which selection works (Kauffman 1995: ch. 4). The first two of these, I take it, are more or less conceptual truths but the third is much more interesting. If a system (organism) is too developmentally robust, mutations do not induce changes to its developmental end point, hence there are no adaptive novelties to pass on. If a system is too developmentally labile, the introduction of mutations causes the entire developmental system to collapse; hence whatever 'novelties' may arise are neither adaptive nor inherited. So, the capacity of organisms to mount adaptive responses to both internal and external perturbations is a *prerequisite* for populations of organisms to undergo adaptive evolution by natural selection. But, to mount adaptive responses of this sort—to attain and maintain these poised, persistent stable states in the presence of unpredictable perturbations—*just is* to manifest goal-directed behaviour. The goals are those of maintaining metabolic processes and leaving/developing high-fidelity copies: in short, the goals of surviving and reproducing.

So far all that the systems dynamics approach has established is that lineages or populations accumulate adaptations only if they comprise goal-directed individuals. This falls well short of demonstrating that teleology is irreducibly implicated in the explanation of adaptations. It is consistent with the view that adaptations are the causal consequences of selection operating over populations of goal-directed individuals.[17] After all, so the story goes, goal directedness is itself a consequence of natural selection (Mayr 1988). There is a common supposition among those who ply this strong selectionist line that without selection there would be no adaptation at all. But, as an empirical matter of fact, that is just not true. The modern study of developmental biology suggests a radically different picture. Kauffman (1993) demonstrates that adaptations arise naturally as a consequence of the dynamics complex systems—with or without selection. Given a few very simple epigenetic rules (laws of form), there is spontaneous development of complex, self-ordering, self-replicating, stable forms. These systems exhibit fantastic capacities to maintain homeostasis despite the vagaries of their environments and to mount hugely plastic adaptive responses to perturbations. 'We appear to have been profoundly wrong. Order, vast and generative, arises naturally. . . . I propose that much of the order in organisms may not be the result of selection at all, but of the spontaneous order of self-organized systems' (Kauffman 1993: 25). Natural selection is not necessary for biological adaptation. All that is required are the sorts of processes that go on in the struggle for life, just as Darwin supposed.

Perhaps this merely suggests an alternative aetiological account of adaptation: to be an adaptation is to be the consequence of certain developmental processes. But this proposal would not serve the purposes of those intent on reducing out biological teleology. Simply being such a consequence is not sufficient for being an adaptation. The vast majority of developmental responses to mutations have either neutral or deleterious effects on survival and reproduction (Lewontin 1974). These are not adaptations. Those responses that *are* adaptations are distinguished from those that are not in that the former contribute positively to the goals of survival and reproduction whereas the latter do not. According to the picture emerging from modern developmental biology, teleology is implicated twice over in the account of adaptation. Adaptive novelties are the causal consequences of developmental processes pursuing the goals of survival and reproduction. Furthermore, a novel trait is an adaptation only if it contributes to the attainment of these goals.

We have an alternative conception of biological adaptation and function—I shall call it the 'teleological conception'—that contrasts sharply with the aetiological theory. According to the former, adaptations are the causal consequences of development. An adaptation is a trait that positively and significantly contributes to the goals of survival and reproduction. The functions of a trait are the effects it has that make this contribution. The aetiological approach, in contrast, casts adaptations as the causal consequences of selection. No unreduced appeal to goals or to any teleological processes is required to account for the phenomenon of adaptation. Certainly, the teleological conception does more justice to

[17] A position similar to this is put forward by Dawkins (1983, 1986).

the pre-theoretic notion of a function. Even setting pre-theoretic intuitions aside, the teleological picture has a certain conceptual primacy. This becomes apparent when we consider the concomitant view of adaptive or functional explanation for each.

According to the teleological conception, a functional explanation explains the systematic positive contribution of a trait to the goals of survival and reproduction. According to the aetiological view, an adaptive or functional explanation explains of a particular trait what it has been selected for in the past. For a trait to have been selected for some effect *f* in the past is just for it to have made a (sufficiently strong) systematic contribution to survival and reproduction in the past by doing *f*. The very notion of systematic contribution to the goals of survival and reproduction is at the heart of the aetiological picture, just as it is in the teleological picture. Functional explanations, then, on *any* account, simply explain a trait's contribution to survival and reproduction. To be sure, there is a debate to be had about *when* these contributions must have occurred—aetiologists insist that, given the role of function explanations in evolutionary biology, only *past* contributions are relevant, whereas an advocate of the teleological approach accepts that either past or present contributions are explanatorily relevant. This disagreement, however, should not disguise an oft-overlooked but fundamental point of agreement: *on either account functions and adaptations are individuated by systematic contribution to survival and reproduction with respect to some environment or other.* On either account, distinguishing functions from mere effects invokes the systematic positive contribution to survival and reproduction. On either account, a functional or adaptive explanation explains the systematic contribution of a trait to survival and reproduction. Once it is conceded that survival and reproduction are goals it must also be conceded that, on either account, adaptation and function are genuinely teleological phenomena.

Those with naturalist scruples will probably balk at this claim. Why should we accept that survival and reproduction are goals? The natural world has no goals, so any explanatory strategy that invokes them, or any categories that presuppose them, must be reduced to—or eliminated in favour of—those that require only efficient causes. I disagree. To think this way is to miss the philosophical windfall afforded by the complex systems dynamics approach to development. That approach tells us that the goals of survival and reproduction are simple causal consequences of a few laws of self-organization operating during development. Nothing could be more natural. In turn, the categories of adaptation and function in which evolutionary biology trades are defined in terms of contribution to the goals of survival and reproduction. Granted, we do not have a *reduction* of the teleological categories of evolutionary biology, but we do have a causal account of how they are realized.

Peter Smith (1992) has outlined a particularly weak form of theoretical reduction—he calls it a 'weak explanatory interfacing'—which I think is particularly germane to this issue. A weak explanatory interfacing occurs when a lower-level theory (T_2) explains why some higher-level theory (T_1) works as well as it does, *without* reducing the categories of T_1 to those of T_2. This is precisely the relation that holds between the complex systems analysis of development (T_2) and evolutionary theory that appeals to adaptations (T_1). The complex systems dynamics approach to development (T_2) shows us how the teleological categories of evolutionary biology (T_1)—survival, reproduction, adaptation, and function—fit into the causal order of the world. What better naturalist credentials could a category have? Yet, it does not provide a means for reducing the teleological categories of T_1 to those of T_2.

So Phase 3 of the teleosemantics programme is misguided. There is no reduction from function (adaptation) to causal history of selection to be had in evolutionary biology. More importantly, none is needed. Evolutionary biology is soaked in the unreduced teleology of goals, adaptations, and functions, and none the worse for it.

3. Brentano's Problem and Darwin's Solution

There is available to us an alternative conception of the ways in which Darwin's theory of evolution informs Brentano's problem. In common with the teleosemantics programme, it holds that intentional mental categories are teleological/functional categories (Phase 1). Beyond this the two part company.

The teleosemantics programme looks to Darwin's theory of evolution as a way of eliminating the appeal to teleology inherent in the notion of intentional function, supplanting it with an appeal to the causal history of natural selection. I have argued against this approach on two grounds. First, it requires that the functions that demarcate intentional mental states are *evolutionary* functions, functions that have contributed to selective success in the past, a position I find extremely implausible. Secondly, it misconstrues the nature of biological teleology in general. One cannot give a reductive account of the teleological categories of biology—*function, adaptation*—in terms of the causal consequences of natural selection.

The alternative picture offers evolutionary biology as a model for naturalized intentionality, but not as a reducing base. I can give only the vaguest of sketches here. Darwin's theory, together with recent work in developmental biology, demonstrates how teleological processes and goal-directed entities arise naturally in nature. They are the simple consequences of a few laws of self-organization manifested in highly complex, dynamic systems. The goal-directed processes of survival and reproduction can be explained by, but not reduced to, the non-goal-directed processes taking place in development. The biological function of a trait is determined by its systematic causal contribution to the natural biological goals of survival and reproduction. Biological functions, then, constitute a natural, irreducible *sui generis* kind of teleological property.

The model is easily applied to the naturalization of intentionality. Psychological agents are themselves highly complex, goal-directed entities. In fact, it seems plausible that one cannot even interpret an entity as an agent unless one construes it as capable of pursuing a distinctive class of goals. If, as the study of complex systems tells us, goals are a natural consequence of complexity, there is no reason to suppose that the goals that characterize agency are *non*-natural. Intentional states—beliefs and desires—are individuated by the causal/functional roles they play in the attainment of an agent's goals: intentional categories are teleological categories, just as the functional categories of biology are. But intentional categories need not be *biological* categories, as the teleosemantics programme supposes. An agent's *psychological* goals need not be *biological* goals. They are, after all, constrained by different norms. The norms of survival and reproduction constrain the functions of biological traits and the norms of rationality constrain the functioning of mental states. If mental states are individuated by their intentional functions and intentional functions are not reducible to *biological* functions, then intentional states may well constitute a *sui generis* realm of wholly natural, irreducibly teleological categories too. Aboutness is real, but it is not really something else.

References

Agar, N. (1993), 'What do Frogs Really Believe', *Australasian Journal of Philosophy*, 71: 1–12.

Amundson, R. (1996), 'Historical Development of the Concept of Adaptation', in M. Rose, and G. V. Lauder (eds.), *Adaptation* (San Diego, Calif.: Academic Press), 11–53.

——and Lauder, G. V. (1994), 'Function without Purpose: The Uses of Causal Role Function in Evolutionary Biology', *Biology and Philosophy*, 9: 443–69.

Ariew, A. (1996), 'Innateness and Canalization', *Proceedings of the Biennial Meetings of Science Association*, supp. vol. 3: 519–29.

Bedau, M. (1992), 'Naturalism and teleology', in S. J. Wagner and R. Warner (eds.), *Naturalism: A Critical Appraisal* (Notre Dame, Ind.: University of Notre Dame Press), 23–52.

——(1996), 'The Nature of Life', in M. Boden (ed.), *The Philosophy of Artificial Life* (Oxford: Oxford University Press), 332–57.

Brentano, F. (1995), *Psychology from an Empirical Standpoint*, 2nd edn., trans. A. C. Rancurello, D. B. Terrell, and L. L. McAlister (London: Routledge; 1st edn. 1874).

Buller, D. I. (1999) (ed.), *Function, Selection and Design: Philosophical Essays* (Albany, NY: SUNY Press).

Carroll, R. L. (1988), *Vertebrate Paleontology* (Chicago: Freeman).

Cummins, R. (1975), 'Functional Analysis', *Journal of Philosophy*, 72/20: 741–65.

Darwin, C. (1996), *The Origin of Species* (London: Penguin Classics).

Dawkins, R. (1983), 'Universal Darwinism', in Bendell Darwin (ed.), *Evolution from Molecules to Men* (Cambridge: Cambridge University Press), 403–25.

——(1986), *The Blind Watchmaker* (London: Longman).

Dennett, D. C. (1987*a*), 'Evolution, Error and Intentionality', in D. C. Dennett, *The Intentional Stance* (Cambridge, Mass.: MIT Press), 287–321.

——(1987*b*), 'Intentional Systems in Cognitive Ethology: The "Panglossian Paradigm" Defended', in D. C. Dennett, *The Intentional Stance* (Cambridge, Mass.: MIT Press), 237–86.

——(1996), 'Granny versus Mother Nature—No Contest', *Mind and Language*, 11: 263–9.

Dretske, F. (1986), 'Misrepresentation', in R. Bogdan (ed.), *Belief: Form, Content, and Function* (Oxford: Oxford University Press), 17–36.

Fodor, J. (1987), *Psychosemantics* (Cambridge, Mass.: MIT Press).

——(1990), *A Theory of Content and Other Essays* (Cambridge, Mass.: MIT Press).

——(1996), 'Deconstructing Dennett's Darwin', *Mind and Language*, 11: 246–62.

Godfrey-Smith, P. (1994), 'A Modern History Theory of Functions', *Noûs*, 28: 344–62.

Griffiths, P. E. (1993), 'Functional Analysis and Proper Functions', *British Journal for the Philosophy of Science*, 44: 409–22.

Hankinson, R. J. (1998), *Cause and Explanation in Ancient Greek Thought* (Oxford: Oxford University Press).

Kauffman, S. A. (1993), *The Origins of Order: Self-Organization and Selection in Evolution* (Oxford: Oxford University Press).

——(1995), *At Home in the Universe* (Oxford: Oxford University Press).

Kitcher, P. (1993), 'Function and Design', in P. A. French, T. E. Uehling, and H. K. Wetstein (eds.), *Midwest Studies in Philosophy XVII* (Minneapolis: University of Minnesota Press), 379–97.

Lauder, G. V. (1994), 'Homology: Form and Function', in B. K. Hall (ed.), *Homology: The Hierarchical Basis of Comparative Biology* (San Diego, Calif.: Academic Press), 151–96.

Lewontin, R. C. (1974), *The Genetic Basis of Evolutionary Change* (New York: Columbia University Press).

McGinnis, N., Kuziora, M. A., and McGinnis, W. (1990), 'Human *Hox* and *Drosophila Deformed* Encode Similar Regulatory Specificities in *Drosophila* Embryos and Larvae', *Cell*, 63: 969–76.

McGinnis, W. (1994), 'A Century of Homeoesis, a Decade of Homeoboxes', *Genetics*, 137: 607–11.

Mayr, E. (1988), *Toward a New Philosophy of Biology* (Cambridge, Mass.: Harvard University Press).

Mellor, D. H. (1988), 'I and Now', *Proceedings of the Aristotelian Society*, 89: 79–94.

Millikan, R, Garrett (1984), *Language, Thought, and Other Biological Categories: New Foundations for Realism* (Cambridge, Mass.: MIT Press).

——(1986), 'Thoughts without Laws: Cognitive Science without Content', *Philosophical Review*, 95: 47–80.

——(1989), 'Biosemantics', *Journal of Philosophy*, 86: 281–97.
——(1990), 'Compare and Contrast Dretske, Fodor and Millikan on teleosemantics', *Philosophical Topics*, 18: 151–61.
——(1993*a*), 'In Defense of Proper Functions', repr. in R. G. Millikan, *White Queen Psychology and Other Essays for Alice* (Cambridge, Mass.: MIT Press), 13–29.
——(1993*b*), "Truth Rules, Hoverflies, and the Kripke–Wittgenstein Paradox', repr. in R. G. Millikan, *White Queen Psychology and Other Essays for Alice* (Cambridge, Mass.: MIT Press), 211–39.
——(1995), 'A Bet with Peacocke', in C. MacDonald and G. MacDonald, *Philosophy of Psychology: Debates on Psychological Explanation* (Oxford: Blackwell), 285–92.
Monod, J. (1967), *Chance and Necessity* (London: Penguin).
Nagel, E. (1977), 'Teleology Revisited', *Journal of Philosophy*, 74: 261–301.
Neander, K. (1991), 'The Teleological Notion of "Function"', *The Australasian Journal of Philosophy*, 69: 454–468.
——(1995), 'Misrepresenting and Malfunctioning', *Philosophical Studies*, 79: 109–41.
Papineau, D. (1993), *Philosophical Naturalism* (Oxford: Blackwell).
——(1994), 'Content (2)', in S. Guttenplan (ed.), *A Companion to the Philosophy of Mind* (Oxford: Blackwell), 225–30.
——(1998), 'Teleosemantics and Indeterminacy', *Australasian Journal of Philosophy*, 76: 1–14.
Peacocke, C. (1995), 'Concepts and Norms in a Natural World', in C. MacDonald and G. MacDonald (eds.), *Philosophy of Psychology: Debates on Psychological Explanation* (Oxford: Blackwell), 277–84.
Price, C. (1998), 'Determinate Functions', *Noûs*, 32: 54–75.
Rowlands, M. (1997), 'Teleological Semantics', *Mind*, 106: 279–303.
Shapiro, L. (1992), 'Darwin and Disjunction: Foraging Theory and Univocal Assignments of Content', *Proceedings of the 1992 Biennial Meeting of the Philosophy of Science Association*, 1: 469–80.
Smith, P. (1992), 'Modest Reductions and the Unity of Science', in D. Charles and K. Lennon (eds.), *Reduction, Explanation and Realism* (Oxford: Oxford University Press), 19–44.
Sterelny, K. (1990), *The Representational Theory of Mind: An Introduction* (Cambridge, Mass.: Blackwell).
Stich, S. and Warfield T. (1994) (eds.), *Mental Representation* (Oxford: Blackwell).
Walsh, D. M. and Ariew, A. (1996), 'A Taxonomy of Functions', *Canadian Journal of Philosophy*, 26/4: 493–514.
Whyte, J. (1990), 'Success Semantics', *Analysis*, 50: 149–57.
Zubay, G. (1993) (ed.), *Biochemistry*, 3rd edn. (Dubuque: Wm. C. Brown and Company).

Part Four

Methodological Issues

13. Human Rationality and the Unique Origin Constraint

Mohan Matthen

ABSTRACT

More than one philosopher has hypothesized that the function of rationality is to lead us to true beliefs. In the context of evolution, this means that human rationality was selected for this capacity. If so, human rationality ought to approximate to the best systems of deductive and inductive logic, a consequence that many philosophers have accepted on independent grounds. However, there are two problems with the above hypothesis concerning function. First, empirical research indicates that human reasoning systematically falls short of 'logic'. But, the hypothesis about function implies that all systems of rationality would converge upon logic. Secondly, this convergence claim is methodologically suspect because it violates the Unique Origin Constraint on function attributions, a constraint formulated and supported in this chapter. If this is right, rationality must be an accidental product of natural selection; a suggestion is made about how it could have originated.

1. The Function of Reason: A Methodological Problem

Ever since Ernest Nagel's path-breaking discussion (1961), philosophers have engaged in a great deal of discussion about the meaning of the word 'function', and how it is to be extrapolated from attributions to artefacts to descriptions of biological organs. The problem we have been trying to address is this: since function talk seems to presuppose the purposes of a Creator, it seems to violate naturalistic methodology in biology. Yet, biologists do use functions. How can we make biology safe for naturalism while still making a place within it for functions?

This worry has been ameliorated somewhat by a new philosophical analysis of the meaning of the word 'function'. Though the details may still be hotly

This chapter was incubated in a discussion group on rationality, of which the other members were Reneé Elio, Jeff Pelletier, and Catherine Wilson. I am grateful to them for inspiration, and to Ronnie de Sousa, Karen Neander, Elizabeth Preston, Larry Shapiro, Elliott Sober, Kim Sterelny, and the editors (in particular André Ariew) for detailed comments on earlier versions.

contested, it is now widely accepted that we may say that a biological organ has function *F* if it has been naturally selected *for*[1] doing *F*.[2] This analysis would legitimize function attribution in the natural realm: an appeal to natural selection is obviously compatible with naturalism. Consequently, one now finds philosophers confidently applying the newly rehabilitated concept of biological function in a large number of different contexts: they posit functional categories,[3] make assertions about health-related norms, use functions to ground attributions of mental content, and so on. Indeed, one might even say that recent philosophical discussions of function have rehabilitated an old way of thinking about norms in nature that had long been abandoned by those who suppose themselves to be right thinking heirs to the Scientific Revolution. Aristotle and Aquinas thought that certain rights and wrongs were a part of the way in which the world is constructed; Aquinas (and arguably Aristotle) thought that these rights and wrongs arose from the purposes of the Creator. The more recent friends of natural norms suggest that we can have norms and functions without a Creator,[4] and even that we have, in many areas, close to the very same norms as used to be attributed to the Creator's rational agency.

As much as the new analysis of function might have quietened the controversy about the place of functions in naturalistic science, the tendency to use the new selection-based conception of function to underwrite old Creationist norms is an unresolved problem. For it does not seem legitimate to think that

[1] The term 'selection for' is due to Elliott Sober (1993: 83).

[2] There are other analyses of function in play in the modern discussion, and though the consensus appears to be that they should coexist with selection-based functions, some still think that what are known as 'causal role functions', first introduced by Robert Cummins (1975, 1983), should displace the selection-for analysis.

To mark my own place in the contest of details concerning selection-based functions, let me say that, while I accept a close relationship between selection-for and function, I do not accept that it is either a necessary or a sufficient condition thereof. I maintain moreover that different analyses are needed in the realm of artefacts, and in possible worlds where biological or other entities are subject to Lamarckian evolution. Even within the actual biological realm, the phenomenon known as exaptation, or 'pre-adaptation', causes difficulties for the standard view (Preston 1998), for here the presence of an item is explained, not by what it now does, but by what it once did: the item was selected for *F*, and then adapted to *G*—at this point it may no longer be any good for *F*. I favour a 'pluralist' account of functions—I hold that each of these different kinds of function is supported by a different set of underlying causal structures—though I hold that an 'analogy' with human products supplies a unifying thread in the *meaning* of the term 'function'. For details, see Matthen (1997).

[3] The defence of functional categories that allow for the inclusion of anomalous instances is perhaps the leading motivation of defenders of the new teleology. The argument of Sections 2–5, taken in conjunction with Matthen (2000), is intended to serve as the foundation for a sceptical enquiry into the use of such categories. Karen Neander, however, defends them in her contribution to this volume.

[4] Mark Bedau (1991: 654) actually suggests that we should adopt 'a broader view of nature, perhaps roughly Aristotelian in outlook, [that] could reckon objective standards of value as part of the natural order'.

evolution could create the same sorts of structures as an omniscient Creator.[5] Nowhere is this problem more bothersome than in the consideration of human cognitive rationality, the mental capacity we use to arrive at new beliefs and abandon old ones, and to make and revoke action plans and decisions. How are we to characterize the function of cognitive rationality, the advantage it gives us in the struggle for survival? In other words: How did it evolve? What was it selected for? One familiar way of answering these questions goes, in outline, and without appropriate cautions and qualifications, like this.

> *R* Start with a simple observation: access to the truth is advantageous to us in the struggle for existence. From this observation, it seems to follow that (as Dan Dennett (1987: 75) puts it) 'natural selection guarantees that *most* of an organism's beliefs will be true, *most* of its strategies rational'.[6] Now one might ask: what sort of belief creating mechanisms would we need in order to maximize our access to the truth? The answer seems obvious: we would need reliable perceptual systems, and mechanisms that would create further reliable beliefs starting from those delivered by perceptual systems. These mechanisms of belief creation would conform to procedures recommended by deductive and inductive logic and decision theory: for these are designed to describe truth preservation and rational action strategies. Thus, we would expect that natural selection would implant in us cognitive procedures that correspond to the best available logical systems. This is why we are rational; conversely, this is why we should assume that our innate reasoning procedures do in fact correspond to logic.[7]

R attributes to human rationality exactly the same divinely bestowed function that Descartes attributed to it in *Meditation IV*—the acquisition of true beliefs.[8] It attempts to use an *a priori* normative science—logic, as we are calling it—to shed light on an issue concerning functions in *descriptive* psychology. Logic provides a hypothesis about function, and *R* is an attempt to adopt the *Design Stance* with respect to this function: the idea is that, if you

[5] That evolution does not mimic rational creation is the main point of Elster (1984: ch. 1). It is also central to Elliott Sober's argument (1993) that well-adapted organisms are empirical support for evolution, not Creation.

[6] Quoted by Christopher Stephens (2001), which I made use of in writing this.

[7] Compare *R* with the 'argument from natural selection' in Stich (1985). It may well be that, as Patrick Rysiew reminds me, there is a difference between designing a reliable system and designing a rational one, in the sense that a rational system might eschew reliability in certain circumstances in order to preserve other values. I shall disregard this point for present purposes: it would complicate matters to build other norms into *R*, but the basic point that I am making would remain the same.

[8] Of course, Descartes disavows teleological attributions, since he claims that one cannot divine God's purposes. Nevertheless, he thinks it impossible that God should deceive us, and claims to derive the characteristics of human cognition from this.

want to know how human cognitive abilities got to be the way they are, you should try to figure out how to design a system that calculates in accordance with logic.

The function attribution and Design Stance application embodied in *R* fails, I shall argue. The problem, initially, is an empirical one—that, as is becoming increasingly clear from the work of social psychologists,[9] human cognitive processes are not, in fact, good approximations to the best systems of deductive and inductive logic and decision theory. As these social psychologists show, humans as a species depart conspicuously from the practices prescribed by the best systems of logic: we tend *systematically* to affirm the consequent, commit the Monte Carlo fallacy, ignore statistical base rates, and violate the principles of Bayesian reasoning concerning expected utilities.[10] This leads to a question. Why, in view of *R*, do we fall so woefully short of perfection? Why is natural selection unable to deliver the goods? If *R* leads to a false conclusion, where does it go wrong?

The standard answer to these questions goes something like this: the cognitive failures of humans result from evolutionary satisficing—given the real-world costs of developing and running an adequate reasoning machine, we have to make do with something short of perfect reason. Thus it has been argued that it is too expensive to develop a perfect reasoning machine given the declining marginal utility of improvements beyond a certain point. This is supposed to explain why evolution did not achieve improvements in reasoning beyond a certain point. Again, some claim that, when we have to respond to environmental challenges in real time, it takes too long to reason perfectly, and so we are forced to use fallible ('quick and dirty') short cuts. These short cuts fall short of the standards of logic, but they are good ways of dealing with the limitations of our intellectual resources. Human reasoning *tries* to instantiate logic, but, because of the regrettable necessity of making do in the real world, it falls somewhat short. In this it is something like human virtue as Aristotle describes it—a second-best life imposed on us by the exigencies of the human condition. God does it the way it should be done; we just do the best we can.

This way of limiting the adaptive value of logic accepts the basic methodology of *R*—that is, the formulation of design goal in terms of the ideals of logic—but argues for the necessity of variances from the design goals posited in that account. In this chapter, I take a different line: I contest the conception of evolutionary end points embodied in *R* itself. (Note: I do not contest the Design Stance *per se*, rather a particular method of formulating design goals.) In Sections 2–5, I argue that *R* is flawed because it is committed to a little-noticed

[9] Nisbett and Ross (1980); Kahneman *et al.* (1982); Gigerenzer *et al.* (1999).

[10] Since these departures are systematic, they cannot be explained in terms of mere performance, as opposed to competence, errors in the manner of Cohen (1981).

form of *adaptationism*. The kind of adaptationism most discussed in the literature consists in exaggerating or overestimating the power of adaptation: this is not my target here. The form of adaptationism that I shall be concerned with consists in the employment of a *universalistic* conception of utility, a conception in which certain characteristics are held to be useful *per se*, to enhance the evolutionary fitness of *any* organism in which they may occur, regardless of its environmental situation or evolutionary history. I shall argue in Section 5 that the avoidance of this universalism entails adopting a general constraint on reasoning about function, which I shall formulate and discuss there. The formulation of this *Unique Origin Constraint*, is, as I see it, the main contribution of this chapter.

I take up rationality again in the final section. The problem with *R* is that its reliance on *logic*—an abstract science that is supposed to be universally valid—violates the Unique Origin Constraint. *R* implies that human cognitive rationality was selected for because it conforms to the laws of logic; since the laws of logic are universally valid, one should conclude that *any* cognitive system will evolve toward the same rules and procedures. Thus *R* should be rejected.

This creates a difficult problem. The argument just outlined gives us reason to doubt absolute or universal conceptions of reason. At the same time, we cannot just ditch reason and logic as absolute values. In the first place, logic really does tell us why reasoning is universally a good thing. Moreover, the above-mentioned work of social psychologists notwithstanding, humans are really pretty good at reasoning in accordance with logic. Thus, the standard kind of pessimism about human rationality (as found, for example, in postmodernist views of reason as parochially imposed by power elites) is not a viable option. The problem, then, is not just to explain why humans fall short of perfection in reasoning—but to do so in a way that explains why they are as good at it as they are. This demands that we reconstruct the very notion of rationality found in *R*.[11] This is obviously too large a task even to begin in this chapter, which after all is primarily concerned with the methodology of function attributions. I shall, however, attempt to establish some ground rules based on the Unique Origin Constraint.

2. Adaptationism: The Need for Positive Alternatives

In a famous and much-cited article, Stephen Jay Gould and Richard C. Lewontin (1978: 581) complain that 'an adaptationist programme has dominated evolutionary thought in England and the United States during the past

[11] Christopher Cherniak (1986) is sensitive to this problem, though he operates within the cost–benefit satisficing tradition sketched above.

forty years'. They allege that the practitioners of this programme adopt what amounts to a teleological view of evolution, and analyse traits solely in terms of the good that each supposedly does. Adaptationists ignore the non-teleological, non-advantageous aspects of biological traits (except with regard to cost–benefit trade-offs) and ignore causes of evolution other than adaptation measured by cost–benefit analyses. Gould and Lewontin think that, much like the creationism that it is supposed to replace, adaptationist thinking is infected by subjective conceptions of value held by individual scientists, and that it is, as such, a fundamentally wrong-headed approach to the theory of evolution.

Despite the fame of the Gould–Lewontin attack on adaptationism, there is no general agreement about its content: there is still considerable controversy in the literature about how exactly we should characterize the sin of which adaptationists stand accused. In Sections 3–5, I shall attempt to define one kind of adaptationism particularly relevant to the assessment of *R*. In the present section, I attempt to set the stage by making a few preliminary points and clearing away some misconceptions.

2.1. *Two Kinds of Anti-Adaptationism*

One aspect of Gould and Lewontin's critique is easy to grasp and has been quite influential. Adaptationists, they allege, overlook causes of evolution other than adaptation. This charge has occasioned a new methodological self-consciousness concerning the ways in which adaptation is used in evolutionary theorizing. Whether or not earlier models really ignored the importance of such evolutionary factors as pleiotropy, drift and random fixation, and developmental and 'architectural' constraints,[12] it is clear that these factors are more readily and more explicitly taken into account than they once were, certainly by philosophers, perhaps even by biologists (though most would plead innocent of having committed any offence in the first place). In recognition of this, the salutary influence of Gould and Lewontin's paper is now widely acknowledged. One self-confessed adaptationist, John Maynard Smith, allows that 'the effect of the Gould and Lewontin paper has been considerable, and on the whole welcome' (quoted in Dennett 1995: 278). (He goes on, however, to say: 'I doubt if many people have stopped trying to tell adaptive stories. Certainly I have not done so myself.' I shall be arguing that, even in the light of the–Gould Lewontin critique, there is no need to stop telling adaptive stories: properly

[12] Gould and Lewontin sometimes suggest, as many others do, that many of these factors—particularly those connected with Mendelian particulate genetics and the 'neutral theory'—are 'non-Darwinian'. What is true is that Darwin did not know of some of these factors. If, however, the claim is that they cannot be fitted into the Darwinian framework of natural selection, I disagree (Matthen and Ariew, 2002).

conceived, their point has to do with the kinds of adaptive stories one tells.) I shall call this *corrective* aspect of the Gould–Lewontin critique *critical anti-adaptationism.*[13] Peter Godfrey-Smith (1996: 22) characterizes adaptationism like this: 'When an evolutionary explanation is given which simply mentions certain alleged benefits associated with a trait, rather than a detailed array of evolutionary forces, constraints and initial conditions, the explanation is often referred to as "adaptationist".' *Critical* anti-adaptationism is what he has in mind: it tells us simply that we should avoid such an unhealthy concentration on 'benefits', especially those merely 'alleged'.

Many philosophers and biologists identify adaptationism with this over-dependence on the 'alleged benefits' of a trait in evolutionary accounts. And they associate Gould and Lewontin simply with the stern injunction that one should cast one's net more widely. This is a mistake. However influential the Gould–Lewontin critique might have been with respect to the practice of evolutionary biology, critical anti-adaptationism does not constitute an original contribution to knowledge: it is simply a call to pay attention to causes of evolution that were antecedently known to exist. Nor does it exhaust what Gould and Lewontin wanted to say: for, both together and separately, they suggest that adaptationism is not just a mistake made by overzealous evolutionary theorists writing about particular cases, but rather a tendency of thought implied by the neo-Darwinian synthesis. To this tendency, they claim to propose an alternative. In other words, their anti-adaptationism is presented, not merely as a critique, but as a positive understanding of evolution that is somehow antithetical to the classic model and the gradualism with which it is associated. Let us call this positive aspect of their thesis *constructive anti-adaptationism.* If Gould and Lewontin have made a novel contribution to knowledge, it is by identifying elements of the neo-Darwinian synthesis that must be purged from any genuinely non-adaptationist perspective. The question is: what is their critique? And what is the positive content of a non-adaptationist approach to natural selection?

2.2. *Cartoon Adaptationism*

Many philosophers of biology attempt to reconstruct adaptationism by trying to figure out, by a kind of 'inference to the best explanation', what sort of belief would justify the practices that *critical* anti-adaptationism excoriates. They think of adaptationism simply as *the theory that underlies adaptationist practice.*

[13] I use the term 'anti-adaptationism' to refer to a critique of adaptationism. A '*non*-adaptationist' perspective is one that does not fall afoul of such a critique. What I shall be calling 'constructive anti-adaptationism' falls in between. It is (by my own stipulation) a positive thesis held by Gould and Lewontin, but it is buried in their attack on adaptationism, and hardly ever, if at all, stated as a positive view.

What does adaptationist practice look like? Here is an especially silly example, in the words of a recent author paraphrasing Galen.

> Women . . . don't need facial hair as men do . . . beards would lend women an air of augustness inappropriate to their natural condition: 'since I have shown many times, indeed throughout the work, that Nature makes for the body a form appropriate to the character of the soul'. None the less, it is clearly a good thing if women are ornamental: and so nature generously provides them with long hair for this purpose. (Hankinson 1988: 138–9)

Translated into a selectionist idiom, this is an extreme example of the sort of fanciful tale that Gould and Lewontin would have us eschew. The idea one often finds in philosophical treatments of the concept is that adaptationism is the theory that underlies this kind of function attribution.

Now, as long as the issue is just the *habit* of adaptationism, that is, the habit of telling stories such as the one Galen offers us above (translated into a selectionist idiom), one might think that there is no compelling reason to diagnose it in terms of any underlying belief or theory. One tells children who are making an intolerable noise: 'You think you are the only one in the world!' Perhaps such a belief *would* explain their inconsiderate actions, but it seems unlikely that they really subscribe to it, or any other discrete and well-articulated self-justification for self-centred behaviour. So also with the adaptationists. Some have attributed to them a 'fundamentalist' thesis (the term is from Gould 1997) that the process of adaptation *alone* accounts for evolution. But no serious student of evolution actually believes in any such fundamentalist thesis. However adaptationist their practice might be, you can bet that nobody will ever pin on John Maynard Smith or E. O. Wilson a myopic ignorance of genetic drift, pleiotropy, architectural constraints, and so on. Dennett (1995: 276) protests against this, and he is completely justified: 'The thesis that every property of every feature of everything in the living world is an adaptation is not a thesis anybody has ever taken seriously, or implied by what anybody has taken seriously, so far as I know. If I am wrong, there are some serious loonies out there . . .'.

Now, why do Gould and Lewontin (and especially Gould—see Gould 1997) suggest that anybody is a fundamentalist 'loony' in the sense mocked by Dennett? To put this into context, let us first acknowledge that today there is nobody at all who thinks that the biological function of beards is to lend men an air of augustness. On the other hand, *everybody* agrees that the wings of both birds and insects are extremely well adapted for flight, and that the aerofoil structure of these wings evolved for precisely this reason. Adaptationists and anti-adaptationists are not going to disagree about these extremes. Presumably, however, there are controversial cases, cases in the middle. *Social behaviour* is an example (the kind of behaviour in which an individual appears to give up its own advantage in order to benefit a group of which it is a part); *sexual aggression* is another. There is a tradition in evolutionary biology that tends to analyse the latter phenomena in terms of adaptive behavioural strategies—that is, behavioural strategies that result in a reproductive advantage for those who practise them. Gould and Lewontin are hostile to such analyses: these analyses are, in fact, at the heart of their hostility to adaptationism. The question is: what, precisely, is the point of disagreement in these contested cases?

There is a rhetorical tendency in Gould and Lewontin to pin on those who disagree with them about, for example, the adaptive value of social behaviour, a line of thought that mimics Galen. That is, they often talk and write as if the *only* reason anybody would think that cooperation or sexual aggression evolved as adaptive strategies is that, like Galen, they think that *every* trait is adaptive, and (like Galen again) they are willing to essay absurd 'just-so stories' about how this is so. Distancing himself from such methodological looseness, Gould professes to be a 'pluralist' because he weaves non-adaptive evolution into his own narratives. This, he suggests, is more responsible than simply telling unverifiable adaptive tales (though, as Orzack and Sober (1994) observe, a *non*-adaptive just-so story is just as easy to invent as an adaptive one). Adaptationists are, by contrast, narrow-minded fundamentalists who rejoice in one thing only, adaptation. So Gould says.

This strategy of creating cartoon adaptationists has not served the subject well. The problem is that, as I suggested above, nobody owns up to, and indeed nobody (since Galen) is guilty of, being an adaptationist in the sense of simply ignoring the plurality of causes of evolution. No ground for serious scientific disputation can be defined in terms of the alleged confrontation of fundamentalism and pluralism. I propose, therefore, to decline Gould's rhetorical gambit. Forget about the neglect of drift and architectural constraints. Forget about the presuppositions of Galenic teleology. These are mistakes of method, no doubt, but the fact that somebody makes such a mistake does not imply that they have a false underlying belief. What is the *positive* thesis? Do Gould and Lewontin present a genuine alternative to the conventional wisdom (*anybody's* conventional wisdom) concerning the nature of evolution? I believe they do. In the next three sections I want to explore *one* strand of their constructive critique of adaptationism.

3. Constructive Anti-Adaptationism: Two Key Ideas

I now introduce two ideas that will (I hope) help us gain a better understanding of constructive anti-adaptationism, and lead to an independent positive conception of functional attribution that does fall foul of the Gould–Lewontin critique.

3.1. *Maxima, Local and Non-Local*

Richard Lewontin (1985: 67) talks about adaptation in the following way:

> The concept of adaptation implies that there is a preexistent form, problem, or ideal to which organisms are fitted by a dynamical process. The process is adaptation and the end result is the state of being adapted. Thus a key may be adapted to fit a lock by cutting and filing it, or a part made for one model of a machine may be used in a different model by using an adaptor to alter its shape. There cannot be adaptation without the ideal model according to which the adaptation is taking place. Thus the very notion of adaptation inevitably carried over into modern biology the theological view of a preformed physical world to which organisms were fitted.

There are two ideas that Lewontin discerns here as characteristic of the concept of adaptation. The first is that adaptation presupposes a 'preexistent form, problem, or ideal' to which organisms are fitted by a dynamic process. The second is that the environment is a fixed and given matrix to which organisms are fitted by the process of adaptation much as a key is fitted to a lock by filing; this process of fitting the organism to the environment proceeds according to criteria generated by the 'preexistent problem'. The problem is for the key to open the lock: the environment files away at the key blank until the lock opens. Ultimately (in Section 5 below), I want to argue that there is a kind of evolutionary process that does chip away bit by bit at 'pre-existent' problems, and another kind that does not. But first, let us try and figure out what Lewontin is getting at in this important passage.

The idea can be made more concrete by means of the following (unrealistic) example.

> *F* A bird's capacity for flight solves a problem that existed before the bird did, namely the problem of aerial transport. It is the result of a long series of evolutionary changes. Over many generations, imperfectly flying birds got better and better because there were always mutant varieties that did the job a little better. In the end, we get something like an *eagle*, which represents the closest existing approximation of perfection with respect to flight, the best solution yet to the problem of flight. All other birds are poised on an upward fitness slope directed towards that optimum solution.

Of course, nobody believes anything like this. Notice, however, the structural similarity with *R*, the evolutionary story about rationality. In both stories an 'ideal' is defined by an independent analysis of the trait, and posited to be the end point of evolution. (In the case of *F*, unlike *R*, the 'independent analysis' is pretty laughable, but let us just go with it.) Though no biologist in her right mind would ever have advanced *F* as an account of flight, a consideration of this account will help us appreciate some of the weaknesses of *R*.

Let us call the kind of 'ideal' posited in *F* (and *R*) a *global fitness maximum.* Global maxima are associated with patterns of explanation that are essentially non-contextual and non-historical in conception, in the following sense. When you attempt to explain why an organism has a particular characteristic *simply* by the observation that the characteristic in question is adaptively perfect *tout court* (and omitting, as Godfrey-Smith says, a 'detailed array of evolutionary forces, constraints and initial conditions'), you are committed to the idea that evolutionary pathways—historical environments and processes of selection—do not matter as far as this particular trait is concerned. Wherever the path to this global maximum may have started out, and whatever terrain it may have traversed, it was going to end up with this characteristic, for no other reason than that this characteristic is the best. This is illustrated by *F* (as well as by *R*). The claim is that the eagle has a certain mode of flight because this mode of flight is the highest point on the entire fitness landscape, the best possible solution to the problem of flight. It makes no difference what assortment of avian flight styles there might have been in the beginning. One way or another, the environment would have chipped away at these original flight styles and achieved eagle flight. Like an enormous star, a global maximum exerts its gravitational influence across the entire adaptive landscape and brings all into conformity with it sooner or later. A global maximum is a lot like Teilhard de Chardin's much reviled *omega point.*

Global maxima contrast sharply with *local* maxima, which are equilibrium points in natural selection, not because they are the best *tout court*, but because they are better than the points surrounding them in the adaptive landscape. The organism got to this particular point because it is fitter here than it was at nearby points. It cannot evolve away from this point because to do so would be to descend the surrounding slope, thereby incurring a loss of fitness. An analysis in terms of local maxima presupposes comparisons with closely similar genotypes. Thus the present constitution of a population will be viewed as a product of where it was just before it got here (that is, on this local slope), and of its being in equilibrium now. Had it started somewhere else, it would have ascended some other adaptive slope, ended up somewhere else, and been in equilibrium there.[14] When you say that an organism has a certain characteristic because that characteristic is a local maximum, you imply that history and context matter.[15]

[14] Local and global fitness maxima are invoked by Jon Elster (1984: ch. 1) in an argument that evolution is not 'rational' because it cannot achieve global maxima. (Note, Elster's argument is not about the evolution of rationality, but about the rationality—or rather the *ir*rationality—of evolution.)

[15] Kim Sterelny has forcefully made the point to me that Lewontin would object to the very idea that adaptive landscapes can be constructed independently of considering the kinds of organisms and lineages that inhabit them. Such a construction, an evaluation of all possible phenotypes independently of the actual array of organisms that inhabit the earth, presupposes that there are 'pre-existent ideals' and values, though it acknowledges that these values may not

Earlier, I said that the notion of a global maximum was extremely implausible. Having defined local maxima, we can now present adaptationism in terms of a less extreme notion. A *conditional fitness maximum* is the highest point in a fitness landscape, *given certain assumptions* and *certain considerations of costs and benefits*. On closer examination, the story *F* does not tell us why the eagle must *fly*: it simply tells us (or purports to tell us) why it flies like this, *if* it is going to fly at all. But why should it fly rather than swim? Another kind of story is required if we are going to understand this. Further, *F* does not take costs and benefits into account. Once you introduce these considerations, you may find an explanation for why a sparrow, say, does not fly like an eagle. A cost–benefit analysis might show that it is wasteful for a sparrow to soar like an eagle. It may be that sparrow flight is the best kind of flight sparrows can afford. Or it may be that, given the sparrow lifestyle, there is no need to indulge in this improvement. This may be why sparrows do not get more like eagles: the benefits of doing so are not worth the cost. These considerations parallel those mentioned in connection with *R* in Section 1: the appeal to costs, benefits, and strategies to cope with error proneness to explain observed human departures from perfect rationality. Perfect rationality is not possible, it was claimed, because it costs too much to achieve the marginal utility of making improvements beyond a certain point. In other words: human cognition, imperfect as it is, constitutes a conditional, rather than a global, fitness maximum.

Like global maxima, conditional maxima are non-historical in their mode of explanation. Conditional maxima are derived from global maxima, by fixing certain lifestyle choices and making adjustments for certain additional constraints. The claim about sparrows is not that they are the way they are because of their evolutionary history. Rather the claim is that, when we analyse sparrow flight not just in terms of perfection, but add in considerations of costs and benefits associated with flying, we see why it is optimal. The sparrow is not solving the problem of flight *regardless* of cost; it is solving the problem of flight *given* costs and benefits. (*Mutatis mutandis*, this is the cost–benefit approach to explaining why human rationality departs from the model predicted by logic.) Similarly, the eagle is not solving the problem of transportation as such; it is solving the problem of *flight*. Let us subsume both global and conditional maxima under the idea of *non-local* maxima, maxima specified independently of genotype space, environmental limitations, and historical

always be combined additively. (Thus adding perfect eagle wings to a bipedal human may not improve him, the Icarus fantasy notwithstanding.) A proper understanding of adaptation demands that we take *competition* into account, not just intrinsic capacity, and this is what 'adaptive landscapes' fail to do. Acknowledging this important point, I shall persist with the fiction of an adaptive landscape: it is a convenient way of showing that, even if one can initially think about fitness and adaptation independently of the organisms that have (contingently) evolved, one will still end up with a lineage-relative conception of adaptation (n. 17 marks the place where this occurs).

starting points. I propose to take from Lewontin's remarks about pre-existent ideals the following characterization: the belief that evolution is directed towards non-local fitness maxima is one source of adaptationism.

3.2. Universalism versus Relativism

We can now introduce two models of evolutionary change, using the contrasting kinds of fitness maxima just discussed.

A *universalist* model supposes that a population is under constant selective pressure in the direction of improvement with respect to non-local maxima. (This implies, implausibly, that every population of living organisms, from bacteria to humans, is constantly under pressure to 'improve', and leaves it as something of a mystery why bacteria still exist—Gould (1996) makes a point of this. In Section 5, I will make an important distinction that accommodates improvement while demonstrating how it is that bacteria continue in existence.) According to universalism, the evolutionary history and prospect of organisms can be regarded as a gradual progress toward non-local maxima. Jon Elster (1984) argues that evolution *cannot* attain such non-local maxima because a population that occupies a local maximum would have to take a temporary loss of fitness in order to reach a distant non-local maximum. In fact gradualists have devices to circumvent such difficulties. Some argue that in the multidimensional space of an adaptive landscape, a winding path can often be found that sticks to a ridge, avoiding fitness valleys. Others argue that, as environmental conditions change, so also does the adaptive landscape, with the result that populations can be ferried across valleys by temporary upsurges in the landscape. However one assesses the validity of such arguments, the point to make here is that universalism posits that *in the end* a population will achieve a 'pre-existent ideal', a non-local maximum.

A *relativist* model supposes that most evolutionary change takes place, not under the attractive force of some non-local maximum, but in response to some immediate and transitory environmental challenge.[16] A relativist holds that most populations are at, or oscillate around, a local fitness maximum and that adaptive evolution occurs only when such an equilibrium is disturbed. This implies that non-local maxima such as those alluded to in *R* and *F* do not exert a constant attractive influence—if they did, populations could not be in equilibrium at a local maximum. The disturbances responsible for evolutionary change are contingent historical events that create a *new* local maximum close by the one hitherto occupied by the population. Such new maxima must be understood by a comparison with erstwhile equilibrium points—that is, in local, or *relative*, terms.

[16] It is important to note that the relativism that is involved here is not a relativism of an ontological or epistemological variety.

To illustrate the difference: in the relativist conception, one canonical form of the evolutionary disequilibrium is *niche differentiation.* Here, two populations come into competition for the same resources because their lifestyles too closely resemble one another's. Because of this similarity, each is obliged to find a way of using environmental resources in a way that is different from the other, thus relieving the competitive pressure on both. The evolutionary change that results is *small* and relatively *fast.* Let us suppose that a population of birds is competing for flying insects. By virtue of a small morphological modification and a correspondingly small change in its predatory habits, a subpopulation might become especially good at catching some subclass of flying insect. Then, it has improved access to this subclass, and this may help not only the new subpopulation, but the ancestral one also, to survive. This form of disequilibrium and response depends for its content on the pre-existing character of the two populations, and their particular environmental circumstances. It cannot be understood in terms of 'pre-existing ideals' or constant pressure on each of two populations to realize a non-local maximum. Catching the subclass is not good in itself, nor was this the problem the population was trying all along to solve. The advantage of catching this subclass emerged only because of the utility of specialization in the face of certain competitive pressures. There is no one thing that each of these populations got better and better at (except for some uninformatively general thing, like surviving). They specialized, and they created a value by so doing.[17] (And note that the ancestral subclass also creates a new value, simply by staying the same.)

Niche differentiation is an example of a process driven by historical contingency. The changes that result can be small and just as contingent as the circumstances that caused them. As Lewontin (1985: 79) insists, evolutionary changes are 'very small changes in a character [that] result in very small changes in the ecological relations of an organism'. Large evolutionary developments are seen as an accidental accumulation of small local changes brought about by constant and recurring ecological stress. Each such change depends cumulatively on the contingencies that preceded, because the starting point of each change is determined by previous change. But there was no universal value aimed at all along: the accumulation of small changes should not be regarded as small steps aimed at such a value. The relativist does not see natural selection as a gradual progress towards a non-historically specifiable value. As Lewontin (1985: 79) puts it:

> It seems pure mysticism to suppose that swimming was a major 'problem' held out before the eyes of the terrestrial ancestors of all these animals before they actually had

[17] Thus, as promised in n. 15, I end up with the idea that you cannot construct an adaptive landscape without reference to the lineages that occupy it.

to cope with locomotion through a liquid medium. It must be that the problem of swimming was posed in a rudimentary and marginal form, putting only marginal demands on an organism, whose minor adaptive response resulted in a yet deeper commitment of the evolving species to the water.

The advantage of the relativist's position is that a series of small changes, each occasioned by a local pressure arising as much from the character of the organism and of the competitors it happens to have encountered as from that of the environment, is inherently more probable than a gradual big change that takes place under a single influence. (This principle is subject to an important qualification, which is articulated in Section 5.)

As remarked before, it is a consequence of relativism that a population is in stasis most of the time. Non-local maxima are not exerting an influence constantly.[18] Note also that the relativist model is essentially historical. It is not enough, in explaining a trait, to show that it is advantageous *tout court*. It is necessary to show why it was advantageous given historical circumstances.

4. Relativism in Practice

4.1. *The Virtues of Relativism*

Now let us revisit *F*, the story about the evolution of flight, in terms of the dispute between universalists and relativists. Clearly *F* is a universalist story: it views the evolution of flight as a gradual ascent to a non-local maximum, to which the closest approximation so far is the flight of the eagle. A relativist tells the story of flight differently. A relativist might start combating the gradualist by contrasting the eagle with the swallow. The flight of the swallow is very different from that of an eagle, it does not soar in straight lines or majestic loops, or achieve high altitudes, but darts quickly up and down and side to side and stays close to the ground. The relativist is apt to start from the observation that the swallow's flight is just as perfect as the eagle's, but perfect with respect to a *different* ideal of flight. It does not make sense to ask why the swallow is not more like the eagle: if it were, both would be under pressure to differentiate

[18] Notice that this model is friendly to the notion of a punctuated equilibrium, the idea that a population is in stasis most of the time, interrupted by short periods of rapid change. (Punctuated equilibrium theory is often thought to be the result of the sudden emergence of new genotypes or structures, and this is quite a different matter from what I am talking about here.) Dennett (1995) pokes fun at the idea, suggesting that it is merely a matter of scale. If change is represented on a graph in which hundreds of millions of years are represented by a single centimetre, change will look jerky; if the temporal axis has a smaller scale, change will look gradual and continuous. Punctuated equilibrium occurs, Dennett suggests, when the tempo of evolution varies and is represented on a small temporal scale. I disagree with this: the relativist model I have just presented *posits* that populations are in equilibrium most of the time. It does not merely rely on the notion of punctuations in the pace of change that go undetected over long periods of time.

their niches, and thus to become more unlike each other. One might not have predicted flight like that of the swallow until one had actually seen it. But obviously this does not mean that the swallow is less perfect. The swallow is as perfect, *relative to an insect-catching way of life*, as an eagle is relative to its search for larger prey. Further, the insect-catching way of life is not a 'pre-existent ideal'. It is rather a way of life that was *created* by historical circumstance—by niche differentiation, for example. By catching insects in this particular way, the swallow may have been able to exploit a new food resource, thus enabling it to coexist with an erstwhile competitor to which it now bequeaths its former resource. The value of the swallow's lifestyle is historically determined.[19]

Now, what about the modes of flight instantiated in the evolutionary predecessors of eagles and swallows? The universalist protests that *they*, at least, must be less perfect with respect to eagle flight or swallow flight than eagles or swallows. Suppose that the golden eagle is an older form than the smaller, sleeker bald eagle: does this not imply that the latter is *better*, with respect to the eagle way of life, an improvement over the ancestral form? Again, the relativist disagrees. The relativist says that the characteristics that differentiate the bald eagle from the golden are the end points of a series of contingent evolutionary changes. These changes might have occurred simply because they gave the bald eagle a distinct niche. For instance, the bald-eagle form of life may have developed because there was too much pressure on the small land-mammal population, and there was thus some advantage in taking up a fish-catching form of life. She is apt to emphasize that every stage in such niche-vacating, new niche-making moves does not have value because of its intrinsic character, but gets value from the competitive situation. Of course, bald eagles do a certain thing very well, and golden eagles do not do that particular thing as well. However, the relativist thinks it a mistake to suppose that this should be understood in terms of the *absolute* fitness values of these end points.[20] What might well be true is that, when the bald eagle first became distinct from the golden eagle, it may not initially have been as good at doing what it now does. But its adaptation to its present role must have been pretty fast. After all, we do not observe directional evolutionary changes occurring in the wild.[21] It is a condition of new lifestyles that a subpopulation can adapt in a few generations.

Relativism is the centrepiece of the non-adaptationist perspective I want to develop in this chapter. Note that it does not have to deny that traits are *in some*

[19] Many readers have pointed out to me that nobody is a universalist in the sense attacked in this paragraph. I acknowledge this (of course). So I have been asked: why spend so much time on a straw man? The answer is: because there are (as I shall show) actual people who are universalists concerning things like social cooperation and rationality. The straw man of the text is a simple way of understanding one feature of their approaches to these other traits.

[20] After all, as Beth Preston pointed out to me, many flightless birds had flight in their histories: what does this say about the value of flight?

[21] We do, however, observe oscillations around equilibrium points (see Grant 1986).

sense optimal. Relativism allows us to suppose that the bald eagle's flight is of value *to the bald eagle.* It even allows that this relativistically understood value might be the whole explanation of why it evolved: there is no insistence on drift or heterozygote superiority having contributed. What relativism forbids is the idea that the flight of the bald eagle has universal value, that this form of flight would have adorned the golden eagle as well, or that it had been exerting an influence on the evolutionary history of the bird all along. When the bald eagle's form of flight is cited by itself, and without specifications of starting points and specific ecological challenges, it is implied that, to the extent that its flight fails to resemble this eagle, the golden eagle, or the swallow, *lacks fitness.* This is what the relativist objects to.

4.2. Adaptationism and 'Reverse Engineering'

Dan Dennett (1995) characterizes adaptationist thinking in terms of 'reverse engineering', which is the practice of analysing artefacts as if their features make a contribution to some design desideratum. However, if I am right, there is no reason why adaptationists and their opponents need to disagree about reverse engineering. The non-adaptationist perspective detects nothing wrong with the idea that it is part of a biologist's job to reverse engineer the flight of a swallow—that is, to discover what it is good for. (Famously, Gould (1985: 23–9) reverse engineers the 'flamingo's smile'.) Naturalized teleology—in the form of 'selected effect', 'etiological', or 'normal' functions—has a place in the non-adaptationist's view of the world: as I interpret it, she simply insists that you derive your conception of design from the swallow's actual practice and the history of this practice. You must resist the temptation to analyse it in terms of pre-existent ideals or non-local maxima. Given an adequate *local* conception of actual practice, you figure out what purpose or end the swallow's flight mechanisms are modified for and adapted to. It is reverse engineering in terms of non-local maxima that the relativist resists.

In fact, localist constraints on reverse engineering correspond very closely to Dennett's own best practice. In his remarks on the spandrels of the basilica of St Mark in Venice (Dennett 1995: 274)—the example Gould and Lewontin used to demonstrate the importance of structural constraints, as opposed to adaptive ones—Dennett rehearses the following *purely local* design considerations in rapid succession: (*a*) St Mark's is 'not a granary', but a church (cf. lifestyle choice), (*b*) 'the primary function of its domes and vaults was never to keep out the rain . . . but to provide a showcase for symbols of the creed' (cf. exaptation), and (*c*) powerful Venetians, with their Eastern vision, wanted to create a local example of Byzantine mosaic iconography (cf. competitive situation). Thus, Dennett says, the dome of St Mark's was designed to solve the 'environmental problem' of how best to 'display Byzantine mosaic images of

Christian iconography'. The domes and vaults of St Mark's are presumably inherited from pre-existent constraints of ecclesiastical architecture. The problem, then, is to adapt these domes and vaults to the local functional considerations just described.

Dennett points out, cleverly, that, though, as Gould and Lewontin insist, spandrels (or 'pendentives', as Dennett calls them) are architecturally required by domes that are supported by arches, this requirement does not fully explain the form of *these* pendentives, for *they* were modified to adapt them to their local function. 'Care has been taken to round off the transition between the pendentive proper and the arches it connects, the better to provide a continuous surface for the application of mosaics.' Dennett thus offers us a perfect example of *relativist* reverse engineering: the environmental problem, showing off Byzantine mosaics, is local—it did not recur in connection with other arch-supported domes in other places and times, and we do not find there pendentives with the peculiarly St Mark's 'rounding off'. Once you have discovered ('reverse engineered') the particular design goals involved in St Mark's, you see how the pendentives are adapted to those goals.

My analysis of Dennett is an attempt to show that the opposition between universalism and relativism does not, or at least need not, hinge on a disagreement about reverse engineering, or about the potency of adaptive peaks as explanatory factors in evolutionary biology. Rather, the controversy is about how such adaptive peaks are to be analysed and identified. The relativist does not object to the Design Stance, functional analysis, or reverse engineering as such. What she opposes is the use of non-local conceptions of adaptive value. For all of Gould and Lewontin's attachment to non-Darwinian factors, they are less relativist in this particular instance than Dennett—for all his attachment to reverse engineering. In this particular instance, Gould and Lewontin assume that 'the dome' is a given, and all else is subordinate to it. Dennett, on the other hand, relativizes to local conditions: the dome in Venice, the dome as a showpiece of Byzantine mosaics, and so on. By taking local considerations into account, Dennett substitutes a plurality of functional kinds where previously there was only one. As I see it, this is the essence of relativism: thinking of small historically determined functions in place of transcendent universalist ones.

5. The Unique Origin Constraint

5.1. *Introducing Phylogeny*

The relativist's emphasis on non-recurring situations as occasions of evolutionary change cause her to be reluctant to allow that the *very same* 'problem' could occur in two different evolutionary situations. (A crucial qualification is needed here in order to accommodate 'convergent evolution', and I shall attend to it in the next subsection.) Different evolutionary paths bring their own unique opportunities and pressures—for example, it is unlikely that the very same niche-differentiation demands should recur in independent lines of descent. Suppose then that traits *T* and *T′* found in two different taxa are adapted to precisely the same environmental problem. Then, since the relativist emphasizes the singularity of such problems, she will infer that *T* and *T′* must have arisen from a common ancestor. Thus:

> A *relativist* analysis is one that applies the same adaptive analysis only to homologous traits, that is, to traits that derive from a single origin.

This is a first pass at the *Unique Origin Constraint*. It constrains relativist attributions of function: to the extent that an analysis of a trait in terms of adaptive function violates this condition, that analysis is committed, to some degree, to universalism.

The contested cases mentioned in Section 2.2 above violate this condition. The analysis of 'cooperation' given by E. O. Wilson in his famous work *The Insect Societies* (1971) has special application to a special chromosomal structure found only in *Hymenoptera*, but in his even more famous *Sociobiology* (1975) he treats cooperation as if it were *one* phenomenon as it occurs in species as widely separated as humans and ants.

> Biologists have always been intrigued by comparisons between societies of invertebrates, especially insect societies, and those of vertebrates. They have dreamed of identifying the common properties of such disparate units in a way that would provide insight into all aspects of social evolution including that of man. The goal can be expressed as follows: when the same parameters and quantitative theory are used to analyse both termite colonies and troops of rhesus macaques, we will have a unified science of sociobiology. (Wilson 1980: 4–5)

Similarly, Barash (1979) treated *all* forced sexual activity, whether in ducks or in humans, as if it were an application of a single behavioural 'rape' strategy. These analyses assume that there is some one optimality analysis that applies equally well to non-homologous instances of cooperative behaviours, instances that can be traced to distinct evolutionary origins. This is what marks them as universalist, and, in this particular way, adaptationist.

Note that this does not necessarily imply that they are wrong. The inefficacy of non-local maxima is not an a priori truth. My own discussion has been sympathetic to the relativists, but, of course, one would not expect Wilson and Barash to be so. However, I have tried to show why independent origins should be reckoned unlikely occurrences, especially in the case of complex capacities such as flight or rationality. (A full treatment, though not in the context of functional attribution, is to be found in Sober 1988.) To the extent that they are methodologically self-conscious, one might expect to find in universalists such as Wilson and Barash a detailed account of how these traits differ from taxon to taxon, and of how different pre-existing structures and traits were modified to the same function. (Unfortunately, one does not in fact find such discussions.)

5.2. Retrodictive Engineering

Convergent evolution is a phenomenon that challenges the anti-universalist line outlined above. Flight occurs non-homologously in birds and in insects. Flight requires aerofoil surfaces. Thus, aerofoil surfaces evolved independently in birds and insects. This is a widespread phenomenon—both sides of the debate admit that it is, and so our definition of universalism should be qualified so that a mere recognition of convergence is not taken to be a sign of universalist thinking.

Before we embark on this task, let us note that many analyses of convergence do not actually require any reference to adaptative value at all. Suppose, for instance, that we *know* (by observation) that a certain species is capable of flight. Then we know it requires aerofoil surfaces. Further, flight control demands devices to effect ascent, descent, aerial changes of direction, and so on. We can therefore infer, via an engineering analysis of flight, that the organism possesses such surfaces and such devices. Let us call this (by analogy with 'reverse engineering') *retrodictive engineering*: it consists in breaking down the performance of a complex task into its component sub-tasks, and figuring out what devices are needed for these sub-tasks. (Retrodictive engineering corresponds closely to Robert Cummins's notion of 'functional analysis' (1975, 1983).) Presumably, retrodictive engineering is a part of all functional attribution, just as reverse engineering is. But, though it presupposes adaptation somewhere in the background—in the example just mentioned, it assumes the adaptive value of flight—it does not make any use of this background fact. All that is needed here is the observation that, if a particular organism is capable of flight, a retrodictive analysis will show what is needed to achieve flight (of a certain type). We can deduce what two organisms that are capable of flight must share in common.

Suppose, then, that a term *T* can be defined in an independent science *S* (for example, aeronautical engineering), in such a way that it is observed to be instantiated in a variety of biological organisms. Suppose further that *S* demonstrates that, if an organism satisfies *T*, then it satisfies $t_1, t_2, \ldots, t_n$. In short, suppose that some science *S* shows that a multi-realizable trait *T* does not come alone, but is part of a collection of properties that we may call the *T*-suite. Then, if we *observe* that *T* (as defined in *S*) occurs non-homologously, we can retrodict by means of *S* that the *T*-suite of properties originates more than once. This is the case for winged flight: that is, it is observed to occur non-homologously, and there is a 'flight-suite' of properties retrodicted by aeronautical engineering.

These reflections should prompt us to make separate inventories of the categories that ground adaptive analyses and those that ground engineering analyses. All organisms are, of course, subject to the physical constraints of their surroundings. They adapt to this common environment in ways that enable them to carry out quite different tasks. Just as a lever, a hammer, a pump, a camera can be used to serve very different purposes, so also limbs that push off against firm land, limbs that grasp small items, mouths that catch liquid at awkward angles, eyes that focus—all of these devices can be used as parts of very different adaptations. The similarity of these devices across different applications may be demonstrated, without reference to adaptation, by the specifications that we get from a retrodictive analysis.

5.3. The Univocity of Categories across Multiple Origins

I am supposing that there will be two levels of analysis appropriate to explaining a trait. The first, an engineering analysis, will tell us what a trait does ('reverse engineering') and how it does this ('retrodictive engineering'). In the case of biological organisms, in which the engineering is the work of evolution, this calls for a second level of analysis. This second level, consisting of *adaptive* analysis, is required in order to demonstrate how a trait contributes to lifestyle, viability, and reproductive success of an organism. It is here that we need to be cautious about attributions that span multiple origins.

Where retrodictive analysis demonstrates the existence of a category that spans multiple origins, that category is defined by the independent science *S*. One should not rely here on observation unaided by theory. Consider flight again. Some organisms float in the air, or are carried by the wind—spiders, for example. These organisms may *look* as if they are flying, but they are not. The flight-suite of properties may not obtain in their case. Here the independent science should be relied upon to determine whether some category genuinely occurs across multiple origins or not. When the independent science is used rigorously in the identification of traits, we need have no worry about univocity.

Now, one problem that plagues *adaptive* analyses across multiple origins is that this kind of test of univocity is hard to come by. Intuitions may tell you that you observe 'rape' occurring non-homologously in a number of different biological taxa—ducks and primates, for instance. But it is dubious that what is actually observed is a robust phenomenon. Apparently, we do observe coerced or physically forced sexual activity in these different species. However, critics have pointed out that the accompaniments of such coercion are different in different species: in primates it might be related to positioning in status hierarchies, in ducks it is not. Thus, it is not clear that there is any one reproductive or behavioural strategy that all cases of coerced sexual activity instantiate. In view of this, one might well conclude that, even if certain salient similarities to *human* behaviour lead us to think that we observe 'rape' in many different species, more is required before we can conclude that it is genuinely a unity across unrelated biological taxa—and one should be cautious in making inferences based on a 'rape-suite' associated with the human variety.[22]

A similar point can be made about cooperation. Observation might lead us to the idea that cooperation exists non-homologously. However, some seemingly cooperative behaviour is apparent only: animals may herd together to reduce their chances of being picked off individually by predators. Even in well-established cases of animals engaging in collective behaviour that invites a 'free-rider' to take advantage of the collateral benefits, cooperation can be for very different purposes. Some organisms work together to reduce the risk of predation, others to build shelters, yet others to nurture and protect the young. Let us assume that all of these phenomena meet the above characterization of cooperation as collective behaviour open to exploitation by free-riders. Even so, it is controversial whether there is an independent science *S* that defines cooperation in a way that enables us (*a*) to recognize it in these multiple realizations (for example, in termite colonies as well as in troops of rhesus macaques), and (*b*) to retrodict a significant 'cooperation-suite' of properties. Actually, it would be astonishing if both of these conditions could be met in the case of cooperation. It might be that one could abstract from these diverse examples some generalizations concerning 'cooperation' (for example, that there needs to be a mechanism to punish or exclude free-riders). But it seems likely that, since the adaptive purposes are highly diverse, most of the explanatory action is likely to take place at the particular rather than the general level. In other words, it is unlikely that in such cases (*b*) above will be satisfied, even if (*a*) is. (Even (*a*) is not achieved in the case of 'rape'.)

Incorporating these considerations into our earlier articulation of universalism, we get the following relativist methodological constraint:

The Unique Origin Constraint. A *relativist* analysis is one that applies the same adaptive analysis only to homologous traits, *except where* there is a property, *F*,

(*a*) defined by an independent science* *S* in such a way that it is recognizable in non-homologous realizations, and

(*b*) actually observed by that definition to have non-homologous realizations, and

[22] For a discussion of the definitional question regarding rape and the like, see in particular Kitcher (1985: ch. 6).

(*c*) such that *S* establishes that *T* is a member of a significant set of properties (the '*F*-suite') necessary for *F*.

*An 'independent science' is one independent of adaptive considerations regarding *F*.

5.5. *Functional Minimalism*

I have been arguing that functional analysis is not likely to furnish us with significant predictions across multiple origins. It seems to me that, if this is correct, then we have little motivation to posit functional categories that span non-homologous instances. This makes me something like a *functionalist minimalist* in the sense of Karen Neander (this volume). However, it is important to note the dichotomy of types of 'functional' consideration that I have noted above. I do not claim that there are *no* categories that span multiple origins: I have conceded that categories defined in terms of an independent science may well admit 'analogies'. (In other words, causal role functions, or Cummins functions as they are sometimes called, admit non-homologous instances.) Even with respect to *adaptive* categories, I cannot claim to have established that there are *no* such categories. However, I do claim that, with regard to the products of cumulative selection, where a present product of evolution is the result of many adaptive changes, one piled on top of another, multiple origins are exceedingly unlikely.

In making this distinction, I am assuming that an important difference may be observed between evolutionary changes that bring along with them a change of lifestyle and those that do not. It has often been observed that the focusing mechanism of the eye, for example, is the product of a series of changes: first a small depression in the skin that is more sensitive to light from one direction than another, then a progressive deepening of the cavity and a narrowing of the aperture, then a membrane over the aperture that at first protects and then acts like a lens as it develops further, and so on. With each such change, lifestyle may remain fixed (though, as the organ becomes more and more developed, it may afford the population opportunities for specialization). Thus it is possible that one might keep the *adaptive* analysis fixed, while improving the engineering aspects of an organ in accordance with the independent science that underwrites its design and performance evaluation. If one wants to trace the history of some organ that is exquisitely shaped to some quite specialized adaptation—a wing or an eye, for instance—one will find in general that the process involves some modifications that lead to an adjustment of lifestyle, and some that lead to an improvement of an antecedently existing lifestyle. My claim is that adaptive changes do not occur in duplicable ways.

It might seem that this is a vague sort of distinction to make. It may be so in conception, but the two kinds of change lead to quite different evolutionary effects. Broadly speaking, the first kind of change leads to populations becoming differentiated from one another: one subpopulation pursuing the new lifestyle, another pursuing the old. Since there is nothing intrinsically worse about the antecedent condition, it will be preserved. Ultimately, this is the kind of change that leads to speciation. The second kind of change—improvement relative to a fixed lifestyle—leads to competition within a population pursuing a common lifestyle, and hence to the extinction of the ancestral variety. Thus, the first kind of change leads to a proliferation of kinds, the second to a development of pre-existing kinds. (Note that many cartoon versions of adaptationism—the pictures of Homo developing an erect gait, for instance, that Gould picks on repeatedly, suggest that evolutionary change is all of the second kind.) My claim amounts to this: the sameness of *adaptive* analyses across multiple origins demands that the first of these kinds of change be governed by universalistic norms.

6. Relativism and Human Rationality

6.1. *Explaining Rationality*

We are now ready to revisit the story *R* with which we started. The special problem here is that *R* can be thought of as an attempt to solve a problem that arises as follows.

NORM. Cognitive rationality is the system of thought *prescribed* by the very best systems of deductive and inductive logic and decision theory. It is good for humans that they should be cognitively rational.

*FACT. Human beings are *naturally* so constituted that they think in ways that closely approximate the patterns of thought prescribed by these best systems.

PROBLEM. We need to provide an account of how the fact came to mirror the norm.

This way of stating the problem is common to creationist and evolutionary treatments of the origins of human cognition. Creationists (for example, Descartes) invoke God's agency to explain the cognitive potency of humans—for instance, they attempt to solve sceptical problems concerning the reliability of perception, memory, induction, etc. by means of the supposition that, in His goodness, God must have created humans in such a way as to instantiate NORM. The question is whether evolutionary accounts can appeal to natural selection to pull off the same trick.

The nub of the difficulty for evolutionists is that, unlike God, natural selection cannot be thought of as doing what is *good*. Thus, NORM has to be rephrased in terms of reproductive fitness. Thus, the problem is taken to be that of showing that *FACT above is fitness increasing. As we have seen, philosophers have sometimes said things that suggest that this is a straightforward transition. (I quoted Dennett in *R* above; here is Quine (1969: 126): 'creatures inveterately wrong in their inductions have a pathetic but praiseworthy tendency to die out before reproducing their kind.') However, others have pointed out that the kind of cognitive rationality that is useful in today's world may not have been particularly fitness increasing in the primitive hunter-gatherer (or scavenger) societies in which it (supposedly) evolved. As Elliott Sober (1981: 95) puts the question: 'How could the fundamental mental operations which facilitate scientific theorizing be the product of natural selection, since it appears that such theoretical methods were neither used nor useful 'in the cave'—i.e., in the sequence of environments in which selection took place?' This is the kind of difficulty that makes PROBLEM more problematic for evolutionists than for their creationist counterparts.

A further difficulty for *R* lies with *FACT. As we said at the outset, recent studies have made it increasingly clear that what we possess by way of instinctive inferential patterns is a collection of ill-assorted 'heuristics' or cognitive tricks that work reasonably well in a limited range of situations, but fall short of anything like a 'best' practice. Further, our cognitive abilities seem not to be systematic in the way that logic is. For example, we are instinctively able to perform *modus ponens* but have considerable difficulty with *modus tollens* until tutored, we instinctively use deficient sampling techniques for purposes of inductive inference, we are better able to calculate frequency-based probabilities than those based on propensities, and we are unable to be consistent in the formulation of action strategies in the presence of uncertainty. These are the sorts of findings that make *FACT not a fact at all.

6.2. *Conceptions of Reason Compatible with Relativism*

These difficulties make it attractive to give up on NORM. And, it might be thought, this is in any case dictated by the relativist perspective that I have been trying to construct. NORM posits rationality as a conditional maximum not only capable of influencing the evolution of human rationality, but applicable to cognitive capacities in any species whatsoever. If one is sympathetic to the relativist approach sketched above, one will need to take a more historically oriented view of human cognitive abilities. Thus, one might substitute for the universalist conception of reason found in NORM, a more historical conception, such as the one hinted at above: human cognition arises out of highly contingent circumstances grounded in our history as hunter-gatherers or scavengers.

Now, some ways of thinking about human rationality are quite compatible with a relativistic conception. Ironically perhaps this can be confirmed by examining the assumptions of those who have argued that the assumption of human rationality is conceptually *necessary*. Some argue in support of this position on the grounds that, if one is going to attribute beliefs to others at all, one has to do so on the assumption that they are rational—if someone is not sufficiently similar to us in respect of their cognitive processing, we will not be able to figure out *what* they think, much less how rational it is. Others, particularly L. J. Cohen (1981), argue that our conception of rationality is nothing other than a codification of human practice. Both of these arguments presuppose that, if we can adjudge *x* to be rational, then *x* must be similar in cognitive processes to ourselves. According to the first group, to adjudge *x* to be rational is to attribute some beliefs to *x*, and one cannot do that unless one assumes that *x* is similar to us. According to Cohen, rationality just is human best practice; thus, any creature that is rational follows human practice. Both approaches presuppose a descriptive conception of human rationality. Hence, both are compatible with the idea that some unrelated species ('Martians') might possess cognitive capacities that are fundamentally different in kind from our own. In so far as these capacities differed from our own, we might be obliged to conceptualize them in terms fundamentally incommensurable with those that we use in our conceptualization of human rationality. But this is a consequence of *a priori* (or transcendent) conceptions of human rationality. As Kant was well aware, all such conceptualizations are grounded on particularities of the human condition.

Another approach to human rationality might consist in supposing it to be instrumentally good for something. It might be held that logic is an independent science that provides us with engineering specifications for any cognitive system as such, whether or not such a system is adapted to the particular circumstances of a hunter-gatherer society. This would support the idea that, despite the peculiarities of a system equipped to deal with a highly specific situation, our cognitive system would still have to conform to standards of cognitive rationality that are applicable across the range of such systems. And there is some justification for such a conception. One might say that reverse engineering shows us that our cognitive systems do actually provide us with the capacity for truth maximization. So, if one were to accept the proposition that logic is the independent science of truth maximization, it would follow that all such systems would be governed by logic. Of course, it is possible that some cognitive systems are not truth-maximizing machines. One *might* say: a crow or a dog is completely a-rational, the lifestyle that evolution has provided these creatures does not require them to perform in accordance with logic at all. (Some of the things these creatures do might have been done also by a rational creature, but these creatures do not calculate or reason.) A Martian, on the other hand—that fictitious alien who can do all or most of the things humans can do—has to be comparable, at least in terms of logic. The Martian's capacity might be better than ours, or the same, or worse—but since there is only one benchmark for cognitive rationality as traditionally conceived, namely the maximization of truth content, it cannot have capacities that are fundamentally different in character. The Martian has converged on the same cognitive values as we have. This approach too is compatible with the Unique Origins Constraint.

6.3. The Trouble with a Relativistic Conception of Human Rationality

The accounts of rationality we have just considered are not open to criticism on the grounds that they sin against relativism. But they run into another kind of problem, namely that (*a*) traditional conceptions of rationality have considerable justification, combined with the fact that (*b*) we humans are reasonably good at following the traditional norms of rationality, even after our cognitive deficiencies have been taken into account.

We can put the difficulty in this way: we seem to possess the capacity rationally to assess, criticize and correct our own irrationalities and mistakes. Look at it this way. Suppose that there is a coherent account of why a certain false belief is valuable. For example, imagine that I want (*G*) to persuade Canadians of the badness of destroying the temperate rainforest on Vancouver Island. Let us suppose that the likelihood of my achieving *G* is just about zero. Then, if I form a *correct* estimate of the likelihood of achieving *G*, I will not embark on the project. That, of course, would destroy any chance that I will achieve *G*. So, as far as achieving *G* is concerned, it is better that I should form a falsely inflated estimate. Now, this particular goal is, of course, irrelevant to my evolutionary fitness. Still, some might argue that *in general* I will get nothing done except by overestimating my chances of success. Thus they argue that it is better that I should systematically overestimate my chances of achieving my goals than that I should estimate them correctly.

In the light of such an argument, what are we to make of the fact that we actually possess the capacity to assess our estimates by means of 'logic'. After all, humans are capable of reasoning correctly and consistently. To be sure, this is not always easy. And it sometimes needs considerable education and training. Nevertheless, we are capable of internalizing and employing some advanced (if not perfect) systems of logical inference. These systems are capable of detecting our overestimation of our prospects. In short, my overestimate of *G* is correctable. The question is this: if evolution got us, for some tangible advantage, systematically to overestimate our prospects, then why did it compromise this very advantage by endowing us with the capacity to detect our errors? Relativistic conceptions of the value of rationality run into difficulties if they are inconsistent with the fact that we do actually possess some pretty powerful epistemic capacities.

I do not know how to reconceptualize rationality in a way that makes it consistent with relativism. But here, to conclude, are two observations that I hope are relevant to the general problem.

6.4. Progress can be illusory

Suppose that the agglomeration of cognitive abilities that constitutes human 'rationality' is in fact a closer approximation to the best inferential systems prescribed by logic than the abilities that an ape possesses. Does it follow that the difference should be explained by progress towards perfect rationality?

Consider a cognitive trick learned by an ape. Suppose that an ape acquires the ability to represent and add any two numbers not exceeding five. This ability, which amounts to a fragment of arithmetic, constitutes an improvement in the ape's rational performance. Now, if it could add numbers up to *six*, it would be doing even better. When it can add numbers up to and including six, it possesses a larger fragment of arithmetic than when it was able to add numbers only up to five. If the ape learns to perform in this enhanced way, should we explain its improvement by reference to its closer approximation to arithmetic? No, the improvement may have a much more local significance. For example, it may be the case that the ape learns this larger fragment of arithmetic in order to play some game with its human keeper. There is no reason to think that it was progressing towards a systematic appreciation of a branch of mathematics.

In exactly the same way, it may well be that many increases in the complexity of cognitive behaviour acquired during the course of evolution constitute progressively closer approximations to NORM. An organism with thirty specialized cognitive tricks may be closer to possessing logic than one with only twenty out of that thirty. However, from

A. Evolution will generally result in an increase in cognitive powers because it will fashion an increasing number of 'cognitive tricks'.

and

B. Many increases in cognitive powers result in a closer approximation to NORM.

it does not follow that

*C. Evolution is driven by the optimality of NORM.

The increase in cognitive power noted in B may amount to a closer approximation to NORM without being explained by that fact. It may well be that many forms of complex cognitive behaviour are advantageous, though not the kind of systematic rationality that the best systems of logic prescribe. In particular, there may be nothing privileged about the particular grab-bag of cognitive tricks and heuristics that constitutes human rationality. A 'Martian' might display equal complex cognitive behaviour without necessarily sharing much with us in terms of the actual instinctive inference patterns she displays. Assume that our cognitive abilities evolved to meet the demands of a hunter-gatherer society. Assume that the Martian's abilities evolved in a more solitary environment. There is no reason *so far* to think that there will be significant convergence between the two. That is the relativist's position.

6.5. But even Illusory Progress can Lead to the Real Thing

Nevertheless, let us assume that an organism continues, by the evolutionary process, to accumulate cognitive tricks. It might undergo a concomitant increase in representational capacity—that is, an increased capacity for the manipulation of symbols. Now it is possible that, at a certain point in this process, the increase in symbol-representational power brings with it the capacity to represent logic, or some significant fragment thereof. In other words, it could be the case that logic, or the capacity to learn it, is the consequence of an increase in representational capacity, despite there being no lifestyle relative advantage in possessing logic.

Demonstrating anything like this requires more by way of the theory of symbols than I can muster. Nevertheless, it does suggest a way in which rationality could have multiple origins. It could be that Martians too started accumulating cognitive tricks. It could be, as I suggested, that their tricks are entirely different in purpose from ours—this would have the consequence that Martian Kahnemans and Tverskys detected quite a different set of 'cognitive illusions' than those found on Earth. Still, it might be that the tricks that got us going on the representational path had nothing to do with logical inference, or even anything to do with cognitive content. Some say, for example, that the crucial development that leads to representational capacity is the emergence of the visual cortex. It would follow that the cerebral cortex developed to accommodate *vision* not thought. If this is right, a crucial part of the evolutionary build-up to rationality has nothing to do with inference as such. Nevertheless, if, with the accumulation of a critical quantity of representational capacity, whatever its origin, Martians too developed the symbolic capacity to represent logic, then they too would ultimately develop logic, or a significant fragment thereof.

In Section 5, we envisaged an independent science that shows, for an adaptive trait T, the existence of a T-suite of properties instrumentally necessary for T. But there is another way to establish the existence of a significant T-suite of properties. As an alternative to showing that a particular suite of properties is instrumentally *necessary* for *T*, one might show that there is a significant suite of properties necessarily associated with *T*. This is, in essence, the kind of demonstration that might be possible for the case of human cognitive rationality. The traditional conception of reason supposes that truth maximization is the favoured activity of cognitive systems, and that rationality is instrumentally necessary for truth maximization. But it might also be that rationality is the necessary consequence of a certain level of symbolic complexity. This level of symbolic complexity could arise out of all sorts of performance capabilities other than truth maximization. Thus, it might well be that increasing symbolic complexity brings an increasing conformity with absolute norms of rationality, without being driven by those norms. This would also show why human cognitive deficiencies coexist with the capacity for considerable cognitive sophistication. These deficiencies should not, therefore, be taken as showing either (*a*) that reason is not a universal value, or (*b*) that humans fall hopelessly short of that universal value, or even (*c*) that we need to reassess the content of the universal value that humans in fact conform to. The fourth way that I am proposing is this. Our capacity for reason is dictated by symbolic complexity required for tasks other than truth maximization. Therefore, it is coherent to suppose that we might have evolved to have become quite good at truth maximization without there being a great deal of pressure to become better. Our excellence at truth-maximizing reason does not imply a drive to perfection in this regard.

These are merely speculations. What we need is an independent science to establish the content of a rationality-suite of properties in the sense of the last section. If we could come by a theorem about representational adequacy relative to, say, second-order logic, then we would be getting somewhere. Until then, all that we can do is avoid the idea that evolution is driven by the universal value of rationality.

References

Barash, David P. (1979), *The Whisperings Within* (New York: Harper & Row).

Bedau, Mark (1991), 'Can Biological Teleology be Naturalized?', *Journal of Philosophy*, 88: 647–55.

Cherniak, Christopher (1986), *Minimal Rationality* (Cambridge, Mass.: MIT Press).

Cohen, L. Jonathan (1981), 'Can Human Irrationality be Experimentally Demonstrated?', *Behavioral and Brain Sciences*, 4: 317–70.

Cummins, Robert (1975), 'Functional Analysis', *Journal of Philosophy*, 72/20: 741–65.

——(1983), *The Nature of Psychological Explanation* (Cambridge, Mass.: MIT Press).

Dennett, Daniel C. (1987), *The Intentional Stance* (Cambridge, Mass.: MIT Press).

——(1995), *Darwin's Dangerous Idea: Evolution and the Meanings of Life* (New York: Simon & Schuster).

Elster, Jon (1984), *Ulysses and the Sirens: Studies in Rationality and Irrationality*. 2nd edn. (Cambridge: Cambridge University Press).

Gigerenzer, Gerd (1991), 'How to Make Cognitive Illusions Disappear: Beyond Heuristics and Biases', *European Review of Social Psychology*, 2: 83–115.

——Todd, Peter M., and the ABC Research Group (1999), *Simple Heuristics that Make us Smart* (New York: Oxford University Press).

——Todd, Peter M., and the ABC Research Group (1999), *Simple Heuristics that Make Us Smart* (Oxford: Oxford University Press).

Godfrey-Smith, Peter (1996), *Complexity and the Function of Mind in Nature* (Cambridge: Cambridge University Press).

Gould, Stephen Jay (1985), *The Flamingo's Smile: Reflections in Natural History* (New York: W. W. Norton).

——(1996), *Full House: The Spread of Excellence from Plato to Darwin* (New York: W. W. Norton).

——(1997), 'Evolution: The Pleasures of Pluralism', *New York Review of Books*, 44/11 (26 June), 47–52.

——and Lewontin, Richard (1978), 'The Spandrels of San Marco and the Panglossian Paradigm: A Critique of the Adaptationist Programme', *Proceedings of the Royal Society of London Series B: Biological Sciences*, 205: 581–98.

Grant, Peter (1986), *Ecology and Evolution of Darwin's Finches* (Princeton: Princeton University Press).

Hankinson, R. J. (1988), 'Galen Explains the Elephant', in Mohan Matthen and Bernard Linsky (eds.), *Philosophy and Biology: Canadian Journal of Philosophy*, supp. vol. 18 (1988), 135–57.

Kahneman, D, Slovic, P., and Tversky, A (1982), *Judgement under Uncertainty: Heuristics and Biases* (Cambridge, Cambridge University Press).

Kitcher, Philip (1985), *Vaulting Ambition: Sociobiology and the Quest for Human Nature* (Cambridge, Mass.: MIT Press).

Lewins, Richard, and Lewontin, Richard (1985), *The Dialectical Biologist* (Cambridge, Mass.: Harvard University Press).

Lewontin, Richard (1985), 'Adaptation', in Richard Lewins and Richard Lewontin, *The Dialectical Biologist* (Cambridge, Mass.: Harvard University Press, 1985), 65–84. Originally appeared in Italian translation as 'Adattamento', in Guilio Enaudi (ed.), *Enciclopedia Einaudi*, i (Turin: Enaudi, 1977).

Matthen, Mohan (1997), 'Teleology and the Product Analogy', *Australasian Journal of Philosophy*, 75: 21–37.

(2000) 'What is a Hand? What is a Mind?', *Revue internationale de philosophie*, 214: 123–42.

——and Ariew, André (2002), 'Two Ways of Thinking about Fitness and Natural Selection', *Journal of Philosophy*, 99: 55–83.

Maynard Smith, John (1995), 'Genes, Memes, and Minds', *New York Review of Books*, 42/19 (30 November), 47–51.

Nagel, Ernest (1961), *The Structure of Science: Problems in the Logic of Scientific Explanation* (New York: Harcourt, Brace & World).

Nisbett, R. E., and Ross L. (1980), *Human Inference: Strategies and Shortcomings of Social Judgment* (Englewood Cliffs, NJ: Prentice-Hall).

Orzack, Steven Hecht, and Sober, Elliott (1994), 'Optimality Models and the Test of Adaptationism', *American Naturalist*, 143: 361–80.

Plantinga, Alvin (1991), 'When Faith and Reason Clash', *Christian Scholar's Review*, 21: 8–32.

Preston, Beth (1998), 'Why is a Wing Like a Spoon? A Pluralist Theory of Function', *Journal of Philosophy*, 95/5: 215–54.

Quine, W. V. O. (1969), *Ontological Relativity and Other Essays* (New York: Columbia University Press).

Sober, Elliott (1981), 'The Evolution of Rationality', *Synthese*, 46: 95–120.

——(1988), *Reconstructing the Past: Parsimony, Evolution, and Inference* (Cambridge, Mass.: MIT Press).

——(1993), *Philosophy of Biology* (Boulder, Colo.: Westview Press).

Sterelny, K., and Griffiths, P. E. (1999), *Sex and Death: An Introduction to Philosophy of Biology* (Chicago: University of Chicago Press).

Stephens, Christopher (2001) 'When is it Selectively Advantageous to have True Beliefs? Sandwiching the Better Safe than Sorry Argument', *Philosophical Studies*, 105: 161–89.

Stich, Stephen F. (1985), 'Could Man be an Irrational Animal?', *Synthese*, 64: 115–34.

Wilson, E. O. (1971), *The Insect Societies* (Cambridge, Mass.: Harvard University Press).

——(1975), *Sociobiology: The New Synthesis* (Cambridge, Mass.: Harvard University Press).

——(1980), *Sociobiology: The Abridged Edition* (Cambridge, Mass.: Harvard University Press).

14. Real Traits, Real Functions?

Colin Allen

ABSTRACT

Discussions of the functions of biological traits generally take the notion of a trait for granted. Defining this notion is a non-trivial problem. Different approaches to function place different constraints on adequate accounts of the notion of a trait. Accounts of function based on engineering-style analyses allow trait boundaries to be a matter of human interest. Accounts of function based on natural selection have typically been taken to require trait boundaries that are objectively real. After canvassing problems raised by each approach, I conclude with some facts that satisfactory notions of trait must respect.

In the extensive literature on what it means for a biological trait to have a function, philosophers and biologists have given the notion of function star billing. They have paid little attention to the notion of trait appearing in the supporting role. My goal in this chapter is to turn the spotlight onto this fundamental notion and demonstrate that we lack and need an analysis of trait. I use 'analysis' here with a broad meaning that encompasses traditional conceptual analysis, theoretical definition, and other ways of specifying meaning.

Before embarking, it is useful to address another terminological point. Many biologists prefer the term 'character' to 'trait'. There are a couple of reasons for this. First, it appears that Darwin himself did not use the term 'trait' to refer to the properties of organisms, writing instead of characters. Secondly, 'character' has found favor with systematists, who have introduced a technical distinction between characters and character-states (explained below). Thus provenance and technique combine to invest 'character' with a cachet that the more homely 'trait' lacks. Nevertheless, 'trait' is ubiquitous and I, like many

The receipt of a Big 12 Faculty Fellowship from Texas A&M University enabled me to visit the University of Colorado, where the resulting conversations with Marc Bekoff were very helpful for getting a draft of this chapter started. Audiences at Sam Houston State University and Texas A&M University, and several subsequent discussions with Glenn Sanford, were very helpful for getting the draft completed. Both individuals provided helpful comments on a very early draft. I wish also to thank Fred Adams, Scott Austin, Marc Bekoff, Sophie Dove, Peter Godfrey-Smith, Karen Neander, Eric Saidel, Glenn Sanford , Elliott Sober, and the editors of this volume for helpful comments on the manuscript.

others, will use the term freely and interchangeably with 'character' to refer to properties of organisms that interest biologists.

Although my route to the topic of traits goes through theories of function, questions about the delineation of traits have broad significance for biology and the philosophy of biology. For example, the boundaries of the traits or structures identified by morphologists and paleontologists are important for phylogenetic inference. Questions of whether to lump or split behavioral patterns into one action or two for the purpose of developing an ethogram play a similar role in ethology. The question of how much precision is to be expected or desired in the notion of a trait is as fundamental to the philosophy of biology as related questions about notions such as *species* (see contributions to Ereshefsky 1992), *environment* (Brandon 1990), and *niche* (Smith and Varzi 1999). Of course, much useful biological research has proceeded in the absence of completely satisfactory analyses of key terms. Darwin himself wrote on the origin of species, without settling the question 'what is a species?' Likewise, Mendel investigated the inheritance of sweet pea traits without settling the obvious question 'what is a trait?' Scientific success is obviously possible in the absence of answers to such questions, but it does not follow that there is nothing to be gained by raising such questions, and it would be fatuous to suppose that those who raise such questions are dismissing the science as hopeless.

This chapter has the nature of preliminary work. I do not intend to offer a definitive answer to the question 'what is a trait?', for I am skeptical that any single answer will be satisfactory, any more than any single analysis of *species* has proven satisfactory for all biological purposes. It is an important task for the future to try to specify what answers suit which purposes, but that is a much larger project than can be accomplished here. The more restricted project in this chapter is to identify roles that different notions of trait might play in the debate about biological functions.

The chapter is divided into five sections. In the first section I describe the two major approaches to understanding functions that have emerged in both the biological and the philosophical literatures. In the second section I survey the relatively scarce amount that has been written about the notion of trait. The third section introduces a distinction between effectively nominalist approaches to identifying the traits of organisms, which make trait selection a matter of human interest, and effectively realist approaches, which attempt to provide more objective criteria for traithood. In the third and fourth sections I describe some apparent shortcomings in the effectively nominalist approach to traits implied by Cummins's account of functions and some problems with an effectively realist proposal derived from the work of Godfrey-Smith. In the fifth, final section I conclude with some facts that satisfactory notions of trait must respect.

1. Two Accounts of Function

To set the stage for understanding the role of the notion of trait in the analysis of functions, this section contains an overview of the two major approaches to the notion of function. While biologists have tended to focus more narrowly on attributions of function within their science, philosophers have often had an eye on the broader applications of the term beyond biology. Nonetheless, the same two classes of approach have come to dominate both the biological and the philosophical literatures.

- *Etiological* approaches to function look to a causal-historical process of selection; functions are identified with those past effects that explain the current presence of a thing by means of a historical selection process (typically natural selection in the case of biological function).
- *Systems-analysis* approaches invoke an ahistorical, engineering style of analysis of a complex system into its components. Functions of components are identified with their causal contributions to broader capacities of the system.

There are variant views within these two classes, and there are accounts of function that attempt to meld them. There are also many accounts of function that fall outside these general approaches. These details are amply illustrated in the other contributions to this volume so they need not detain us here.

Philosophers have tended (perhaps somewhat parochially) to identify the two kinds of approach with the philosophers who first articulated them in the philosophical literature. Etiological approaches are associated with Wright (1973) and engineering-analysis approaches are associated with Cummins (1975). Indeed it is quite common to see the labels 'Wright functions' and 'Cummins functions' used by philosophers to identify the two approaches. Precedents for each kind of approach can, however, be found independently in the biological literature. (This is not to deny that Wright and Cummins significantly advanced the development and understanding of the respective approaches. But see Pittendrigh 1958, Tinbergen 1963, Williams 1966, Ayala 1970, and Hinde 1975 for examples of the historical approach; Rudwick 1964, and Bock and von Wahlert 1965 for the systems-analysis approach; and see Allen *et al.* 1998 for an overview and representative papers.)

As the preceding citations indicate, much of this work on the notion of function was done during the third quarter of the twentieth century. During the latter part of the century, particularly since the mid-1980s, there has been a resurgence of philosophical interest in analyses of function, driven largely by the development of evolutionary approaches to the phenomena of mind and language. Most prominently, proponents of 'teleosemantics' have constructed naturalistic accounts of the phenomena of meaning and mental content in terms of proper function, while offering solutions to puzzles about cognitive error and misrepresentation by assimilating them to typical cases of biological malfunction. The notion of biological function has, consequently, occupied a central position in the debate about teleosemantics.

Evolutionary approaches to the phenomena of mind and language have borrowed extensively from the pre-existing literature on functions and teleology (with the possible exception of Millikan 1984, who seems to have constructed her theory of proper function *ab initio*—see Millikan 1989 for comparisons). The earlier work was not directly concerned with questions about the functions of mind or language. Indeed (rightly or wrongly) many of the biologists responsible for this early work would be more than a little skeptical of the possibility of justifying claims made about the evolution of such traits. Whether or not such skepticism is justified, it behoves those who would make claims about the functions of mind, language, beliefs, or sentences to say something about the identification of such traits within their preferred approaches to function.

Different accounts of function place different theoretical and practical constraints on an adequate notion of trait. For instance, etiological accounts that center on natural selection require an account of trait that allows the reidentification of traits across generations while providing a suitable framework for the causal requirements of natural selection. Furthermore, the comparative method required to establish hypotheses about etiological function mandates, if the approach is practicable, the identification of traits across taxonomic groups. Systems analyses, as well as being expected to support the comparative method, require that traits be identifiable as components that make a distinctive causal contribution to the system capacity in question.

Criteria for trait identity are important to arguments about human evolution. Some arguments about the uniqueness of the human mind, human language, and human culture depend on the refusal to identify a human trait with certain traits of non-humans or of our ancestors (see Allen and Saidel 1998). Other claims about the evolutionary function of complex features of organisms presuppose that it is proper to classify these complex features as single, real traits. None of these claims or arguments can be properly investigated without a clear conception of what we mean by 'trait'. We cannot persuasively argue that human language is a different trait from vervet monkey communication, or even that it is a trait at all, without first having specified what we mean by 'trait' and whether that meaning is suitable to the task at hand. The next section surveys some attempts to define the notion of trait.

2. Definitions of Trait

Although ubiquitous, the notion of trait is rarely introduced explicitly. Authors typically help themselves to it without comment. There are a few exceptions, but these are typically less than enlightening. For example, in their exposition of evolutionary theory, von Schilcher and Tennant (1984: 31) explicitly introduce 'trait' alongside 'characteristic' as a stylistic variant of 'property', but they remark that the notion of property has itself 'eluded satisfactory analysis by philosophers'. Aside from a remark about Wittgenstein to illustrate their point, they leave it at that.

As mentioned above, biologists typically link the notion of trait to that of character. For example, the entry for 'character' in the glossary of Futuyma's widely used textbook (1998) on evolutionary biology makes this connection. 'Character' is itself defined in conjunction with 'character-state'. The distinction between characters and character-states is explained by Eldredge and Cracraft, who write (1980: 30n.):

> The terms 'character' and 'character-state' merely refer to similarities at two different hierarchical levels. Thus the character 'feathers' is a common similarity of all birds, although specific character-states of the character 'feathers' (e.g., variation in color, texture, and pattern) would be similarities common to various groups of birds. At the same time, it is apparent that even the character 'feathers' could be considered a character-state, say within the vertebrates, if the systematist were considering the 'character' to be the vertebrate integument.

In this passage, Eldredge and Cracraft note that the distinction between character and character-state is relative to the level of phylogenetic analysis. Sober (1988) describes the character/character-state distinction as a special case of the determinable/determinate hierarchy familiar to philosophers. Roughly, a determinable is a more general property (for example, color) that can be specified more determinately (for example, red, green). Typically, the more determinate categories are themselves determinable by yet more determinate properties (scarlet, crimson, . . .), leading to a hierarchy of determinable/determinate relations. This is illustrated by Sober (1988: 35) with the example of bipedalism and quadripedalism as character-states of walking, which is in turn a character-state of locomotion.

Phylogenists are faced with the problem of how to enumerate characters. Cladists, who base their systematics on counting shared characters among members of different species, must be especially careful not to double count by treating two dependent characters as if they were independent, for this can lead to mistaken inferences about the relationship between species. The number of digits on the left forelimb should not, for example, be treated as an independent character from the number of digits on the right (indicating that the relevant notion of independence is not merely logical or nomological). Whether double counting matters for other purposes to which the notion of trait might be put is an open question.

On the same page as the passage quoted above, Eldredge and Cracraft express some embarrassment about their appeal to perceptual similarity as a basis for judgments of shared character. Specifically, they make a remark about the 'infinite regress' in the idea that what is similar is what is judged similar by biologists. Here, perhaps without realizing it, they have hit upon a special case of a difficulty facing any effectively nominalist understanding of properties that is based on a primitive, unanalyzed notion of similarity.

Some explanation of this last remark is in order. Eldredge and Cracraft do not provide a theoretically motivated set of criteria for, nor even constraints upon, the classification of characters or traits. If such criteria or constraints existed and were formulated to be independent of human choice, they would provide a framework for a 'realist' account of characters: that is, an account that provided for the existence of characters delineated by natural processes independent of human theorizing. In metaphysics, nominalism is the view that abstract categories have no real existence. While we should not infer that Eldredge and Cracraft are committed to metaphysical nominalism—for their failure to provide independent criteria or constraints might simply be a symptom of a gap in knowledge rather than a sign of a particular philosophical disposition—nevertheless, their account of characters is *effectively nominalist* (or *effectively non-realist*) because the account itself provides no criteria for character differentiation that would support a realist account.

Bock and von Wahlert (1965: 272) also give an effectively non-realist account of trait—although their preferred term is 'feature', characterized as 'any part, trait, or character of that organism, be it a morphological feature, a physiological one, behavioral, biochemical, and so forth'. They offer a definition of 'feature' as follows: 'Any part or attribute of an organism will be referred to as a feature if it stands as a subject in a sentence descriptive of that organism.' (It is clear from the context that they mean to exclude relational properties by the terms 'part' and 'attribute'.) In contrast to Eldredge and Cracraft, Bock and von Wahlert can be read as endorsing a more explicitly nominalist theory of traits when they embrace the implication that their broad definition commits them to, that there is no privileged way of dividing an organism into features (traits). As they put it: 'The limits of a feature are generally set arbitrarily.' Using the notion of a morphological feature to illustrate, they continue, 'the limits of a feature defined in morphological terms are almost always set in an arbitrary fashion with regard to its development, genetics, and evolution. Hence morphological units should not necessarily be considered as real or absolute units of an organism for other biological studies.'

One might wonder whether effectively nominalist or non-realist conceptions of traits are too shifty to serve as bedrock for function attributions. An alternative, self-avowedly realist conception of trait is suggested by Godfrey-Smith in his book *Complexity and the Function of Mind in Nature.* He writes (1996: 2–23):

[my] thesis refers simply to 'cognition' as if this was a single thing, a single trait. However, the generalized category of 'cognition', even if it picks out a single kind relevant to everyday discussion, may not reflect a single evolutionary reality. 'Cognition' may well be a collection of disparate capacities and traits, each with a different evolutionary history. . . . we will focus on a small and allegedly fundamental set of mental phenomena. We are concerned with a basic apparatus that makes possible perception, the formation of belief-like states, the interaction of these states with motivational states such as needs and desires, and the production of behavior. Even this contracted set may not constitute anything like a single trait, for evolutionary purposes, but we will proceed on the assumption that it does.

Godfrey-Smith does not fully articulate what he means by 'single trait' or 'single evolutionary reality'. Nor is it exactly clear from his discussion whether his intention to apply both the etiological and analytical notions of function requires the assumption that he makes. Nonetheless, this is one of the few intimations in the literature on evolutionary approaches to function that the notion of a trait needs to be analyzed and that a realist analysis might be provided.

3. Nominalism and Realism

Before proceeding to discuss the relationship between the two major approaches to functions and the positions I have labeled 'realism' and 'nominalism', it is important to note that these terms as I have been using them must be treated with care because ultimately there are metaphysical issues that remain after the more empirical issues facing biologists and philosophers of biology are addressed. It is possible to maintain that scientific theories are ultimately neutral on the traditional metaphysical dispute between realists and nominalists. Hence, even if one were to accept something like Godfrey-Smith's construal of a single trait, a philosophical issue of whether to treat any such property as metaphysically real remains. Where appropriate, I use the qualifier 'effective' (or 'effectively') to indicate the more limited distinction between, on the one hand, effectively realist accounts that place relatively strong constraints on the individuation of traits, and, on the other hand, effectively nominalist accounts that do not provide strong constraints.

There is a correlation between positions taken on this more restricted version of the realism/nominalism debate and the two main approaches to functions. Effectively realist views appear more often among those sympathetic to etiological theories of function, while effective nominalism is more common among supporters of ahistorical systems-analysis approaches. Although both approaches appeal to causal mechanisms in their accounts of function, the type of causal processes consistent with etiological accounts, because based on natural selection, is more circumscribed and arguably less relative to researchers' interests. This idea that etiologically characterized functions are objectively real appears to be a factor in the claim represented by Millikan's subtitle (1984): 'New foundations for realism'.

Although different conceptions of trait have different consequences for accounts of function, it is not my objective here to argue for or against any particular approach to function. I hold a pluralistic view about the notion of function in biology, according to which both etiological and systems-analysis notions of function have roles to play (Bekoff and Allen 1995). On such a view, one would not expect to find decisive arguments favoring one type of approach over the other for all biological purposes. Nonetheless, it is quite compatible with such a view that there are reasons for preferring one approach over another for specific purposes. The next two sections are concerned with pointing out some consequences of adopting effectively nominalist and effectively realist accounts of function.

4. Nominal Functions

I shall use Cummins's account of functions to illustrate the consequences of an effectively nominalist approach to traits. Cummins's account notoriously places only very weak constraints on the choice of components and capacities selected for analysis. The account is also explicit about the relativity of function attributions to specific explanatory contexts. Different explanatory aims will yield different function attributions, and, although only some of these will be of interest to biologists, all are examples of the same pattern of explanation. Thus, for example, the beating sound of a heart may have a function relative to the capacity of the body to indicate the presence of heart disease to a physician, but not with respect to the capacity of the body to move around in the world. Again, I emphasize that *the account* is *effectively nominalist* in the sense intended—namely, that the account itself specifies functions only with respect to systems and capacities delineated according to the pragmatic interests of those who employ it. (Of course, if one latches onto systems and capacities deemed real by objective standards, then the functions of system components that explain those capacities may also be real, so those who follow Cummins's account need not be committed to metaphysical nominalism.)

Cummins's formulation of the definition of function employs the phrase 'appropriately and adequately' to characterize the connection between the explaining effect of a component and the explained capacity of a system. This phrase is intended to mark a graded distinction between cases where the systems-analysis approach has a high degree of 'explanatory interest' because the explaining capacity is markedly simpler than the explained capacity and cases where there is little explanatory interest because the two capacities are relatively similar in sophistication or complexity. The question of how to measure relative complexity must be set aside here. But assuming a common-sense notion of relative complexity, it follows from Cummins's account that there would be little explanatory interest in, for example, accounting for an organism's capacity for selecting a fit mate by postulating a cognitive module that detects fitness—the capacity of the postulated detector is no less sophisticated than the analyzed capacity of the whole organism.

One might argue about the extent to which biologists are constrained in their identifications of the components and capacities of the systems they study, but the point here is that any so-called realist conception of those components is external to Cummins's account of function itself and, indeed, to any engineering-style analysis of a biological system into its components. The notion of 'appropriate and adequate' explanation that is at work in Cummins's account provides constraints on the kinds of traits and capacities involved insofar as completely invented traits may not feature appropriately or adequately in causal explanations of system capacities, whether or not those capacities are themselves construed realistically. But, if one uses Cummins's account of functions and chooses to limit one's attention to particular components or capacities, it is not because the account of functions requires one to do so. This is in contrast to etiological accounts of function, which, by appealing to specific historically causal events of natural selection, place stronger constraints on the traits to which one may attribute functions. For example, to be assigned etiological functions the traits must be heritable.

It is worth noting some consequences of Cummins's approach for two current debates. One consequence is for Godfrey-Smith's project of identifying the function of mind. Even given Godfrey-Smith's stripped-down characterization of mind as a perception–belief–desire–action device, the device thus characterized seems complex enough that one might question the explanatory interest in assigning this 'component' of an organism the function to 'enable the agent to deal with environmental complexity' (1996: 3) In fact, Godfrey-Smith seems to recognize this as he moves (1996: 16) 'to drop the requirement that the capacity explained by a function is more complex than the function itself'. This move leaves Godfrey-Smith free to attribute a complex function to the mind in order to explain a similarly complex capacity of the organism, but the result is a notion of function that arguably has lesser explanatory power.

A second debate is over whether human language and primate communication are comparable traits (Pinker 1994; Allen and Saidel 1998). Linguists are fond of pointing out ways in which human language differs from the signalling systems of other species. Ethologists are fond of pointing out the ways in which they are similar. If, given what I am calling an effectively nominalist view of traits, any perceived similarity is a good enough basis for trait identification and if the identified set of communicative behaviors in humans and non-humans can be used to account for the same capacity (also, presumably, determined by perceived similarity), then both human and non-human forms of communication will be attributed the same function relative to that capacity. It is a feature of Cummins's account, about which I make no judgement, that it is neutral with respect to different stances on these issues. Either the traits are identified as similar or not, and either the explained capacities are similar or not. From one perspective, relative to the goal of explaining a given capacity, the functions of human language and another communication system are the same. From other perspectives, relative to other explanatory aims, they are not. Whether this should count as a 'bug' or a 'feature' of Cummins's account is an open question.

Finally in this section, I turn to the issue of matching systems analysis with the hierarchical conception of character and character-state. Systems analyses are based on the identification of concrete parts of a concrete particular system and figuring out the ways in which the operation of the parts contribute to the operation of the whole. It is an oft-mentioned criticism of Cummins's account that a component that fails to contribute to a capacity lacks a corresponding function. Thus, in his account, a severely deformed heart that cannot pump blood therefore does not have pumping blood as its function. Yet, if it does not have this function, it cannot be said to be malfunctioning. This objection could be escaped if the analysis of function allowed a comparison to hearts of other organisms in the same taxonomic group to determine a function for the malformed particular. But Cummins does not take this route, preferring instead to downplay (indeed deny) the significance of a notion of malfunction for scientific purposes.

Taxonomic groupings play no explicit role in Cummins's account of function. Nevertheless, it is an important part of biological practice to explain homoplasy—the existence of non-homologous common features in different taxonomic groups—in terms of common function. In Cummins's account, function attributions are relative to the identified capacity. It may, then, be possible to accommodate the biological practice by identifying capacities at different levels of abstraction corresponding to different taxonomic levels. For example, to return to Sober's examples from the character–character-state hierarchy, bipedal walking and locomotion are both capacities of mine. The contribution made by my legs to my locomotion might be described abstractly enough to match the contribution made by a parrot's wings to his locomotion. So, among the vertebrates for example, one could perhaps end up with the same function relative to the same capacity attributed to different traits.

There are difficulties, however. The level of generality that is required to identify a common function with respect to locomotion for my legs and the parrot's wings seems unlikely to meet Cummins's condition that the explained capacity be of significantly greater complexity than the explaining function. The challenge is to complete the statement that legs and wings both contribute to the capacity for locomotion by ——, where the blank is filled by something that is of considerably less complexity than locomotion while being of sufficient generality to cover the roles of both perambulation and flight in producing movement. As before, I am relying on an intuitive notion of complexity to make this point. It remains to be seen whether this notion can be spelled out adequately. Problems doing this might provide a reason to drop Cummins's greater-complexity requirement on the explained capacity (Godfrey-Smith, pers. comm.).

This is not the only difficulty facing attempts to fill in the blank at such a high degree of abstraction as the walking/flying example entails. The identified function must also be appropriate to feature in a causal explanation. Perhaps, in this case, the blank could be filled by stating that legs and wings both produce forces in the environment that, through the 'mechanism' of equal and opposite reaction, cause an acceleration of the organism. But, even if this 'function' is of sufficiently lower complexity than the capacity it is supposed to explain, it is not clear that it is adequate for a causal explanation of the capacity for locomotion, for it looks much more like a statement of Newton's law rather than a description that specifies actual causes. The more abstractly one describes a capacity, the less likely one is to be able to find a non-trivial analysis of it into components that have the right kind of causal features. What, for instance, do feathers, skin, and scales all do that makes these character-states of the vertebrate integument good candidates for the attribution of function? On a strict application of Cummins's account, Godfrey-Smith's notion that 'response to environmental complexity' is a function of mind would arguably be a victim of this difficulty. This further motivates his dropping the greater-complexity requirement so as to deploy Cummins's definition of function alongside the etiological account.

4. Real Traits, Real Problems

Let us turn now to considering consequences of an effectively realist account of traits for discussions of biological function. The reader will recall that from Godfrey-Smith's brief remarks on the matter we derived the effectively realist suggestion that a single trait is any (possibly complex) feature of an organism whose components have a common evolutionary history (CEH). Under the CEH notion of a trait, the phenotypic expression of a single allele will constitute a single trait as a limiting case, for there can be no parts with different evolutionary histories. Of more interest are traits that are polygenic in origin. If a heart is to count as a single trait, it is because its various parts (chambers, valves, and so on) have a common evolutionary history. If there are to be non-adaptive or maladaptive traits, CEH should be allowed to encompass non-adaptive mechanisms as well as the history of natural selection. The CEH approach can be considered successful insofar as it allows parts of the organism or its behavior to be carved out objectively from a phenotype. I believe, however, that the CEH approach is not adequate to the task. One problem is due to the fact that the extent to which parts of an organism's phenotype share a common history is almost always guaranteed to be a matter of degree. The evolutionary history of a finger is linked to that of other fingers, which is in turn linked to the rest of the hand, the hand to the arm, the arm to the torso, and all of it to the circulatory and nervous systems, and so on. It is this holism that underlies the skepticism of Gould and Lewontin (1979) about dividing an organism into traits for separate analysis.

Gould and Lewontin attack the idea that organisms are composed of isolable traits on the grounds that organisms are tightly integrated systems whose parts cannot be antecedently identified. They write (1979: 217), 'The dissection of an organism into parts, each of which is regarded as a specific adaptation, requires two sets of a priori decisions. First, one must decide on the appropriate way to divide the organism and then one must describe what problem each part solves.' They argue that such decisions cannot be justified aprioristically. Sanford (1997) attempts to undermine their critique, arguing in contrast that a posteriori methods of trait identification are available, based on an understanding of the operative causal mechanisms. I believe that the best way to understand the notions of 'a priori' and 'a posteriori' at work in this debate is not in the traditional epistemologist's sense of prior or subsequent to experience. Rather, at issue is whether biological theory provides any way of making the divisions. In the absence of reasons for dividing up the organism in a particular way that are based on an empirically derived theory, the division is empirically and theoretically arbitrary and in that sense aprioristic.

Even if Gould and Lewontin's charge that such divisions are unacceptably aprioristic can be avoided, I would argue that CEH is problematic in failing to provide a non-arbitrary way of isolating single traits. CEH is a matter of degree, and there is no non-arbitrary way to set a threshold on how much shared history is enough shared history to identify a single real trait, except by requiring complete linkage of the genes responsible for the expression of the phenotypic character. If that threshold is adopted, then there will be relatively few complex single traits, and those that would pass the test are unlikely to correspond to our common (albeit pre-theoretical) ideas about what is to count as a single trait. There is a growing recognition that genes are shared by different systems (Piatigorsky 1998). For example, squid maintain symbiotic, bioluminescent bacteria in structures designated as 'light organs'. Enzymatic assays reveal that these organs contain high densities of proteins that are found in both ocular lenses and ordinary muscle tissue (Montgomery and McFall-Ngai 1992), indicating that lenses and muscle have a common evolutionary history. Yet it is counter-intuitive that they should be counted as the same trait.

It might be argued that such an outcome is the cost of progress—that our intuitions should not count for much and that such a thorough revision of our conception of trait is required—but that is a discussion I do not intend to pursue here. My purpose is served if this gives one more reason to agree that the notion of trait requires further investigation.

If the collection of capacities that constitute cognition, or a mind, is not a single trait by the CEH standard, is this fatal to Godfrey-Smith's project of finding the function of mind? Godfrey-Smith explicitly makes the assumption that mind is a single trait, but need he make this assumption? If the mind fails to count as a single trait, containing parts with separate evolutionary histories, then one might worry that functions grounded etiologically in natural selection could be attributed only to the parts and not to the whole. But is this worry justified? If the worry has purchase anywhere, it would be most acute in what I will call 'gerrymandered' traits: pairings or larger groupings of features that seem entirely *ad hoc*. Consider, for instance, the combination of toes and nose. Does toes + nose have any functions of its own, beyond the functions of toes and the functions of noses?

It is my view that functions of a gerrymandered trait might be objectively attributable and distinct from the simple summation of the functions of each part considered separately. Take objectivity first. Gerrymandered entities or properties can generally be accommodated within an objective framework. For instance, consider the gerrymandered entity consisting of myself (CA) and the Statue of Liberty (SOL). Call it 'CA–SOL'. The combined mass of CA–SOL is an objectively measurable property derived from the masses of each component of CA–SOL. Furthermore, we might define some other quantity, call it 'cohesion', which is a function of the gravitational attraction between CA and SOL. 'Cohesion' is not a property of either part alone, but it is a perfectly objective property of the whole that could, in principle, be measured or calculated at any moment. In a corresponding fashion, then, if toes have a function via their contributions to ancestral fitness, and noses have a function via their contributions to ancestral fitness, the gerrymandered 'trait' of toes + nose has, minimally, the combined functions. But because the toes + nose complex may have had synergistic effects for which there has been selection—in locomoting towards good smelling food, for instance—it may be possible to assign a function objectively to toes + nose that does not belong to either toes or nose alone.

Put differently, the contribution to ancestral fitness made by the complex toes + nose may have been greater than the contributions of each separately. Thus there may have been selection for the complex over and above selection for each component. (Something like an iteration of this argument may provide additional support for Gould and Lewontin's holism about selection on phenotypes.)

Of course the mere fact of objectivity does not entail that these objects or properties are of any scientific or intellectual interest. The 'cohesion' of CA–SOL is hardly of any fundamental interest to physicists, and unlikely even to be of any practical interest to anyone—not even its sentient component. It is perhaps here that Godfrey-Smith's worries about the biological reality of the traits he labels 'mind' and 'cognition' have some purchase. It can reasonably be maintained that 'mind' is not as gerrymandered as toes + nose. But the notion of mind is gerrymandered nevertheless. To see this, consider the role of learning in cognition. Classical conditioning, operant conditioning, and observational learning all have different evolutionary histories. The capacity for classical conditioning is, for instance, sensitive to the kinds of correlations between stimuli that have been salient in the natural environments of the particular species. In other words, the capacity to form a given connection between a given unconditioned stimulus (US)—for example, food—and a given conditioned stimulus (CS)—for example, bell—varies from species to species depending on the evolutionary history of that species. The capacity for operant conditioning is also affected, but by different facts about the evolutionary history. For instance, it is much easier to get pigeons to tap a stimulus with their beaks for a food reward than it is to get them to press a bar with a foot for that same reward, whereas for rats the opposite (*mutatis mutandis*) is true. These differences presumably can be explained in terms of the different natural feeding strategies (for example, pecking versus digging) in each species. Furthermore, within a given species, the capacity for classical conditioning will have been shaped by different selection pressures than the capacity for operant conditioning.

Hence, learning is not a single trait according to the CEH criterion. Since cognition subsumes learning neither is it a single trait. Even so, there may be biological functions that arise from the interactions between the various components that cannot easily be reduced to the functions of the individual system components. And, if so, it may be possible to conduct an inquiry into the biological functions of cognition without Godfrey-Smith's assumption that mind is a single trait.

The possibility remains that a trait such as 'cognition' is too gerrymandered to be of interest to biologists. Without a non-arbitrary way to say how much gerrymandering is too much, we have only a vague conception of trait. Determining the boundaries of this conception is presently a matter of interests and expedience. Consequently, etiological functions are perhaps less sparse and more interest relative than the realists have suggested.

5. What is a Trait?

In this chapter I have raised some questions about how well we understand the notion of trait. I suspect that different notions of trait are needed to serve different purposes for biologists, but here our focus has been on how these questions bear on the accounts of biological function. Even in this narrower context, it may well be that no single definition of trait will be forthcoming.

No matter what account of function biologists explicitly endorse, their approach to function attribution depends, as a matter of practical necessity, upon the comparative method that utilizes comparison of similar structures or behaviors within and outside a taxonomic group. Etiological accounts explicitly identify the functions of trait T in organism O by looking to O's ancestors to find the effects of ancestral versions of T (T*) relevant to the survival and reproduction of O's ancestors. In practice, of course, one cannot literally look to O's ancestors, who are long since deceased. Rather, one usually must apply the comparative method to the traits (T*) of related species to try to draw inferences about the functions of ancestral traits. The issue of cross-taxonomic trait identifications must also be faced by proponents of systems accounts of function if they are to provide an adequate framework for the application of the comparative method.

Attempts to deal with these cross-generational and cross-taxonomic issues must respect the following facts:

(*a*) *Phenotypic variation.* T and T* will typically be distinguishable. My nose need not be identical in shape to any of my ancestors' noses (for which I am extremely grateful in several cases). Nor need it be identical in form to a chimpanzee's nose for the comparative method to be applied.

(*b*) *Genetic variation.* The genetic causes of T will not be identical to T*. Typically, T and T* will be polygenic. The exact set of alleles responsible for my nose need not be the same as those responsible for my ancestors' noses or a chimpanzee's nose. The further back in history or across taxa we go, the less correspondence there will be of alleles, or even of loci on chromosomes.

Consequently the notion of tokens of the same trait cannot be based on genes or phenotype alone. Given the complexity of genotypic and environmental causes of phenotypes, it seems to me that any definition of 'trait' based on these notions will result in a notion of 'same trait' that is inherently vague. (See also Millikan's definition (1984) of 'higher-order reproductively established family' wherein it is a proper function of reproduced genes to produce 'similar' structures. This melding of phenotypic and genetic aspects still relies upon a potentially troubling, unanalyzed notion of similarity. For example, for some purposes squid muscles and the covers of their light organs are similar, as well as having a common genetic cause.)

Of course, biology does not and should not grind to a halt without an answer to the question 'what is a trait?' But one who has physics as a model for scientific inquiry might see the vagueness apparently inherent in the notion of trait as a reason to think that scientific biology cannot preserve such a notion as a basis for its future development. Others might see such vagueness as quintessentially biological. Just as there are no clear boundaries between species, why should there be a clear boundary between traits? Perhaps a hierarchy of trait-types can be mapped onto the taxonomic hierarchy. I argued above that, as traits are identified across broader and broader taxonomic groups, they become less suited for the kinds of causal explanations that are supposed to ground function attributions, especially in systems-analysis approaches. But perhaps this counts in favor of the proposal, for it seems rightly to capture the difference between the rather minimal informativeness of answering the question 'why do vertebrates have limbs?' with 'for locomotion' and the somewhat greater informativeness of answering 'why do cheetahs have long legs?' with 'for speed in catching antelope' without appealing to an unanalyzed notion of complexity.

References

Allen, C., and Saidel, E. (1998), 'The Evolution of Reference', in D. Cummins and C. Allen (eds.), *The Evolution of Mind* (New York: Oxford University Press), 183–203.

——Bekoff, M., and Lauder, G. V. (1998) (eds.), *Nature's Purposes: Analyses of Function and Design in Biology* (Cambridge, Mass.: MIT Press).

Ayala, F. J. (1970), 'Teleological Explanations in Evolutionary Biology', *Philosophy of Science*, 37: 1–15.

Bekoff, M., and Allen, C. (1995), 'Teleology, Function, Design, and the Evolution of Animal Behavior', *Trends in Ecology and Evolution*, 10/6: 253–63.

Bock, W., and von Wahlert, G. (1965), 'Adaptation and the Form–Function Complex', *Evolution*, 19: 269–99.

Brandon, R. N. (1990), *Adaptation and Environment* (Princeton: Princeton University Press).

Cummins, R. (1975), 'Functional Analysis', *Journal of Philosophy*, 72/20: 741–65.

Eldredge, N., and Cracraft, J. (1980), *Phylogenetic Patterns and the Evolutionary Process* (New York: Columbia University Press).

Ereshefsky, M. (1992), *The Units of Evolution: Essays on the Nature of Species* (Cambridge, Mass.: MIT Press).

Futuyma, D. J. (1998), *Evolutionary Biology*, 3rd edn. (Sunderland, Mass.: Sinauer Associates).
Godfrey-Smith, P. (1996), *Complexity and the Function of Mind in Nature* (New York: Cambridge University Press).
Gould, S. J., and Lewontin, R. C. (1979), 'The Spandrels of San Marco and the Panglossian Paradigm: A Critique of the Adaptationist Programme', *Proceedings of the Royal Society of London Series B: Biological Sciences*, 205: 581–98.
Hinde, R. A. (1975), 'The Concept of Function', in G. Baerends, C. Beer, and A. Manning (eds.), *Function and Evolution in Behaviour: Essays in Honor of Niko Tinbergen* (Oxford; Oxford University Press), 3–15.
Millikan, R. Garrett (1984), *Language, Thought, and Other Biological Categories: New Foundations for Realism* (Cambridge, Mass.: MIT Press).
——(1989), 'In Defense of Proper Functions', *Philosophy of Science*, 56/2: 288–302.
Montgomery, M. K., and McFall-Ngai, M. J. (1992), 'The Muscle-Derived Lens of a Squid Bioluminescent Organ is Biochemically Convergent with the Ocular Lens. Evidence for Recruitment of Aldehyde Dehydrogenase as a Predominant Structural Protein', *Journal of Biological Chemistry*, 267/29: 20999–21003.
Piatigorsky, J. (1998), 'Multifunctional Lens Crystallins and Cornea Enzymes: More than Meets the Eye', *Annals of New York Academy of Science*, 842: 7–15.
Pinker, S. G. (1994), *The Language Instinct* (New York: W. Morrow and Co.).
Pittendrigh, C. S. (1958), 'Adaptation, Natural Selection and Behavior', in A. Roe, and G. G. Simpson (eds.), *Behavior and Evolution* (New Haven: Yale University Press), 390–419.
Rudwick, M. J. S. (1964), 'The Inference of Function from Structure in Fossils', *British Journal for the Philosophy of Science*, 15: 27–40.
Sanford, G. M. (1997), 'Explaining Evolution: Genes, Culture, Environment, and Mechanisms', Ph.D. dissertation, Duke University.
Smith, B., and Varzi, A. C. (1999), 'The Niche', *Noûs*, 33/2: 198–222.
Sober, E. (1988), *Reconstructing the Past: Parsimony, Evolution, and Inference* (Cambridge, Mass.: MIT Press).
Tinbergen, N. (1963), 'On Aims and Methods of Ethology', *Zeitschrift für Tierpsychologie*, 20: 410–33.
von Schilcher, F., and Tennant, N. (1984), *Philosophy, Evolution and Human Nature* (London: Routledge & Kegan Paul).
Williams, G. C. (1966), *Adaptation and Natural Selection: A Critique of Some Current Evolutionary Thought* (Princeton: Princeton University Press).
Wright, L. (1973), 'Functions', *Philosophical Review*, 82: 139–68.

15. Types of Traits: The Importance of Functional Homologues

Karen Neander

ABSTRACT

The traditional view was that biology has many functional categories that classify organismic traits. Now, a number of philosophers maintain that there are no important functional categories in biology, with the possible exception of the analogous categories. This chapter explains why function, structure, and homology all have important and complementary roles in the classification of homologous traits.

1. Introduction

Philosophers have long believed that biology is richly endowed with functional categories. The idea has been that a heart counts as a heart because it is for pumping blood and eyes count as eyes because they are for seeing, and so on. More generally, on the traditional view, the traits of organisms are often classified into types at least in part on the basis of their biological function. This view, which extends as far back as Aristotle, has long been entrenched in contemporary philosophy of biology (see e.g. Beckner 1959: 112–18). Indeed, it is still the accepted wisdom in philosophy more generally. For instance, we find Jaegwon Kim (1996: 77) explaining that 'even many biological concepts (e.g., the gene, heart) appear to have an essentially functional component', and Tyler Burge (1989: 312) telling us that 'to be a heart, an entity has to have the normal, evolved function of pumping blood in a body's circulatory system'.

I have maintained the same view myself in earlier work (Neander 1983; 1991a: 180) and I also illustrated the point with reference to the vertebrate heart.

I am grateful to Mohan Matthen, Kim Sterelny and Colin Allen for reading and commenting on an early draft. And I am especially indebted to Michael McCloskey for useful discussion and some very helpful suggestions. Versions of this have been given to the philosophers at Stirling, Syracuse, UC Davis and Boston University and in each case I benefited in significant ways from the discussion. I am grateful to those in the audience who gave me such useful feedback.

For instance, 'heart' cannot be defined except by reference to the function of hearts because no description purely in terms of morphological criteria could demarcate hearts from non-hearts. Biologists need a category that ranges over different species, and hearts are morphologically diverse. . . . Highly significant, moreover, is that for the purposes of classifying hearts, what matters is not whether the organ in question manages to pump blood, but whether that is what it is supposed to do. The heart that cannot perform its function (because it is atrophied, clogged, congenitally malformed, or sliced in two) is still a heart. (Neander, 1991*a*: 180)

There are two claims here. The first is the traditional claim, that functional classification is required due to morphological diversity. The second is the claim that the concept of function that is implicated is the concept of a normal (or proper) function—that is, the concept of what a trait is 'supposed' to do, which is not necessarily what it is disposed to do. These claims were not made in the context of an extended treatment of trait classification but were offered in passing by way of partial motivation for an inquiry into how attributions of normal function are to be understood. However, slight as this discussion of trait classification was, for some time it was hard to find any more elaborate defense of the traditional view. For that reason this brief discussion has served to represent the traditional position and has been a target for those who now want to reject that position.

My guess is that the traditional view was probably so little elaborated because it had for so long gone unquestioned and unchallenged. Now it remains as poorly elaborated as before, but at last it has come in for extended criticism. This criticism begins with two papers published in the same year, one by Ron Amundson and George Lauder (1994) and the other by Paul Griffiths (1994). We owe these authors gratitude for stirring up our stale assumptions. They argue that I did not establish that functional categories are pervasive in biology, and I agree with them on this. They have gone further, however. They have rejected the traditional view and not merely my defense of it, which is where things become more interesting. At a very rough first pass, their view is that the importance of function in the classification of biological traits has been greatly overrated. Actually, I agree that there is some truth to this too, although not nearly as much as they suppose. However, these are vague claims. At the risk of imposing more clarity on the debate than there really is, I suggest that there are some more precise claims that capture our differences. I will explain them, and the terms in which they are expressed, more carefully in the next section. But I will mention them here so that those already familiar with this fledgling debate can have a glimpse of where I am headed. Amundson and Lauder and perhaps also Griffiths seem to maintain that there are no functional categories of any scientific significance in biology, with the exception of the analogous categories, analogous categories being categories of traits that have evolved independently to serve the same function. Others who are now sympathetic to this anti-traditional view have serious doubts about even this much of a concession. However, I am not interested in the status of the concession—the importance or unimportance of analogous categories—in this present chapter. What interests me here is the claim that there are no interesting or important functional categories outside the analogous categories. With or without the concession, this claim involves a dramatic overturning of the traditional view. For even the vertebrate heart, to return to our familiar (not to say hackneyed) example, is a homologous and not an analogous trait. So, on this radical view, not even the vertebrate heart is a functional category.

I suggest that this view, which I will refer to as Functional Minimalism, distorts the truth at least as much as the overly simple traditional view. In what follows I will propose a view that lies somewhere between the two. I will argue that function has an important role in many homologous trait-types, not to the exclusion of but alongside other complementary principles of classification. To some extent, my positive proposal overlaps with some versions of Functional Minimalism. Those who reject the traditional view often claim that the alternative to classification by function is classification by homology and I agree that homology is important in the classification of traits. Where we differ is on the role of function in homologous classifications. In what follows I explain that homology is inadequate as a solitary principle of classification and that it requires assistance from some further complementary principle or principles of classification. This further assistance is not always provided by functional considerations and it is no part of my aim here to contest the importance of structural considerations in trait typology. However, I maintain that function also plays an immensely important role. I will argue that homologous kinds are sometimes differentiated in terms of structure, sometimes in terms of function, and sometimes, and often ideally, in terms of both.[1] I like to think that some Functional Minimalists will agree with my position the moment it is explained. Perhaps this is optimistic, but I suspect that the view I am about to defend has sometimes been rejected or apparently rejected only because it had not been articulated as a possible position.

There are three points of agreement among the parties to this debate that I want to note by way of further preliminaries. One is that the present discussion concerns a certain class of categories that I will call *abnormality inclusive categories.* These are categories of organismic traits (i.e. of traits of organisms) that include both normal and abnormal items. The category of all hearts is an abnormality inclusive category because abnormal hearts count as hearts along with normal ones. Very many categories of traits are abnormality inclusive but

[1] I am using the terms 'kind', type', 'category', and 'classification' interchangeably. Some will object to my use of the term 'kind', but they are free to substitute less onerous terms. See also n. 14 below.

not all of them are. Trivially, the category of normal hearts is not, since it excludes abnormal hearts. Nor are the categories of abnormal livers or cAMP deficient neural pathways, since they exclude normal livers and normal neural pathways respectively.[2] It is clear that the Functional Minimalists also take themselves to be describing abnormality-inclusive categories. As we will see, they are at pains to show that the categories in question, on their own analysis of them, are appropriately inclusive of variation, both normal and abnormal.

The second point of agreement is the second of my two earlier claims, mentioned above, put into conditional form. It is agreed that, *if* these abnormality-inclusive categories are functional categories, then the concept of a *normal function* is implicated. This concept underlies the distinction between normal and abnormal functioning in physiology and goes with talk of traits malfunctioning, of functional deficits, functional impairment, dysfunction, and so on. It contrasts with a concept of function that sometimes goes by the name of a 'causal-role function'—roughly, the concept of an aspect of or subset of a trait's actual causal dispositions. This latter concept cannot play the definitive role for functional categories that are abnormality inclusive. For example, if a heart does count as a heart because of its function of pumping blood, and does so even if it cannot pump blood, it must be that it counts as a heart in virtue of its normal function and not in virtue of its causal-role function.[3] The Functional Minimalists accept this much of my original argument. It is the antecedent of the conditional that they question. In this respect, the title of Amundson and Lauder's paper is misleading. Although it is called 'Function without Purpose: The Uses of Causal Role Function in Evolutionary Biology', it is not a defense of causal-role functional categories. It is an extended argument against classification by function and in favor of classification by some other means (homology or perhaps 'anatomy').

The third point is really only a point of near agreement rather than one of total agreement. I will assume in what follows that something like the account of normal functions that I have defended elsewhere is correct (Neander 1983, 1991*a*, *b*). That is, I will assume that normal functions are (what are elsewhere

[2] It is a good question why biology would want abnormality-inclusive categories. This chapter is extracted from a draft of a book-length manuscript in which that question is addressed, but I will say little about it here. It is plain that biology makes extensive use of abnormality inclusive categories and I hope that this will be sufficient motivation for this chapter.

[3] Christopher Boorse (this volume) claims that malfunctioning hearts do not have the functions that they cannot perform. On such an account, the item that is malfunctioning retains its normal function in a quite straightforward sense. That is, despite the fact that it is malfunctioning, it still belongs to the lineage it belongs to and that lineage was still selected for that which it was selected for. Details aside, this is what confers a normal function upon the item, according to an etiological account. This, of course, is quite consistent with its inability to perform its function. If one reads 'has the function X' as meaning the same as 'performs or can perform the function X', this will seem incoherent, but such a reading is not justified on an etiological account. Nor, as far as I can see, is it the usual reading.

in this volume referred to as) *etiological functions.* I will assume, without any attempt to argue for it here, that the concept of normal function is to be understood in terms of an etiological account. The gist of such an account, at least as I would express it, is that the normal function of an ordinary somatic trait is the disposition for which the relevant lineage of traits was selected by natural selection. Put simply, the function of hearts is to pump blood because that is what hearts were selected for.[4] Some of what follows will hang on this assumption. However, the main proponents of Functional Minimalism are willing to grant it to me. Amundson and Lauder cite my account by way of elaborating the notion of a normal function and Griffiths (1993) has defended an account that is basically the same as my own. In addition, those who reject an etiological account of normal function might be willing to grant me my assumption for the sake of the present argument, if only because they might expect it to be more of a hindrance than a help.

One more preliminary comment. My interest in functional categories is on the present occasion mostly from the perspective of an interest in physiology (and in neurophysiology and neuropsychology), although some of what I have to say is more broadly relevant. In fairness to the Functional Minimalists, I must add that they rarely seem to share this interest. However, I believe they should share it, given that their subject is functional categories, especially given that they draw conclusions regarding the importance of functional categories for biology in general. Some of the Functional Minimalists are more focused on evolutionary theory and others on anatomy or even morphology. Sometimes it is claimed that their discussion relates to some such area only, although this limitation is then forgotten and morals are drawn for biology in general. Sometimes there is no explicit suggestion of any such limitation. But physiology is the discipline devoted to the study of the functional organization of living systems and the importance of the concept of normal function to physiology is undeniable. So it is physiology that is the obvious hard case for the Functional Minimalists and they cannot establish their position without careful consideration of it.

2. Functional Minimalism

There are a number of philosophers who have contributed to a growing skepticism concerning the importance of functional categories in biology, apart from those I have already mentioned.[5] Naturally, there are differences among

[4] There are two recent collections that readers might turn to for related discussion, see Allen *et al.* (1998) and Buller (1999).

[5] Mohan Matthen (2000) argues for views that might be construed as Functional Minimalism. Matthen sees himself as doing so in his chapter in the present volume, but I disagree (see n. 8 below). Walsh and Ariew (1996) also express support for the view, though with little discussion.

these philosophers with respect to the positions they maintain and the arguments they give in their support. Nonetheless, I think it is useful to identify a negative core view for the purpose of discussion and it is this negative core view that I am referring to as Functional Minimalism. This is the view that there are no important functional categories in biology except perhaps for the analogous categories. Or, to put it another way, the view is that there are no intra-homologous functional categories of any scientific significance. This seems to be the view expressed in the following passage from Amundson and Lauder, for instance. They say,

> there is indeed a set of important biological categories which group organic traits by their common biological roles or functions. The most general of these apply to items which have biological roles so broadly significant in the animal world that they are served by analogous structures in widely divergent taxa. . . . Narrower function categories occur also (e.g., kneecap and ring finger) but they are of limited scientific interest. . . . The importance of the above function categories comes from the fact that they all apply to features which result from evolutionary convergence—the selective shaping of non-homologous parts to common biological roles. (Amundson and Lauder 1994: 455–6)

Griffiths (1994: 213–14) independently reaches what appears to be the same conclusion: 'If functional classifications are to be of value in biology it must be because of their superior generality—the fact that they unite disjunctions of cladistic homologues.'

Let us take a moment to review the standard definitions of 'homology' and 'analogy', since these will be important for the discussion that follows. The concept of homology has had a long and interesting history in biology (see Matthen 2000). As is only to be expected, different periods of biological thought have conceived homologous traits differently, but in modern biology the concept of homology is an evolutionary concept. The standard definition of the term tells us that the same trait (or similar traits) in two or more separate species are homologous when that trait (or those traits) have been derived without interruption but with or without modification from a common ancestor. So the wings of pigeons and parrots count as homologous given that there is a common ancestral species whose members were winged and given also that all of the intervening ancestral species on both sides were also winged. Another instance of homology is the vertebrate forelimb; the forelimbs of bats, birds, humans, horses, and so on. Biologists sometimes speak of a continuity in the flow of genetic information and common embryological and developmental pathways in this context (see e.g. Roth 1984). The first is generally considered to be constitutive whereas the second is generally considered to be, not constitutive, but strongly evidential (that is, confirming or disconfirming) of homology, given that embryological pathways are highly conservative in evolution.

Homology is contrasted with homoplasy. The standard definition of 'homoplasy' again defines it in terms of a relation between the traits of two or more separate species. In this case, there is the possession by two or more separate species of similar traits that are not derived in an uninterrupted line from a common ancestor. The evolution of the two traits might be entirely independent, in which case it is referred to as a case of convergent evolution. Or the development of the two traits might be due to a reversion to or a re-evolution of a more primitive form. In either case, if the homoplasy is due to the two traits being shaped by similar selection pressures for the same function, the homoplastic traits are then said to be analogous. Wings are an example of an analogous classification, because bird wings, bat wings, and insect wings have each evolved separately.

Simply put, homology is resemblance that is due to descent and analogy is resemblance that is due not to descent but to the separate evolution of similar traits for the performance of the same function. These are admittedly only rough and ready definitions of the two concepts, but they are in keeping with those commonly found in biology texts as well as those found in the philosophical literature of interest to me, including the writings of the Functional Minimalists. The definition of homology, especially, strikes me as being problematic in a number of ways. For instance, it does not comfortably accommodate iterative homologies (for example, our fingers, which are said to be iterative homologues of each other insofar as the genetic information for one finger shares a common ancestry with the genetic information for the other fingers[6]). It does not seem to accommodate them, because it seems to require that homologous traits be in separate species. For the same reason, this standard definition of homology does not allow a homologous grouping of traits to be restricted to a single species. This is at least a questionable restriction. It is also a restriction that one would not want to insist upon if one was advocating that non-analogous categories are essentially homologous categories, since some types of traits are species specific. However, it will be more straightforward in this context to work with the standard definition, simply noting that we have some reason to think that the standard definition might be improved upon. We might also say instead that a broader notion of a lineage of traits might better suit our purpose.[7]

One thing is quite clear from these standard definitions. The analogous categories are functional categories as a matter of definition (that is, as a matter

[6] Given the complexity of the interactions between genes in the genome and also between the genome and the environment, it is admittedly crude to speak of the genes for such and such a trait. However, this kind of simplification is harmless in this context I think. As far as I can see, to speak otherwise would merely complicate the expression of the present discussion without changing its substance.

[7] For discussion of the concept of homology, see Roth (1984, 1988, 1991), Hall (1994), and Lauder (1994).

of the definition of the term 'analogous'). Analogous categories are, by definition, groupings of diverse homologous types that have a common function. So there is no choice but to agree that function plays a role in classifying traits in the analogous categories. There is room for disagreement concerning the value to biology of such analogous categories, but it is not open to dispute that they are functional categories of some sort. Griffiths (1994) has some discussion of their value, in which he argues rather tentatively for their importance. Matthen (this volume) also has some discussion of them, in which he expresses skepticism about their use.[8] But, as I said in the Introduction, I plan to leave this issue to one side. I doubt that they are as important as the homologous categories. If that is correct, when it comes to assessing the traditional view of the importance of functional classification in biology, the importance or unimportance of analogous categories is a secondary issue. The first question to ask is, what role does the concept of function play in homologous classifications? None, as the Functional Minimalists seem to claim? Or a significant role, as I maintain?

Judging from conversations that I have had with some philosophers on this issue, Functional Minimalism can also seem to be a simple definitional matter, at least at a superficial glance. Just as the analogous categories are, by definition, functional categories, so too, some want to say, the homologous categories are, by definition, not functional categories. However, this is just a mistake. There is nothing in the definition of 'homology' that rules out the possibility that some homologous categories are also functional categories. It would, of course, be a simple logical error to think that, because analogous categories are functional categories and homologous categories are not analogous categories, it follows that homologous categories are not functional categories. But there is apparently a temptation to equate analogous categories with functional categories and to contrast homologous categories with analogous categories and hence with functional categories. As we will see, analogous categories and functional categories cannot be simply equated. We could, if we chose, simply stipulate that the term 'functional category' is to be treated as synonymous with the term 'analogous category'. But, while that would render Functional Minimalism true, it would also render it trivial, and we would have to find other terms in which to ask the interesting question. If it is a matter of stipulation that the only functional categories are the analogous categories, then of course there can be no functional categories of any importance aside from the analogous categories. But the question would remain, does the concept of function play a role in some important homologous classifications?

[8] Matthen says he is arguing against Functional Minimalism (as I define it) in his chapter in this volume. However, I don't see it. Perhaps my definition of it was not clear in the earlier version of this chapter that Matthen read. There is nothing in what I say that involves a rejection of the idea of local fitness peaks, for instance.

Amundson and Lauder do not make either the simple logical error or the trivializing move mentioned above. However, for instance in the following passage, they sometimes stress the definition of 'homology' as if they think that those who disagree with them cannot have understood it and as if their conclusion somehow follows fairly immediately from this definition.

> Whatever the favored definition of homology, one feature of the concept is crucial: *the relation of homology does not derive from the common function of homologous organs.* Organs which are similar in form not by virtue of phylogeny but because of common biological role (or SE function) are said to be *analogous* rather than homologous—they have similar SE[9] function and so evolved to have similar gross structure. . . . The fact that anatomical or morphological terms typically designate homologies shows that they are not functional categories. (Amundson and Lauder 1994: 455, emphasis in original)

Note that they are saying that it is *because they are homologous categories that they are not functional categories.* They are not saying, as one might expect them to, that it is because they are anatomical or morphological categories that they are not functional categories.

The definitional point that Amundson and Lauder stress is perfectly correct, but their conclusion does not follow from it. Or not without further argument. Certainly, homologous relations can and do extend across radical changes in function. This is uncontroversial. As Amundson and Lauder mention to illustrate their point, certain of the mammalian inner ear bones and certain of the reptilian jawbones and portions of the gill arches in fish are homologous. As are the vertebrate forelimbs, such as the wings of birds, the front legs of horses, the arms of humans, and the fins of fish. These two broad homologous groupings include functionally diverse items. But the most that could follow from this is that some homologous categories are not functional categories. We have to be cautious about drawing any further and stronger conclusion.

Indeed, the very examples that Amundson and Lauder use to illustrate their position can as well be used in support of my own. While the two larger categories of traits that they mention are functionally diverse, the various smaller categories within these two larger categories are functionally uniform. These small homologous groupings within the two larger homologous groupings seem to be sensitive to functional similarities and differences. These same mammalian inner ear bones, for instance, seem to form a grouping that is distinct by virtue of a conjunction of both homologous relations and specialized function. As do these same reptilian jawbones, and these same portions of the gill arches of fish. Perhaps the functional uniformity of the group is incidental,

[9] What they refer to as an 'SE function' is what I am referring to as a 'normal function'. 'SE' is short for my 'selected effects' (Neander 1991*a*). That is, it is short for 'effects for which there was selection'.

but my present point is just that they could be functional categories, and at a glance they appear to be so. At any rate, it is perfectly consistent with the definition of 'homology' that these are homologous groupings of traits that we differentiate from related homologous groupings of traits on the basis of functional considerations.

One of the problems with the terms in which the issue has been discussed is that the possibility of mixed criteria is too often ignored. The most obvious instance of this is in a brief argument sketch given by David Walsh and André Ariew (1996) following Lauder (1994). They argue that, since similarity of function is only defeasible evidence for sameness of biological trait-type in Cladistic systematics, function cannot be the criterion of sameness of trait-type (homology is instead). Putting aside the use of the definite article, the inference is a bad one, unless we assume the hidden premise that there can be only one criterion *per* category. Suppose there were two. Then function can be defeasible evidence of sameness of trait-type and criterial as well. If being A and being B are both required for being C, then x's being A is criterial for its being C but it is also defeasible, since x's being not-B can still make it the case that it is not-C. If we allow that there can be mixed criteria, we can stress the importance of homology all we like without necessarily saying anything to effect against the importance of function. For that reason, I find myself agreeing with much that the Functional Minimalists say in defense of their positive proposal but without being in the least persuaded of their negative conclusion. We can agree that homology is important without concluding that function is thereby unimportant.[10]

Although I am going to say little about the role of functional characters in phylogenetic inference, since a serious discussion of it would take me too far afield, a brief aside on cladogram construction is in order, given that the matter has been raised. Homology has a special role in Cladistics, but it is not true that Cladistics decrees that traits be typed on the basis of homology. A cladogram is a kind of diagram that illustrates a hypothesis concerning phylogenetic relations between extant species. And what biologists do, in creating a cladogram, is feed into a computer a lot of data about what traits the individuals in the various species have. The computer program is designed to determine, on the basis of this information, which traits should be considered homologous and which traits should be considered homoplastic. It does this by determining which phylogenetic hypothesis would require the least evolutionary

[10] It might be helpful in this context to speak of 'pure functional categories' and 'mixed functional categories'. A pure functional category is a category for which function is the only criterion. (Even analogous categories are not pure functional categories. The biologist is not referring to the wings of a jet when she talks about wings.) It might be thought that some of the 'Functional Minimalists' meant only to argue that most trait types are not purely functional. But they do not *say* this. And they contrast classification by function with classification by homology and urge upon us the importance of the latter and the redundancy of the former.

change, if it were true. By maximizing homology and minimizing homoplasy, it determines which evolutionary pathway is most economical. Not that evolution is necessarily economical, but this is thought by the Cladists to be our best guess as to which historical hypothesis is true. Notice that some pre-sorting of traits—some identification of them as similar or different—has already occurred prior to the decision to treat them as homologous or not. This in itself suggests that whether or not two traits are homologous cannot be *all* there is to trait classification.

3. Homologous Categories

Homology cannot do the work of classification on its own. There are no pure homologous kinds. To see why, let us take a look at Griffiths's positive proposal (1994), which attempts to describe homologous kinds.

Griffiths suggests that, since taxonomies of organisms into species and higher taxa are cladistic it would be fitting if taxonomies of traits were too. 'If the natural kinds of organisms are cladistic, one might well suppose natural kinds of organismic traits are cladistic homologues. Cladistic traits, like cladistic species, can be seen as the natural kinds for which biology has always been looking' (Griffiths 1994: 211).

Griffiths variously refers to his favored kinds as 'trait clades', 'cladistic traits', 'cladistic homologues', or 'homologous traits'. In the usual way he notes that, 'since Darwin, homology can be identified with resemblance due to descent from a common ancestor, and analogy with resemblance due to convergent evolution' (Griffiths 1994: 212). However, his concept of a cladistic homologue, as the name suggests, also involves the concept of a clade. A clade is a monophyletic group—that is, a group of organisms that consists of all and only the descendants of some ancestral population. Cladistic homologues are, he says, traits that 'unite monophyletic groups' in the sense that 'every species in the clade has the trait or is descended from a species that has it' (Griffiths 1994: 212). The idea seems to be that we identify the first population in which a given trait occurs and then define a trait clade with respect to that population, such that it consists of all and only the descendants of that population.

Notice that such trait clades clearly embrace diversity across species. A clade includes all of the descendants of the designated ancestral population, no matter what morphological modifications (or changes in the 'species design') occur along the way. And, as Griffiths remarks in response to my earlier comments on this topic (as quoted in the Introduction to this chapter), pathological variation will also be included in the trait clade because pathological individuals are just as much descendants of the original population as normal individuals are (Griffiths 1994: 213). So we might be tempted to conclude with

Griffiths that trait clades are appropriately inclusive of variation, for they do include divergent designs across species as well as normal and abnormal variation within species.

The idea of a trait clade seems clear enough, but, as it stands, it is hard to make sense of it as a complete proposal for the classification of traits. A trait clade includes *all* of the descendants of the original population, no matter *what* morphological changes have occurred. Consider, for instance, the fact that descendant species that lack the trait may nonetheless be members of the trait clade. There is, of course, a difference between a lineage losing a trait and a lineage never having possessed it. But there is also a difference between a lineage still possessing a trait and its not possessing it any longer, and it is this latter distinction that is obscured in Griffiths's proposal. The problem that he fails to see is that, as he defines a trait clade, belonging to a trait clade does not guarantee that the relevant trait is possessed. For example, consider the category of gills. The idea seems to be that we identify the population in which gills first occurred and then we define the trait clade for gills as the clade that descends from that population. The problem is that we are members of that 'trait clade' and yet we do not have gills, so it is hard to see how this concept of a trait clade has in any sense captured the category of gills. Therefore, while it is true, as Griffiths claims, that a trait clade embraces inter-species and intra-species variation, his trait clades are much too liberal. There are other problems with this proposal too. One is that the trait clade does not distinguish between different traits that unite the same monophyletic group. If two traits arise in the same founding population, then they will share the same trait clade. Since there can be two distinct types of traits and only one trait clade between the two, a trait-type cannot be identified with a trait clade. Another problem is that the proposal does not explain how we identify when the trait first occurs. The proposal presupposes that we can identify sameness and difference of trait-type. Of course, we can do so intuitively. That is, biologists *do* do so intuitively. But the point of the exercise is to make the principles explicit.

No criticism of Cladistics as a system for classifying organisms follows. However, we have to conclude that Griffiths's proposal fails as a basis for classifying organism traits. What we need is a suitable way to differentiate one homologous category of traits from other related homologous categories of traits, and Griffiths seems to appreciate that fact. But the concept of a trait clade does not do this in an appropriate way.[11]

[11] Matthen has suggested that an inverted clade might do better. The idea here is that we designate a population that possesses the trait in question and then work backwards, including all of its ancestors, as far back as some designated stopping point. However, this proposal seems seriously incomplete too. Suppose that the group of creatures that possesses the trait forms a paraphyletic group (i.e. the group includes only but not all of the descendants of the population in which the trait originally arose). Then in that case some branches of the trait homologue will be left out. The last two objections raised against Griffiths's proposal can also be raised against Matthen's proposal.

To return to one of our earlier examples, we need to know how to differentiate between the mammalian inner ear bones and the reptilian jawbones and the portions of the gill arches of fish that are their homologues. Given that all three categories are homologous, the relation of homology cannot on its own suffice to sort traits into one or another of the three smaller classifications. Homology is a relation of degree, somewhat akin to the relation of resemblance or genetic relatedness, and it is too open-ended to serve as a solitary principle of trait taxonomy. On the assumption that life originated just once, and allowing for unlimited modification, all of our traits derive, more or less directly or indirectly, from the traits of a common ancestral population—that is, the first forms of life. Indeed, the idea that trait-types are purely homologous runs the risk of reducing all trait-types to just one: The Trait. Clearly, we need and have ways of differentiating one type of trait from ancestral, descendant, and cousin traits, in the face of our mutual relatedness. We have ways of distinguishing smaller homologous groupings within larger homologous groupings.

We can come at something like the same point from another direction by considering the question, 'Are these two traits homologous?' Suppose that we have two species of bird, both of which have adult males with long red tail-feathers. And suppose someone points to a tail of a male of each species and asks, 'Are these homologous?' Assuming one might know the answer when the question is better specified, the appropriate response is to ask for more information. This is a version of the familiar problem of ostension—that is, pointing does not tell us what feature is being indicated. Are we being asked if the length of their tails is homologous, or if their color is, or if their getting wagged in a winning way during courtship is? Their length, for instance, might have derived from a common ancestor, while their similar color might have arisen independently. So, before two traits can be identified as homologous with respect to each other, we need some *specification of the traits* in question.

My final point in this section concerns taxonomic restrictions on trait-types. Someone might agree with me that homology is not sufficient as a solitary principle of classification. He might agree that lineages of traits need to be constrained by some further principle or principles of classification. However, he might also think that neither functional nor structural considerations of the kind to be discussed in the next few sections are required for this. There is at least a question as to whether homologous traits can be differentiated from related homologous kinds of traits solely on the basis of taxonomies of creatures, such as the species and higher taxa. Perhaps the problem of trait taxonomy at this point simply reduces to the problem of organism taxonomy. Taxonomies of creatures are certainly used to constrain categories of traits. For instance, we speak of *human* ovaries, the *barn owl's* optic tectum, the *primate* immune system, and so on. However, this strategy of taxonomic restriction clearly overlays other classificatory practices. There remains the question of which species have *ovaries*, or *optic tectums*, or *immune systems*. In any case, according to Cladistics, which perhaps deserves to be considered the dominant school of thought on the issue of the nature of species and higher taxa, the right way to taxonomize organisms is in terms of clades. And, as we have just seen, the concept of a clade does not seem to do the work that we want done in the case of *trait* taxonomy.

4. Functional Homologues

If we conceive of the phylogenetic trait tree as a branching flow of (genetic and other) information, the issue is how to draw conceptual lines in this flow. Clearly there will be few if any sharp boundaries. Nonetheless we must distinguish one trait from another, for physiology requires such distinctions. My suggestion, the central suggestion of this chapter, is this. One main way in which this is done is by drawing conceptual lines at those places where there is a significant change in what there was selection for. (For example, from selection for respiration to selection for jaw support, and from selection for jaw support to selection for audition.) Talk of what there was selection for is usually regarded as talk about function. However, there is also a structural component to what there was selection for, since there is selection for both dispositions and the structural features that are their categorical bases. We will return to this point towards the end of this chapter. First, however, I want to make some remarks about the functional side of things.

We have seen that homologous classification requires assistance. At present it is an open question whether that assistance is sometimes provided by function, although I am suggesting that this is the case. But now let us note that it is an implication of an etiological account of (normal) function that classification by (normal) function also involves classification by homology, or at any rate by lineage.[12] Remember that, on an etiological account, the function of a token trait depends on the selective advantage of past traits of the type in ancestral individuals. Its function is to do what traits in the relevant lineage were selected for. So, before we can determine the function of a token trait, and therefore before we can classify it according to its function, we need to be able to locate it in a lineage. In general, to determine that a token trait, x, has the function, Z, we need to determine that x belongs to a lineage of traits that was selected for Z-ing.

Thus the lack of synonymy between 'functional categories' and 'analogous categories'. On an etiological account of functions, functional categories are

[12] Griffiths (1994) also sees this. If I read him correctly, he regards this as evidence that functional classification is redundant.

homologous categories in the first instance. Functions are assigned to traits on the basis of their belonging to a lineage of traits. And we differentiate related lineages on the basis of interesting differences in the dispositions for which there was selection. Such homologous-cum-functional categories are like species insofar as they are both spatio and temporally connected historical 'individuals'.[13] They (the homologous-cum-functional categories) are necessarily unified segments of the phylogenetic trait tree. Such categories are not analogous categories, for analogous categories include discontinuous segments. Analogous categories are second-order groupings of first-order functional-cum-homologous categories. They are groupings of distinct functional homologues.

Assuming an etiological account of normal function, any category that implicates normal function will also implicate homology. So my earlier brief remarks on the importance of function to trait typology should not have been interpreted as a rejection of the importance of homology. On the contrary, they entail that homology is important also. Notice also that there is no reason to conclude from the fact that functional classification implicates homology that functional classification is less 'fundamental' than homologous classification, or that it is rendered unnecessary by prior homologous classification. If, as I have argued, classification by homology also implicates some further principle of classification, that conclusion is premature. For, if the further principle is sometimes function, then these principles are complementary.

Notice that, if a homologue is distinguished in terms of functional considerations, the resulting category need not be functionally uniform. So far, I have spoken as if functional categories must be functionally uniform, but this was just a convenient simplification. The most straightforward kind of functional category is one in which a requirement on membership is possessing the defining function. This will be the result if we choose to begin a homologous kind with the onset of selection for a particular disposition and choose to end it at cessation of selection for it. But other kinds of functional categories are also possible. Mohan Matthen (pers. comm.) has claimed that ostrich wings, although vestigial, are avian wings nonetheless. I am not sure if this is true (or, if it is true, if it is of any theoretical importance). But let us suppose it is true for the purposes of illustration. Ostrich wings are vestigial; they have lost the function of flight. So, if they are really wings, some avian wings will not have the function of flight and therefore the category of avian wings will not be a

[13] Of course, traits in the same lineage do not remain spatially connected to the other traits in their lineage. But they connect with each other at the point of genetic replication. So, too, individual organisms are all spatially connected to other individuals of their species at the points of replication and reproduction. The idea that trait types are historical 'individuals' might seem problematic for talk of kinds, but see Griffiths (1994) for discussion of this and origin essentialism in this context.

functionally uniform category. It will include items that have the function of flight and also items that do not themselves have the function of flight but are descendants of items that had that function. Function can still delineate the category. We might start the relevant type with the start of selection for flight but continue it past cessation of selection for flight, motivated by the fact that some of the original wing design remains and by the fact that what remains has not been exapted for another purpose.[14]

What happens in practice is that biologists give descriptions of traits, sometimes in terms of their function, sometimes in terms of their structure, and often in terms of both. In addition they will often identify the taxonomies of creatures that possess the traits or they will restrict the category under discussion to some subsection of them. These functional and structural descriptions are revisable (as are the claims about taxonomic restrictions) but they serve as practical definitions of category terms. I am suggesting that, in the case of homologous traits, the kinds to which biologists refer, and the kinds that they are attempting to describe, are lineages of traits that are differentiated from related lineages in theoretically interesting ways, and (most?) often with respect to interesting changes in what there was selection for.

5. Form and Function

My position here is a pluralist one. I am arguing that function is important in classifying homologous traits. I am not arguing against the importance of other criteria, such as homologous relations or structural specifications of the traits in question. The Functional Minimalists, however, have an exclusionary claim. Their position is that function is not involved, whatever else may be. Once they acknowledge that homology cannot do the work of classification on its own, there are (at least) two positions Functional Minimalists could adopt. They could claim that classification is by homology plus structure or that classification is by structure alone. But why think that form and function are competing criteria or that, if they are, form is to be preferred to function?

Philosophers are apt to talk as if our only reason for wanting functional categories is our wanting categories that are multiply physically realizable. This is the standard motivational story for functional categories. It is, for instance, the standard story that we hear when Functionalism is introduced in introductory

[14] Notice that, if Matthen were right about avian wings, this would entail (in my view) that token avian wings do not possess the function of flight by virtue of their belonging to the category of avian wings. For the purposes of function attributions, types must be identified in the way that results in functionally uniform categories: that is, lineages must start around about where selection for the relevant disposition begins and end around about where selection for the relevant disposition ceases.

and not so introductory texts in philosophy of mind. According to this story, we want functional categories, if and when we do want them, because we want to abstract away from underlying differences in the structural details of items that have or perform the same function. In doing so, we make salient the similar capacities or dispositions of functionally equivalent but structurally diverse items. To a significant extent, this standard motivation for having functional categories is undermined in the case of homologous traits. The closer the genetic relation between two homologous traits, the more they will share by way of structural details. A lot could be said about this, but I hope it is obvious enough to need no further elaboration here. I certainly do not want to dispute it. Suffice it to say that the usual reason for wanting functional categories is at least seriously weakened when the relevant traits are homologous. We might, therefore, think that structure rather than function is the relevant complementary principle in such cases. Analogous categories, in contrast to homologous categories, are structurally diverse. The wings of bats, birds, and insects, for instance, vary greatly in the details of their structural design. So, in that kind of case, the standard motivation for having functional categories definitely holds. But it is far from obvious that it holds in the case of homologous traits.

What reason can there be for wanting classification by function if there is substantial structural similarity among all of the traits of the type? Someone sympathetic to Functional Minimalism will answer that there is no reason for wanting functional categories apart from the analogous categories. Indeed, the Functional Minimalists often speak as if virtually the only context in which functional categories could have a significant place in biology is in broad comparative studies across widely divergent species.

Now there are a number of ways one might choose to reply to this line of argument. One might claim that homologous traits that share the same function are not so structurally uniform that there is *no* need to abstract away from structural differences. After all, it is of the essence of neo-Darwinian biology that much variation exists even within a species. Most organic traits are so complex that there is almost never precise structural sameness (Rosenberg 2000: 59). And there are, after all, significant structural differences between the wings of hawks and hummingbirds. I think there is something to this line of reply, but I also think that a more fundamental response is possible. The main reason why we want functional categories has little to do with wanting to abstract away from differences in structural design or material realization.

I will say why in a moment but, before that, I want briefly to comment on the role of functional categories outside broad comparative studies. Even if our wanting multiply realizable categories were our only reason for wanting functional categories, this would not make them redundant outside broad comparative studies. Obviously enough, there are the polymorphic traits (such as different blood types). But there are also many other multiply realizable categories within a single species. There are many categories of traits whose members have the same function, broadly construed, and yet whose members have different structural designs, owing to the fact that they have different functions, more specifically construed. For instance, there are the many different *digestive enzymes* that catalyze different chemical reactions. There are the different *hormones* that circulate in body fluid, coordinating the operation of various parts of the system by interacting with various different target cells. There are also such things as *regulator genes*, the various *synaptic enhancers* and *inhibitors*, the different *sensory receptors*, and so on. Perhaps some of these have enough structural similarity, despite their significant differences, for us to wonder if functional classification is required, but this will not be so in all cases. Since such broad functional categories are useful, functional categories are useful within species-specific studies, even if the rest of what I have said (and am about to say) turns out to be incorrect.

But now, the more fundamental reply. Abstracting away from underlying structural difference is not our only reason for wanting categories that are sensitive to functional considerations. It is not even our main reason. The main reason is a terribly obvious one, but it seems that it needs to be said. The main reason why we want categories that are sensitive to functional considerations is that we are interested in what things do. Or, to put it in less subjective terms, the main reason why we should want such categories is that what things do is immensely important for the business of describing and explaining the operation of living systems. The key question is this, what in nature should our categories track? What are the theoretically important similarities and differences or theoretically important groupings? And there is no question that functional similarities and differences are at least as important as structural ones. Indeed, structural similarities and differences are significant largely because of their functional implications. In physiology, in particular, we seek classifications of traits that are illuminating for describing and explaining the functional organization of the system (see Cummins 1983 for helpful discussion). I think that those who seek to diminish the importance of function for the classification of traits must have lost sight of this.

The conceptual decomposition of the system that is useful for the purposes of describing and explaining the functional organization of a system is sometimes hard to discover. While it is true that a surgeon can remove a heart whole from the chest and that the bones of the long dead may fall apart on the paleontologist's table, the conceptual decomposition of an organism is far from always so easy. At the other end of the spectrum, for instance, there is the ongoing attempt to fathom the working of the human visual cortex and the concomitant struggle to identify its meaningful components. There have been and still are competing and coexisting systems for its compartmentalization. What makes certain components 'meaningful', I suggest, is not mere structural distinctiveness. There is a confusing, dismaying, plenitude of mere structural distinctions that might be drawn. What makes certain components meaningful is instead structural distinctiveness combined with functional significance.[15]

Consider some recent research into the classification of muscle fiber types. A team of researchers began by identifying types on the basis of their distinctive functional properties. Three distinct types and one less distinct type were identified on the basis of their different contractile properties. 'Slow types' were identified as having long twitch times, low peak force, and resistance to fatigue; 'fast fatigue resistant types' were identified (this will come as no surprise) as having a fast contraction rate and resistance to fatigue, and so on. It was only with the assistance of these distinctions that histochemical assays could then reveal molecular similarities among units of the same type. For instance, the slow types were found to be high in oxidative enzymes but low in glycolytic markers and ATPase activity, and the fast fatigue resistant types were found to be high in all three. We do not need the full details. Now, assuming that these classifications are useful and will remain in use, we can ask, which kinds of criteria—the functional or the structural—will become constitutive in this case? But this sets up a false dichotomy. The two are intimately interwoven and are more or less two aspects of the same thing. Insofar as the biologists are seeking structural similarities within types and structural dissimilarities between types, they are seeking those similarities and dissimilarities that have functional significance. They are seeking the structural basis for interesting functional properties.

There are, moreover, two ways in which structural heterogeneity can be an issue. One is the standard and much discussed way in which the same function might be performed by mechanisms that vary in their structural design. This is what is ordinarily referred to as 'multiple realizability'. But there is also another kind of 'heterogeneous realization'. In this case, a component of a given type can be structurally heterogeneous, not in the sense that the type comes in different designs, but insofar as the design of each particular instance is complex such that each instance consists of structurally heterogeneous subcomponents. In large part, such complex components are meaningful components, not because they are tied together into a bundle with something like structural string, but instead because they are functionally coherent units. The various heterogeneous parts may collaborate closely in the performance of some overall function or small set of functions. For example, genes are individuated in part in structural terms on the basis of being molecules of DNA with contiguity on a chromosome, but they are also individuated in part on the basis of the proteins that they specify, which is a functional consideration. That

[15] On these last few points, see Zeki (1993).

is, which gene a molecule of DNA belongs to can depend in part on the protein that it helps to specify. At a far higher level of analysis, the immune system is another example, for it too is in part individuated with respect to the overall function to which its various parts contribute—that is, immunity. Whether or not some item belongs to the immune system depends in part on whether it contributes to immunity.

6. Narrow and Normal Structure

Amundson and Lauder are usually interpreted as defending the idea of classification by homology, and I have joined in this interpretation up until this point. However, their joint paper can as well be read as defending the idea of pure structural categories—that is, categories of traits where structure is the sole consideration. For one thing, they claim that the style of classification that they are advocating is 'less inferential' and 'more observational' than functional classification. This is an odd claim to make on behalf of homologous categories. It would be hard to choose between function and homology as to which of them was more or less inferential or observational. In order to determine homologous relations, we have to make inferences about evolutionary history, just as we do when we determine normal functions. Moreover, in some systems of systematics (systems that are designed to assist us in making inferences about phylogenetic relations), evidence concerning function is taken into account in determining homology. For example, the presence of function may be counted as weak evidence against homology on the grounds that similar traits that lack a function are less likely to have resulted from convergent evolution. Thus, sometimes an inference to function is prior to an inference to homology. The point is that there is little between function and homology when it comes to a contest over which is the most inferential and least observational.

The structural interpretation of Amundson and Lauder is also suggested by a number of more direct statements. For instance, having noted that kidneys, properly so-called, exist only in vertebrates where they are all co-homologous, they continue,

> But isn't kidney a function category? Well, kidneys do all perform the same function (in vertebrates). But they are also homologous. This means that we could identify all members of the category 'kidney' by morphological criteria alone . . . So, at least in that sense, 'kidney' is not a functional category, or at least not essentially and necessarily a function category. (Amundson and Lauder 1994: 456)

What they seem to be saying here is that, if a group of traits are co-homologous, then they can be classified on the basis of their structural similarities. They do not seem to be saying that if a group of traits are co-homologous then they can be classified on the basis of homology specified in structural terms.

Plainly, there is a difference between (*a*) criteria by which we *recognize* and in that sense *identify* that something is of a given kind and (*b*) criteria that *constitute* something's being of a given kind, or that *individuate* that kind from other kinds. Acidity can be identified by its turning litmus paper red but that does not make turning litmus paper red constitutive of acidity. We can identify water by its look and taste, by its quenching our thirst and its coming from our faucets, and so on, but that does not make these features constitutive of water. Perhaps we almost never recognize that water is water by testing for its molecular structure. Nonetheless, this is consistent with the claim that what makes water water is its being H_2O. Along similar lines, biologists may in practice recognize that a token trait is a trait of a certain type by reference to features that are not necessarily criterial. For instance, many a histological marker—for example, cytochrome oxidase staining—can be simply that, a marker. For that reason, we have to be wary about what conclusions we draw from the kinds of facts that Amundson and Lauder appeal to above, and also in this passage: 'Even a severely malformed vertebrate heart, completely incapable of pumping blood (or serving any biological role at all), could be identified as a heart by histological examination' (Amundson and Lauder 1994: 457). Cardiac tissue can occur in tumors and abnormal growths in the oddest places in the body and such tissues do not amount to extra hearts. So the histological markers of cardiac tissue do not provide the criteria for being a heart. If someone smashed my television set and left the pieces where they lay, I could identify it by the color and texture of its plastic, but the color and texture of its plastic are irrelevant to its classification as a television.

What remains of the point being made by Amundson and Lauder? Their point is presumably that there is enough structural similarity among co-homologous tokens for them to be classified on that basis. Little detail is given to support this claim, but let us suppose that this is true for the sake of the argument. Curiously, they also claim that all of the tokens of these types—the vertebrate heart and kidney—also have a common function. So it might seem that they should claim only that structural and functional considerations are *on a par* in these cases. Nonetheless, they conclude against function. Why? Recall that Amundson and Lauder's view is an exclusionary one: outside the analogous categories, function has no significant role in the classification of traits. So we must ask, what gives their alternative the edge, on their view? Part of the answer may lie with the standard motivational story for functional categories that I discussed above. That is, their thinking might be that, in the absence of underlying structural difference, there is just no need for functional classification. But is there another more positive reason on offer? We find another reason that makes sense if we interpret their positive proposal as being in favor of structure rather than homology. For, so interpreted, their claim that their preferred alternative is less inferential and more observational makes sense. Another claim sometimes made in this context is that classification by structure is more predictive than classification by function. Although this chapter is already long, some thoughts on these claims are in order.

Let us begin with the one about prediction. Now you might think that a trait's structure is more predictive than both its normal function and its homologous relations. You might think so because you might well think that the structure of a trait is intrinsic to it whereas its normal function and its homologous relations are extrinsic, since both depend on its history and the history of its ancestors. Since the causal dispositions of a trait depend on its intrinsic physical properties, similarities and differences in intrinsic structure are a better reflection of similarities and differences in the causal powers of traits. The historical precedents relevant to assigning normal function and homologous relations cannot affect the present causal powers of my eyes—their degree of myopia, for example—except insofar as they have left their trace on their present structure.

Here we run some risk of becoming enmeshed in what is known as the Methodological Individualism debate, but we can avoid that fate. As I remarked at the outset, all parties to this present debate agree that we are dealing with abnormality inclusive categories. By hypothesis, these include both normal and abnormal items of the relevant type. Thus they will include items with very different causal dispositions, no matter what their criteria happen to be. Whatever the criteria are for being a heart, it is a given that some hearts can pump blood and some hearts cannot. So it seems to follow that tight predictions expressed in terms of abnormality-inclusive categories are not to be had, and that opting for structural as opposed to functional criteria cannot change that fact without, as it were, changing the subject. If the structural criteria are suitably abnormality inclusive, then they will result in categories that are just as poor or just as good for the purposes of prediction as the ones that we would get if we captured the same extension but appealed to function or homology instead. Of course, someone might want to change the subject at this point and urge upon us the need to purge physiology of abnormality-inclusive categories. But that is a radical revisionist option that the Functional Minimalists do not seem to be entertaining and it is outside of the scope of the present chapter.

I have not discussed the rationale for abnormality-inclusive categories in this chapter. To do so will involve a long discussion of the nature of functional explanation and I want to leave that for a later occasion. But I can say something brief to some effect here. Although abnormality-inclusive categories embrace traits with very different causal dispositions, we can use them to generate more fine-grained categories with greater predictive power. Knowing that something is a heart does not tell us what its actual causal capacities are. It may or may not be able to pump blood. But knowing that it is a normal heart or that it has ventricular fibrillation (a rapid and fluttering beat) or coronary artery occlusion (blockage) or stenosis (partial blockage) or myocardial infarction (heart muscle cell death) tells us much more. The complaint that abnormality-inclusive categories are not predictive is short-sighted. We need to look a little beyond them at those more fine-grained categories that are based upon them.

There is a further problem with the claim that structural classifications are more predictive than functional or homologous classifications. This is also a problem for the other claim, that structural criteria are less inferential and more observational. The problem is that both claims assume that the relevant notion of structure is that of actual, intrinsic structure. However, this cannot simply be assumed, for there is a notion of *normal structure* that parallels the notion of normal function. Some tokens are structurally normal but others are malformed or structurally abnormal owing to injury, disease, congenital abnormality, or misuse. And a trait's actual structure may not be what is normal for traits of its type. Although this concept of normal structure receives little attention in the philosophical literature, it is as much in common currency in physiology as is the concept of a normal function. A description of the normal system is a description of both normal function and normal structure. It is far from obvious that normal structure is more predictive or less inferential or more observational than normal function is. One highly plausible suggestion is that, if normal functions are dispositions for which there was selection, then normal structures are structures for which there was selection.[16] I will not try to defend this analysis of normal structure here, nor will I attempt to develop it in more detail. But I believe that something along these lines is correct, and that, in any case, any plausible analysis of the concept of normal structure will render the concept extrinsic. This makes normal structure and normal function comparable in terms of the kinds of concerns that have been raised (predictive powers and so on).

Finally, it does look as if it is the notion of normal, not narrow, structure that is, strictly speaking, criterial for abnormality-inclusive categories. Consider, for example, the anaerobic bacteria that have an arrangement of magnetite particles that respond to the local magnetic field and thereby direct the bacteria down deep and away from the toxic oxygenated waters on the surface. We will suppose that the system of the northern-hemisphere bacteria is a simple inversion of that of the southern-hemisphere bacteria and vice versa. Perhaps this inversion can be achieved by a single point mutation, but in any case we will suppose that such inversions do sometimes accidentally occur. Now let us consider an individual northern-hemisphere bacteria that has such an accidentally

[16] In my Ph.D. dissertation (Neander 1983), my view was that normal structure was defined in terms of normal function. Because of this, I did not regard normal structure as an independent notion.

inverted magnetotactic system relative to its lineage. Its magnetesome will probably be lethal, because it will lead the bacteria upwards instead of downwards if it remains in its ancestral environment. By hypothesis, this system is structurally identical to that which is normal for the southern-hemisphere bacteria. So, let us suppose, for the sake of the argument, that the trait in question is classified on the basis of structural as opposed to functional criteria, or at least on the basis of both structure and function if function is also involved. Our question is this: how is the trait to be classified with respect to the relevant abnormality-inclusive categories? Does it belong with the northern-hemisphere type or with the southern-hemisphere type? If we keep in mind that we are speaking of abnormality-inclusive categories, I think it is clear enough that this is an abnormal northern-hemisphere type, not a normal southern-hemisphere type, although it is structurally identical to the latter and not the former. If so, and if it is classified on the basis of structure, it must be classified on the basis of normal structure rather than actual structure. While it might be rare in nature, a given trait could be structurally more like most of the members of one kind than most of the members of another kind and yet belong to the latter and not the former, where the relevant kinds are abnormality inclusive. This does not rule out the possibility that the relevant classification involves structural considerations. But it does rule out the possibility that belonging to the relevant classification is a simple matter of sharing the same intrinsic structure. Of course, we *can* classify traits on the basis of their intrinsic structure. However, the abnormality-inclusive categories do not do so. They do not do so because they are essentially historical categories.

It is time to conclude. First of all, the claim that many or even most types of traits are homologues turns out to be entirely consistent with the traditional claim that functional classification is important in biology. Analogous categories are one kind of functional category, but they are not the only kind. Homologous categories can also be functional categories, as we can see if we allow, as we must, for the possibility of mixed criteria. Homology is clearly inadequate as a solitary principle of classification, because some further principle of classification is needed to differentiate one homologous category from other related homologous categories. Homologous groupings of traits are segments of the phylogenetic trait tree, and what we have to consider is how this conceptual segmentation is achieved, in the face of the continuity in the flow of information. I have suggested that this is a matter of drawing conceptual lines at those places where there is a significant change in what there is or was selection for. I have also suggested that there is both a functional and a structural component to this, for there is selection for both dispositions and the structural features that are their categorical bases. This results, in effect, in our classifying traits on the basis of their normal structure or their normal function, or both. Function is of central importance in physiology, since physiology is just the study of the functional organization of living systems. Thus classifications that are sensitive to similarities and differences in function, or that are sensitive to functionally significant groupings of structurally heterogeneous components, will be interesting and important from the perspective of physiology. Structural similarities and differences are also important, but we need to keep in mind that they are largely important because of their functional implications. Indeed, it seems somewhat strained to oppose structure to function as a principle of classification. One can emphasize one aspect or another, but they are so intimately interwoven in classificatory practice that it is a mistake to insist on choosing between the two. Furthermore, functional and structural criteria are comparable with respect to such issues as predictive powers and inferential or observational status. For the relevant notions are both 'normative', in the sense that they are both notions of the normal, in the teleological as opposed to the statistical sense of the term, if we assume an etiological account of each of them. Abnormality inclusive categories involve a notion of structure and function that is, to recall the title of Amundson and Lauder's paper, with, not without, a purpose.

References

Allen, Colin, Becoff, Marc, and Lauder, George V. (1998) (eds.), *Nature's Purposes: Analyses of Function and Design in Biology* (Cambridge, Mass.: MIT Press).

Amundson, Ron, and Lauder, George V. (1994), 'Function without Purpose: The Uses of Causal Role Function in Evolutionary Biology', *Biology and Philosophy*, 9/4: 443–69.

Beckner, Morton (1959), *The Biological Way of Thought* (New York: Columbia University Press).

Buller, David J. (1999) (ed.), *Function, Selection and Design: Philosophical Essays* (Albany, NY: SUNY Press).

Burge, Tyler (1989), 'Individuation and Causation in Psychology', *Pacific Philosophical Quarterly*, 70: 312.

Cummins, Robert (1983), *The Nature of Psychological Explanation* (Cambridge, Mass.: MIT Press).

Griffiths, Paul E. (1993), 'Functional Analysis and Proper Functions', *British Journal for the Philosophy of Science*, 44: 409–22.

——(1994), 'Cladistic Classification and Functional Explanation', *Philosophy of Science*, 61: 206–27.

Hall, Brian K. (1994) (ed.), *Homology: The Hierarchical Basis of Comparative Biology* (New York: Academic Press).

Kim, Jaegwon (1996), *Philosophy of Mind* (Boulder, Colo.: Westview Press).

Lauder, George V. (1994) 'Homology: Form and Function', in B. K. Hall (ed.), *Homology: The Hierarchical Basis of Comparative Biology* (San Diego, Calif.: Academic Press), 151–96.

Matthen, Mohan (2000), 'What is a Hand? What is a Mind?', *Revue internationale de philosophie*, 214: 123–42.

Neander, Karen (1983), 'Abnormal Psychobiology', Ph.D. dissertation, La Trobe University.

——(1991*a*), 'Functions as Selected Effects: The Conceptual Analyst's Defence', *Philosophy of Science*, 58: 168–84.

——(1991*b*), 'The Teleological Notion of "Function"', *Australasian Journal of Philosophy*, 69: 454–68.

Rosenberg, Alexander (2000), *Darwinism in Philosophy, Social Science and Policy* (Cambridge: Cambridge University Press).

Roth, V. Louise (1984), 'On Homology', *Biological Journal of the Linnean Society*, 22: 13–29.

——(1988), 'The Biological Basis of Homology', in C. J. Humphries (ed.), *Ontogeny and Systematics* (New York: Columbia University Press), 1–26.

——(1991), 'Homology and Hierarchies: Problems Solved and Unresolved', *Journal of Evolutionary Biology*, 4: 167–94.

Walsh, Denis M., and Ariew, André (1996), 'A Taxonomy of Functions', *Canadian Journal of Philosophy*, 26/4: 493–514.

Zeki, Semir (1993), *A Vision of the Brain* (Oxford: Blackwell).

BIOGRAPHIES

Colin Allen

Colin Allen is Professor of Philosophy at Texas A&M University. His work focuses on the epistemological and methodological problems involved in the scientific study of the minds of non-human animals, particularly in the area of cognitive ethology. He collaborates extensively with biologists and psychologists, and is co-author (with ethologist Marc Bekoff) of *Species of Mind* (MIT Press, 1997). He is co-editor with Marc Bekoff and comparative morphologist George Lauder of *Nature's Purposes* (MIT Press, 1998), an anthology of classic papers on teleology in biology, and he is co-editor with psychologist Denise Cummins of *The Evolution of Mind* (Oxford University Press, 1998), a collection of original essays in evolutionary psychology. Allen and Bekoff, together with comparative psychologist Gordon Burghardt, are currently editing a collection of fifty new essays about animal cognition under the title *The Cognitive Animal* (MIT Press, 2001).

> My initial interest in functions came about from thinking about the meanings and functions of vervet monkey alarm calls and some of my earliest work argued that functional description of animal communication requires specification of the content of their signals in intentional terms. In 1994 came an invitation to write a review piece with Marc Bekoff on function, natural design, and animal behavior for volume 11 in the annual *Perspectives in Ethology* series, which enabled us to bring the ethological and philosophical literatures into contact. As a consequence of writing this piece we saw the need for a comprehensive collection of articles that became *Nature's Purposes* and we were very fortunate to get George Lauder, a comparative morphologist, to help us bring a third strand of the literature into the mix.

André Ariew

An assistant professor of philosophy at the University of Rhode Island, André Ariew specializes in the philosophy of biology. He has written or co-written papers on functional explanation, probabilities in evolutionary theory, fitness, and innateness.

> I came to be interested in teleology and functional explanations when I was a graduate student at the University of Arizona. I was first interested in understanding how Aristotle's teleology fit in his theories of nature. Fortunately I had the opportunity to work with Julia Annas on this project. Then I became interested in the intersection of philosophy of biology and philosophy of mind. Robert Cummins (at Arizona) and Elliott Sober and Denis Walsh (at the University of Wisconsin-Madison) all tutored me through that project. I consider the chapter I have contributed to this volume as a homage to the teachings of Julia, Rob, Elliott, and Denis.

Christopher Boorse

Christopher Boorse attended Oberlin College in 1963–7, graduating with a bachelor's in philosophy; he attended Princeton University in 1967–72, getting a Ph.D. in philosophy in 1972; he has taught at Delaware since 1971. He has written papers on philosophy of medicine, philosophy of biology, and ethics. He currently teaches logic, decision theory, and medical ethics.

I got interested in functions because I was developing an analysis of medical concepts of health and disease and became convinced that, at least in Western scientific medicine of the past two centuries, health is conceived as species-normal biological function of the parts of the organism.

David J. Buller

David J. Buller was an undergraduate student of Larry Wright's, though he learned nothing about functions from Wright, being primarily interested in Nietzsche and Heidegger at the time. After meandering through numerous subdisciplines in philosophy, and numerous periods in the history of philosophy, he dedicated himself to the philosophy of psychology, picking up on interests that led him into his original undergraduate major. Through literature in the philosophy of psychology, which he now recognizes to be deeply misguided, he became interested in the topic of functions. That moved him into the philosophy of biology, and he has never looked back, becoming what Kim Sterelny has so poetically called 'a refugee from the philosophy of psychology'. He wrote some articles on the topic of the biological concept of function and also edited an earlier anthology, *Function, Selection, and Design* (SUNY Press, 1999). In recent years he has been working on evolutionary psychology, and he is currently writing a book (under contract with the MIT Press) on everything that is wrong with evolutionary psychology as we know it today. He is Associate Professor of Philosophy at Northern Illinois University, and he lives with his wife and son where he hopes no one will be able to find him.

Robert Cummins

A graduate of Carleton College and the University of Michigan, Robert Cummins specializes in the philosophy of psychology and cognitive science, with a special emphasis on mental representation. His books include *The Nature of Psychological Explanation* (MIT Press, 1983), *Meaning and Mental Representation* (MIT Press, 1987), and *Representations, Targets and Attitudes* (MIT Press, 1996). He is co-editor of *Philosophy and AI* (MIT Press, 1991) (with John Pollock), and of *Minds, Brains and Computers* (Blackwell, 2000) (with Denise Dellarosa Cummins). He is currently Professor of Philosophy at the University of California, Davis.

In my second year in graduate school—1968—I wrote 'Functional Analysis,' a paper that was turned down by every major journal in philosophy in 1970. I put it away in a drawer, and resubmitted literally the same piece of paper (no word processors then) to the *Journal of Philosophy* in 1975. It has since been reprinted more times than I know, and is in four languages (that I know about). I revised and expanded the material in that article for the first chapter of *The Nature of Psychological Explanation* in 1982. Although 'Functional Analysis' and its sequel in the book were not intended as analyses of the concept of function in biology or psychology, but rather as part of an attempt to distinguish analytical-decompositional explanations from explanation by nomic subsumption, they eventually found their way into the functions literature proper, a literature I have not contributed to since. 'Neo-Teleology' in this volume, is, therefore, the only thing I have written on this topic in thirty-three years. I offer it with some trepidation. I have done well in this literature by staying out of it.

Berent Enç

Berent Enç is Professor of Philosophy at the University of Wisconsin-Madison. He is author of numerous articles on intertheoretic reduction, philosophy of the mind, explanation of behavior, action theory, and functions, as well as on Hume. He is currently working on a book manuscript on the causal theory of action.

My interest in functions was triggered by general concerns over Functionalism in the philosophy of Mind. In the mid-1970s the consensus view at Wisconsin was that the identity of mental states partly derives from the function of the systems that generate these states. Sober's adjunction to put functions back into functionalism, Stampe and Dretske's work involving semantic contents that later got to be known as 'semantics Wisconsin style', as well as my efforts in finding intentional content in the states of mechanical devices, were all manifestations of this consensus view. Since these efforts were all premised on the presumption that teleology could ultimately be naturalized, I took it upon myself in my 1975 *Philosophy of Science* paper on functions to build on the Cummins and Wright analyses of functions and to provide a set of necessary and sufficient conditions for function attributions. Later, starting with Millikan's seminal book, studies of functions much greater in depth and detail appeared in the literature. But one idea that I carried with me from that time on until recently was that our understanding of functions and of certain types of counterfactuals went hand in hand. In fact I used such counterfactuals to offer a solution to the problem of deviant causal chains that plague all causal theories. Only recently I woke up to realize that the semantic evaluation of these counterfactuals was not fully determinate. And it is that rude jolt that resulted in the present chapter on the indeterminacy of function attributions.

Valerie Gray Hardcastle

Valerie Gray Hardcastle (Ph.D. Cognitive Science and Philosophy, University of California, San Diego) is currently the Director of Science and Technology Studies and Associate Professor of Philosophy at Virginia Tech. The main focus of her research is to map out the theoretical relationships between psychology, psychiatry, and neuroscience, particularly where consciousness and other human experiences are located.

She is the author of four books and numerous articles and is putting the finishing touches on her latest book, *Constructing Selves.*

I think I got interested in functions by being in a bad mood one day and believing that only dreck had been written on the subject. In a fit of hubris, I thought I could do better. It turns out that not only dreck has been written on the subject, but I did like what I wrote, so perhaps the bad mood was worth it.

Mohan Matthen

Mohan Matthen's first degree was in physics from the University of Delhi. He studied philosophy at Stanford University, and has taught at numerous universities in California, Alberta, Quebec, and British Columbia. Currently, he is Professor of Philosophy at the University of British Columbia. His research has been about Aristotelian metaphysics and science, philosophy of perception, and philosophy of biology.

I first got interested in teleology by puzzling over what it could mean to say, as immunologists often seem to say, that a human organ can make 'errors'. In 'Teleology, Error, and the Human Immune System' (*Journal of Philosophy*, 1984), Ed Levy and I argued that, if we take this seriously, we would be committed to an irreducible form of goal attribution. This thesis was refined in 'Biological Functions and Perceptual Content' (*Journal of Philosophy*, 1988), in which I introduced the notion of 'normal error', a kind of perceptual error that is not due either to maladaptation or to malfunction. In the meanwhile I worked on Aristotle's notion of teleology, publishing 'Four Causes in Aristotle's Embryology' (Apeiron, 1989). Up to this point, my work on teleology suggested a realistic interpretation of functions. This attitude was founded on 'scientific realism': if good immunological and psychological theories imply the truth of teleology, then it is true. But in 'Teleology and the Product Analogy' (*Australasian Journal of Philosophy*, 1997), I argued, on the basis of the plurality of contexts in which teleology is and can be used, that there is a subjective analogy at its core. This article offers an alternative to the Wright–Millikan view of functions as selected effects, in a domain distinct from that of Cummins' causal-role functions.

Ruth Millikan

Ruth Garrett Millikan is a graduate of Oberlin College and Yale University. She is a Board of Trustees Distinguished Professor and the Alumni Association's Distinguished Professor 2000–3 at the University of Connecticut. She is author of *Language, Thought and Other Biological Categories* (MIT Press, 1984), *White Queen Psychology and other Essays for Alice* (MIT Press, 1993), and 'On Clear and Confused Ideas' (Cambridge University Press, 2000). She has lectured extensively all over the world, including the Gareth Evans Memorial Lecture at Oxford University and the Jean Nicod Lectures in Paris.

'Biofunctions: Two Paradigms' was written in protest against conceptual analysis, against the myth that there exists THE notion of biological function, against the myth that there exist THE current function(s) of a biological trait (e.g. 'exaptations'), and against the ubiquitous but mistaken view that my own 'proper functions' were intended to analyze biological function. Still, I do intend it as a positive contribution!

Karen Neander

Karen Neander completed her Ph.D. from La Trobe University, Melbourne, Australia, in 1984 and has held teaching positions in Australia at the University of Sydney, the University of Adelaide, and Wollongong University. From 1988 until 1995 she held an appointment at the Australian National University in the Philosophy Program of the Research School of Social Sciences, where she was first a postdoctoral fellow and then a research fellow. In 1996 she moved to America and since then she has been at Johns Hopkins University. She has written numerous papers on biological functions, natural selection, mental representation, and consciousness and is presently at work on a book on the representational functions of neural components.

I first became interested in biological functions in the late 1970s, when I was a graduate student at La Trobe University. I had started out with the idea of writing a dissertation on insanity and moral responsibility, which led me to 'the anti-psychiatry debate', a hot debate at the time. It was about the medical status of psychiatric conditions that are not caused by organic lesions such as a stroke or too much or too little neural transmitter. Two candidates for this were thought to be sociopathy (otherwise known as psychopathy) and conversion hysteria (e.g. 'blindness' or 'paralysis' where there is no dysfunction in the visual or motor system). The central question was this: if there is no organic lesion, should it be treated as a medical problem or is some other paradigm more appropriate? Are conversion hysterics better viewed as malingerers? Are sociopaths really sick or do we disguise our moral disgust by dressing it in the costume of objective medical science? These questions, I thought, turned in part on whether there was a sense of function and dysfunction that was univocal for both body and mind. I argued that whether or not mental illness is a 'myth' depends on whether minds can malfunction in the same sense that bodies can. So, I turned to the topic of biological functions. I gave my first conference paper on the topic in 1980—under the title 'Teleology in Biology'—defending the same etiological account I have since continued to defend. I sent the paper off to a journal at about the same time, but after nine long months it was rejected without comment. Alas. I too stuffed it in a drawer and forgot it for many years. Happily, it didn't go entirely unread over the next decade. For an unpublished manuscript it was quite widely circulated and cited. The material in it finally emerged in 1991 as 'Functions as Selected Effects'.

Mark Perlman

Mark Perlman received BA degrees in philosophy and anthropology from Ohio State University in 1987, and an MA in philosophy from Ohio State University in 1989. He also studied for a year at the Ludwig-Maximilians-Universität in Munich, Germany.

He received his Ph.D. from the University of Arizona in 1994, working under Robert Cummins. He taught for five years at Arizona State University before moving to Western Oregon University in 1998. He has written a book on mental representation and misrepresentation, *Conceptual Flux: Mental Representation, Misrepresentation, and Concept Change* (Kluwer, 2000), as well as papers on conceptual role semantics and philosophy of law. He is also a musician, both a string bass player and conductor.

> I became interested in functions from their use in teleosemantics. My major area of research in philosophy of mind is mental misrepresentation, solving the 'disjunction problem', and finding a theory to explain how an idea could represent an object yet represent it incorrectly. Functions are a promising source of this, since a thing can have a function yet fail to perform its function. But I became skeptical of the prospects for finding a way to ground this mismatch between idea and its object, and my book is a sustained attack on all misrepresentation, including the teleosemantics route. The chapter in this volume is a greatly expanded version of the arguments criticizing teleosemantics in my book *Conceptual Flux.*

Michael Ruse

Michael Ruse was for thirty-five years at the University of Guelph in Ontario, Canada. He is now Lucyle T. Werkmeister Professor of Philosophy at Florida State University, in Tallahassee. The author of many books on the history and philosophy of biology, including *The Darwinian Revolution: Science Red in Tooth and Claw* (Chicago University Press, 1979), *Taking Darwin Seriously: A Naturalistic Approach to Philosophy* (Blackwell, 1986), and *Can a Darwinian be a Christian? The Relationship between Science and Religion* (Cambridge University Press, 2001), Ruse is a fellow of the Royal Society of Canada and of the American Association for the Advancement of Science, a former Guggenheim and Killam Fellowship holder, and the recipient of an honorary degree from the University of Bergen in Norway. Ruse was founder and for fifteen years the editor of the journal *Biology and Philosophy.* He has just given up the editorship and his greater pride is that he has helped to make the philosophy of biology the thriving subject that it is today.

> When I first became interested in the philosophy of biology back in the 1960s, the problem of teleology—is there something distinctive about biology because of the functional or purposive language?—was already a big issue, thanks to the writings of the logical empiricists such as Carl Hempel and Ernest Nagel. Although, like them, I was inclined to think that most of the arguments for the autonomy of biology were thin to the point of inadequacy, reading an article by Morton Beckner (a student of Nagel) convinced me that teleology really was something distinctive, and simply could not be eliminated from biology without significant loss of content. At the same time, reading a discussion by the geneticist C. H. Waddington (most of whose thinking I thought verged on neo-vitalism) showed me that the popularity of models of teleology (especially Nagel's) based on goal-seeking mechanisms were fundamentally mistaken, and that the secret lay rather in the concept of adaptation, so central to Darwinian evolutionary biology. At that time, I knew virtually nothing

about Darwinism and its history (and what I did know was wrong), but over thirty years of subsequent study of the history and philosophy of evolutionary thought has simply confirmed for me that my first insights about biological functionality were correct. Indeed, now finally I am writing a book on the subject, looking both at history and at the state of play today, not only in philosophy but in other areas also. *Darwin and Design: Science, Philosophy, Religion*, will be published by Harvard University Press, and will complete a trilogy (the earlier books being *Monad to Man: The Concept of Progress in Evolutionary Biology* (Harvard University Press, 1996), and *Mystery of Mysteries: Is Evolution a Social Construction?* (Harvard University Press, 1999)) on end-directed, value concepts in evolutionary thinking.

Peter Schwartz

Peter H. Schwartz received his MD and Ph.D. in philosophy from the University of Pennsylvania in 1999, with his research interests concentrated in philosophy of biology, philosophy of science, and medical ethics. His dissertation presents an account of teleological-appearing statements in biology and medicine, including function ascriptions, natural selection explanations, and claims about dysfunction and disease. He is a resident in internal medicine at the Brigham and Women's Hospital (Harvard Medical School) and an adjunct assistant professor in the philosophy department at Boston University (starting September 2001).

Like many philosophers of biology, I was led into the function debate by trying to answer the question, 'What is the place of teleological-appearing statements in current science?' But I was also led to examine the concept of function by a route traveled by philosophers of medicine, stemming from their attempts to answer the question, 'What is disease?' Studying the concept of disease leads to an examination of the concept of dysfunction, and then back into the concept of function. Therefore my work on function reflects my interests in both biology and medicine.

My chapter in this volume actually began life as an attack on the etiological approach, rather than a defense. I understood that the Modern History theory was the only version of the etiological approach to solve the problem of vestiges, but I became convinced that it makes untenable assumptions about natural selection. Once the Continuing Usefulness account occurred to me, though, the chapter turned into a defense of this account and thus of the etiological approach in general.

Denis Walsh

Denis Walsh studied biology at the University of Alberta and McGill University, Montreal, and philosophy at King's College Cambridge and King's College London. He is currently Lecturer in Philosophy at the University of Edinburgh. He first encountered the tangled concept of function while doing research in functional morphology of vertebrates and again later in the philosophy of mind. He is currently preparing a manuscript that makes a plea for full-blown Aristotelian teleology in both evolutionary biology and psychology.

William Wimsatt

When I came into philosophy from engineering physics (at Cornell) I took a course in ancient philosophy from Richard Sorabji in the fall of 1963. There smack in my way to being a philosopher of physics lay Aristotle's problem of teleology. Philosophers seemed to want to dispose of teleology, or to translate talk of it into talk that wasn't 'really' teleological, but my father was a biologist, so I was sure that it was real, and no amount of progress in molecular biology would change that. Different kinds of it arose from servomechanisms and cybernetics, or from evolution. The cybernetics was interesting, but when I took a look at evolution, I fell in, and I've been swimming around in it ever since. I discussed both cybernetics and evolution in my undergraduate honors thesis on a selectionist account of function (1965), and my dissertation (finished six years later after a post-doc!) was on that too. In the meantime I began to explore evolution and game theory, complexity, development, mathematical modelling, and a lot of other things that have interested me since. Dick Lewontin and Herb Simon were both major influences on me early on. And Lewontin's pivotal influence continued at Chicago, where I went to work with him, and continue to this day. My father's influence was to get me to take the biology seriously, and not to try to sell biologists a philosophical problem in disguise, but to see how real (and often new) philosophical problems arose out of the science.

INDEX